The Plant Microbiome in Sustainable Agriculture

The Plant Microbiome in Sustainable Agriculture

Edited by

Alok Kumar Srivastava

ICAR-National Bureau of Agriculturally Important Microorganisms (NBAIM)
Kushmaur, Mau, Uttar Pradesh
India

Prem Lal Kashyap

ICAR-Indian Institute of Wheat and Barley Research (IIWBR)
Karnal, Haryana
India

Madhumita Srivastava

Sunbeam College for Women
Varanasi, Uttar Pradesh
India

WILEY Blackwell

This edition first published 2021

Registered Office(s)
John Wiley & Sons, Inc., 111 River Street, Hoboken, NJ 07030, USA
John Wiley & Sons Ltd, The Atrium, Southern Gate, Chichester, West Sussex, PO19 8SQ, UK

Editorial Office
The Atrium, Southern Gate, Chichester, West Sussex, PO19 8SQ, UK

For details of our global editorial offices, customer services, and more information about Wiley products visit us at www.wiley.com.

Wiley also publishes its books in a variety of electronic formats and by print-on-demand. Some content that appears in standard print versions of this book may not be available in other formats.

Library of Congress Cataloging-in-Publication data applied for

9781119505167 (Hardback)

Cover Design: Wiley
Cover Images: Wheat field © Dennis Fischer Photography/Getty Images, Chromosome unraveled © Photon Illustration/Stocktrek Images/Getty Images, Genome data © Tetiana Lazunova/Getty Images, Aspergillus fumigatus fungus © KATERYNA KON/SCIENCE PHOTO LIBRARY/Getty Images, Migraine therapy and CGRP receptor © JUAN GAERTNER/SCIENCE PHOTO LIBRARY/Getty Images, Plant microbiome graph courtesy of Alok Kumar Srivastava

Set in 9.5/12.5pt STIXTwoText by SPi Global, Pondicherry, India
Printed and bound by CPI Group (UK) Ltd, Croydon, CR0 4YY

10 9 8 7 6 5 4 3 2 1

Contents

Preface

Global agriculture is facing a major challenge of ensuring sustainable and healthy food production for an ever exploding human population, while seeking to reduce adverse effects on the ecosystem. Recent reports indicate that factors like soil health (nutrients, water, and pH), vulnerability to diseases and pests, agronomic practices and climate change affect crop growth and yield. These factors are the prime cause of crop failure and decline in average yields.

The quest to harness the potential of useful microbes from different ecological niches have shown interesting outcomes in that microbiome plays an important role in growth and development of other living communities. Plants depend on their microbiome for multifarious life supporting activities including nutrient acquisition and augmentation of the defense system towards biotic and abiotic stresses. However, the process of crop domestication may have negative associations with the composition and function of the host associated microbiota, thereby limiting their advantageous effects on crop health and development. With major emphasis on agriculture, characterizing the plant microbiome and its function could be applied for better crop designing and management to grow crops in resource limited environments, and protect them from intruding pathogens. Unfortunately, at present, most of the breeding programs across the globe have not taken microbial action into account. Therefore, a deeper understanding of the interrelationships of the soil–plant–microorganism system is essential for improving the efficacy and potential applications of plant growth promoting inocula for achieving sustainable food security and development.

Many of the laboratories working on plant growth-promoting rhizobacteria (PGPR) have reported that cocktails of useful bacteria in the form of synthetic communities are better than the single inoculants which face competition from other microorganisms and which could be killed or suppressed under suboptimal conditions. These findings clearly establish the importance of the microbial community in the well-being of crop plants. Now the researchers are focusing towards deciphering the microbial communities using next generation molecular approaches, and they are dominating the conventional methods. Modern molecular tools are utilized to recover the microbial information's links with different ecological niches. This information can be used to establish and maintain plant and human health, and finally to achieve comprehensive information of the plant microbiome that can be helped to improve agricultural production. In the past two decades, the plant microbiome has gained interest and crop performance is increasingly being recognized as the result of multipartite interactions. The huge gene pool of the microorganisms living in close association as endosymbionts and surface colonizers extends to the host genome and contributes to its phenotype. The studies clearly indicate that the totality of this genetic information in the form of the hologenome may allow adaptation of crops to diverse environmental conditions and interactions.

The present book provides a comprehensive review and compiled information on different aspects of plant microbe research with reference to its scope in the agriculture system which can be transformed by a complete understanding and application of the specific microbiota in a holistic manner. In the book, the chapters are contributed by active researchers having expertise in the domain. Following an introduction to the specificities of microbiome research, modern tools and techniques to understand the plant microbiome are described. The updated information on the microbiome of different crops and cropping systems, followed by functional ecology and its potential for abiotic and biotic stress management, crop health and nutrient fortification, has been presented in different chapters. As they are of particular relevance for the future of agriculture in a sustainable manner, the biotechnological and molecular aspects of the translational microbiome are thoroughly covered across the book. Lastly, the relevance of the microbial community as the reservoir of novel genes and metabolites and as the key to green and clean agriculture have been discussed. This book will stimulate the readers to understand this complex subject in a lucid manner. It provides a path to researchers to address some of the contemporary issues before the scientific community, towards development of environmentally friendly and sustainable agriculture to meet the needs of our universe. With great pleasure, the editors acknowledge the efforts and contributions of expert authors, which were crucial for the quality of information provided.

Alok Kumar Srivastava
Prem Lal Kashyap
Madhumita Srivastava

List of Contributors

Akhilendra Pratap Bharati
ICAR-National Bureau of Agriculturally Important Microorgansims,
Mau,
Uttar Pradesh,
India

P. Veera Bramhachari
Department of Biotechnology,
Krishna University,
Machilipatnam,
Andhra Pradesh,
India

Siddhardha Busi
Department of Microbiology,
School of Life Sciences,
Pondicherry University,
Puducherry,
India

Son Truong Dinh
Vietnam National University of Agriculture,
Hanoi,
Vietnam

Nabti Elhafid
Laboratoire de Maitrise des Énergies Renouvelables,
Faculté des Sciences de la Nature et de la Vie,
Université de Bejaia,
Bejaia,
Algeria

Martin Filion
Université de Moncton,
Moncton,
New Brunswick,
Canada

Betty N. Fitriatin
Department of Soil Sciences and Land Resources Management,
Faculty of Agriculture of Padjadjaran University,
Jatinangor,
West Java,
Indonesia

Jose Pedro Fonseca
The Noble Research Institute,
Ardmore
OK,
USA

Moni Gupta
Division of Biochemistry,
Sher-e-Kashmir University of Agricultural Sciences and Technology of Jammu,
Jammu and Kashmir,
India

Sachin Gupta
Division of Plant Pathology,
Sher-e-Kashmir University of Agricultural Sciences and Technology of Jammu,
Jammu and Kashmir,
India

Diyan Herdiyantoro
Department of Soil Sciences and Land Resources Management,
Faculty of Agriculture of Padjadjaran University,
Jatinangor,
West Java,
Indonesia

Cuong Tu Ho
Institute of Environmental Technology,
Vietnam Academy of Science and Technology,
Hanoi,
Vietnam

Long Hoa Hoang
Agricultural Genetics Institute – Vietnam Academy of Agricultural Sciences,
Hanoi,
Vietnam

Abbaci Hocine
Laboratoire de Maitrise des Énergies Renouvelables,
Faculté des Sciences de la Nature et de la Vie,
Université de Bejaia,
Bejaia,
Algeria

Houali Karim
Department of Microbiology and Biochemistry,
Université Mouloud Mammeri de Tizi Ouzou,
Tizi Ouzou,
Algeria

Prem Lal Kashyap
ICAR-Indian Institute of Wheat and Barley Research (IIWBR),
Karnal,
Haryana,
India

Ashutosh Kumar
ICAR-Indian Institute of Seed Science,
Mau, Uttar Pradesh,
India

Dhananjay Kumar
Department of Botany and Microbiology,
HNB Garhwal University,
Srinagar Garhwal,
Uttarakhand,
India

Sudheer Kumar
ICAR-Indian Institute of Wheat and Barley Research (IIWBR),
Karnal,
Haryana,
India

Sunita Kumari
ICAR-Indian Institute of Seed Science,
Mau, Uttar Pradesh,
India

Bensidhoum Leila
Laboratoire de Maitrise des Énergies Renouvelables,
Faculté des Sciences de la Nature et de la Vie,
Université de Bejaia,
Bejaia, Algeria

Van T. Luu
Institute for Molecular Physiology
Heinrich-Heine-Universität,
Düsseldorf,
Germany

Ganugula Mohana Sheela
Department of Biotechnology,
Krishna University,
Machilipatnam,
Andhra Pradesh,
India

Kirankumar S. Mysore
The Noble Research Institute,
Ardmore,
OK,
USA

Xuan Canh Nguyen
Vietnam National University of Agriculture,
Hanoi,
Vietnam

Amy Novinscak
Université de Moncton,
Moncton,
New Brunswick,
Canada

Parasuraman Paramanantham
Department of Microbiology,
School of Life Sciences,
Pondicherry University,
Puducherry,
India

Subhaswaraj Pattnaik
Department of Microbiology,
School of Life Sciences,
Pondicherry University,
Puducherry,
India

A.M.V.N. Prathyusha
Department of Biotechnology,
Krishna University,
Machilipatnam,
Andhra Pradesh,
India

Roxane Roquigny
Université de Moncton,
Moncton,
New Brunswick,
Canada

Manisha Sachan
Department of Biotechnology,
Motilal Nehru National Institute of Technology, Allahabad,
Uttar Pradesh,
India

Pragati Sahai
Amity Institute of Biotechnology,
Amity University,
Noida,
Uttar Pradesh,
India

Vimlendu Bhushan Sinha
Department of Biotechnology,
School of Engineering and Technology,
Sharda University,
Greater Noida,
Uttar Pradesh,
India

Usha Seshachla
Department of Environmental Sciences,
Sri Krishna Degree College,
Bangalore,
Karnataka,
India

Mieke R. Setiawati
Department of Soil Sciences and Land Resources Management,
Faculty of Agriculture of Padjadjaran University,
Jatinangor,
West Java,
Indonesia

Anjney Sharma
ICAR-National Bureau of Agriculturally Important Microorgansims,
Mau,
Uttar Pradesh,
India

Shivesh Sharma
Department of Biotechnology,
Motilal Nehru National Institute of
Technology, Allahabad,
Uttar Pradesh,
India

Irina Sidorova
Lomonosov Moscow State University,
Moscow,
Russia

Tualar Simarmata
Department of Soil Sciences and Land
Resources Management,
Faculty of Agriculture of Padjadjaran
University,
Jatinangor, West Java,
Indonesia

Preeti Singh
Department of Botany and Microbiology,
HNB Garhwal University,
Srinagar Garhwal,
Uttarakhand,
India

Rahul Kunwar Singh
Department of Botany and Microbiology,
HNB Garhwal University,
Srinagar Garhwal,
Uttarakhand,
India

Alok Kumar Srivastava
ICAR-National Bureau of Agriculturally
Important Microorgansims,
Mau, Uttar Pradesh,
India

Madhumita Srivastava
Sunbeam College for Women,
Varanasi,
Uttar Pradesh,
India

Baby Summuna
Division of Plant Pathology,
Sher-e-Kashmir University of Agricultural
Sciences and Technology of Kashmir,
Jammu and Kashmir,
India

Padmavathi Tallapragada
Department of Microbiology,
School of Sciences,
Jain University,
Bangalore,
Karnataka,
India

Shree Prakash Tiwari
Department of Microbiology,
VBS Purvanchal University,
Jaunpur,
Uttar Pradesh,
India

Rishi Kumar Verma
Department of Biotechnology,
Motilal Nehru National Institute of
Technology, Allahabad,
Uttar Pradesh,
India

Elena Voronina
Lomonosov Moscow State University,
Moscow,
Russia

Yuan Wang
The Noble Research Institute,
Ardmore OK,
USA

Antoine Zboralski
Université de Moncton,
Moncton,
New Brunswick,
Canada

About the Editors

Dr. Alok Kumar Srivastava is a dedicated scientist in the area of molecular microbiology, presently working as Principal Scientist, ICAR-National Bureau of Agriculturally Important Microorganisms, India. He has made an outstanding and pioneering contribution in the area of Molecular Plant Pathology, contributed to structural, functional and comparative genomics of agriculturally important microorganisms, pathogens, development of molecular diagnostic tools and biological control of important diseases. He successfully sequenced the whole genome of 15 AIMs, deciphered microbial communities of Leh, mangrove soil of Andaman, landfill sites, and saline soils through metagenomics. Dr. Srivastava is Ph.D. from Banaras Hindu University, India, and completed his post doc at Otto Warburg Centre of Biotechnology with Prof Ilan Chet, at The Hebrew University of Jerusalem, Israel. He has also visited several countries including Hungary, France, The Netherlands, and Norway. He worked as a visiting research scientist in the Department of Plant Sciences, McGill University, Canada in the year 2010. He has 31 years of research experience in the area of biological control of fungal pathogens and plant growth-promoting rhizobacteria (PGPRs) and has supervised eight PhD students. He has successfully completed several externally funded research projects from DST, DBT and other agencies of the Indian Government. He has published about than 130 research papers in journals of international repute, several review articles, edited three books, and has more than 2473 citations to his credit (H index-25, I_{10} index 55). He is also associated with the capacity building program in the area of microbiology and has organized/participated in many training programs as a resource expert. Currently, he is associated with one of the mega network projects, "Application of Microorganisms in Agriculture and Allied Sectors (AMAAS)" of ICAR, India and is performing whole genome sequencing of important microorganisms. He is also providing molecular identification services for the microorganisms to be registered with the Central Insecticide Board and Registration Committee (CIB&RC) Govt. of India. His group is focusing on the management of abiotic stresses and soil health through *Bacillus* and other predominant genera as PGPR, its capabilities related to soil fertility and plant nutrition mobilization, the production of bacterial phytohormones and solubilization of mineral phosphates, allowing them to inhabit diverse niches in agro-ecosystems. For further details of this work visit https://scholar.google.co.in/citations?user=De5ciSsAAAAJ&hl=en

Dr. Prem Lal Kashyap had his agricultural education from Chaudhary Sarwan Kumar Himachal Pradesh Krishi Vishvavidyalaya (CSKHPKV), Palampur and Punjab Agricultural University (PAU), Ludhiana (Punjab). Presently, he is working as Scientist (Plant Pathology) at ICAR –Indian Institute of Wheat & Barley Research, Karnal and has nine years of research experience. He is a dedicated researcher in the area of plant pathology, and has made an outstanding and pioneering contribution in the area of phytopathogenomics, biomarker development, biocontrol and integrated disease management. He identified five new chemical molecules for the management of wheat diseases and pests; reported five new races of yellow rust of wheat and contributed in development of "GEHOON DOCTOR" mobile app. He decoded whole genome of several agriculturally important microorganisms, developed biomarkers for early and precise diagnosis of plant pathogens (*Alternaria, Fusarium, Colletotrichum, Tilletia indica, Urocystis agropyri, Puccinia triticina etc.*) and developed a formulation 'Biogrow' available for commercialization by AgriInovate (ICAR) India. He has three wheat varieties to his credit and has filed two patents. Dr. Kashyap has contributed immensely in resource generation through externally funded projects from DBT and other agencies of Indian Government. He also created a large pool of trained human resource in the area of plant pathology that has enabled successful product development following his approach. He has more than 100 publications to his credit, including 65 peer reviewed research papers, 3 books, 22 book chapters, 24 popular articles and 6 technical manuals. He has visited several countries including USA, Kenya, Mexico, Bangladesh and Bolivia for academic pursuits. For significant research contribution in the field of agricultural sciences, Dr. Kashyap has been recognized with several prestigious awards including NAAS Associateship-2019, NAAS Young Scientist Award, Dr. Basant Ram Young Scientist Award, Prof. Abrar Mustafa Memorial Award, M. K. Patel Memorial Young Scientist Award and Prof. Mahatim Singh Memorial Award.

Dr. Madhumita Srivastava has more than 11 years of teaching experience. She works in the capacity of Assistant Professor in Zoology, Sunbeam College for Women, Varanasi, Uttar Pradesh, India. She has completed a Doctoral Degree from VBS Purvanchal University India on the topic "Studies of cyanotoxins and its effect on reproductive physiology of *Cirrhinus mrigala* in Surha lake" and worked as Wetland Expert for two years in a World Bank funded project on Water restructuring and wetlands conservation for livelihood security in Utter Pradesh. Her main areas of interest are evolution and developmental biology, distribution, microbial ecology and environmental studies. She has published a few research papers and review chapters. She is an active faculty member, and has given an outstanding performance. She has been awarded as a *Teacher with Excellence* and in the recent past she convened two National Seminars on environment and climate change issues funded by scientific agencies of India.

1

Plant Microbiome

Past, Present and Future

Akhilendra Pratap Bharati[1], Ashutosh Kumar[2], Sunita Kumari[2], Anjney Sharma[1], Prem Lal Kashyap[3], Sudheer Kumar[3], Madhumita Srivastava[4], and Alok Kumar Srivastava[1]

[1] *ICAR-National Bureau of Agriculturally Important Microorgansims, Mau, Uttar Pradesh, India*
[2] *ICAR-Indian Institute of Seed Science, Mau, Uttar Pradesh, India*
[3] *ICAR-Indian Institute of Wheat and Barley Research (IIWBR), Karnal, Haryana, India*
[4] *Sunbeam College for Women, Varanasi, Uttar Pradesh, India*

1.1 Introduction

The evolution on earth have revealed the association of microbes to a plant and its specific tissue or organs (Compant et al. 2019; Hardoim et al. 2015; Reinhold-Hurek et al. 2015; Spinler et al. 2019; Uroz et al. 2019; Vorholt 2012). Plant microbiome encloses all associated plant microbes, whether phyllospheric, rhizospheric, or endospheric (all microbial genomes) (Brader et al. 2017; Lemanceau et al. 2017). Revealing of plant microbiome functionality has led to knowledge about plant microbe interactions, which are advantageous for plant growth and its production. The demographic, environmental, climatic, and man-made conditions have made crop production very challenging (Asl 2017; Kanianska 2016; Templeton and Scherr 1999). Microorganisms have been shown very advantageous in sustainable crop production. They have shown potential as biofertilizers, biopesticides, and growth enhancers (Bhardwaj et al. 2014; Cheng et al. 2017; Gopalakrishnan et al. 2015; Goswami et al. 2019; Kashyap et al. 2017b; Kushwaha et al. 2019a; Marrone 2019; Mendes et al. 2013; Mitter et al. 2016; Singh et al. 2019b). They have proven good alternatives compared with chemical products as excessive use of these can affect the harmony and microbiota of plant, which may lead to disintegration of soil fertility and quality (Lemaire et al. 2014; Marenya and Barrett 2009; Vorholt et al. 2017). There are number of inoculants proposed by the researchers, but with limited success in the field (Müller et al. 2016; Souza et al. 2015). Manipulation of the plant microbiome has the potential to improve crop production, reduce plant diseases (Andrews 1992; Bloemberg and Lugtenberg 2001; Hegazi et al. 2019; Singh et al. 2019a) and greenhouse gas emission and chemical inputs (Adesemoye et al. 2009; Cheng et al. 2017; Marrone 2019; Singh et al. 2010). In addition to this, it is important for nutrient cycling (the global biogeochemical cycle) also (Philippot et al. 2009).

The Plant Microbiome in Sustainable Agriculture, First Edition. Edited by Alok Kumar Srivastava, Prem Lal Kashyap and Madhumita Srivastava.

Almost all the plants host a microbial community, including bacteria, archaea, fungi, and blue green algae. The rhizospheric microbiome includes microbes associated with the roots along with the soil. The phyllospheric microbiome includes the plant arial surface microbes, whereas the endospheric microbiome include all microbes associated with the internal tissue termed endophytes. The rhizosphere is very rich in a soil-derived microbial community influenced by plant mucilage and root exudates (Kent and Triplett 2002). The phyllosphere is nutrient poor and subject to extreme temperature, radiation, and moisture (Kushwaha et al. 2019b; Vorholt 2012). The rhizospheric and phyllospheric microbes are closely associated with the plant surface and termed epiphytes, whereas the microbes associated with the internal organs and tissues are termed endophytes. The enrichment of microorganisms in the plant is not random, but a targeted process (Berg et al. 2017). The attraction of microbes to the root by the nutrients and the secondary metabolites has been studied. Chemoattractants and repellants have also been studied in the past few years (Feng et al. 2018; Oku et al. 2014; Pinedo et al. 2015). Although the structure of plant microbiomes is well studied, there are many knowledge gaps because of the plant species-specific components; targeted studies have been performed on crops and model plants such as *Arabidopsis*. The gaps are especially related to plants in natural ecosystems and their relationship to plant health.

So much application and the increasing complexity of the microbiomes have led to the development of many techniques in the field of identification, as well as taxonomy. For the identification of microbes, several classical methods had been used previously which included phenotypical as well as biochemical methods. These methods were only effective with the culturable microbes, so it was mandatory to develop the technology (Jesumirhewe et al. 2016). Nowadays, there are many more sequencing technologies which have been further improved. Besides 16S ribosomal RNA and internal transcribed spacer (ITS) sequencing, the use of MALDI-TOF-MS (matrix assisted laser desorption ionization-time of flight-mass spectroscopy) have proven to be useful in microbe identification (Adekambi and Drancourt 2004; Chen et al. 2000; Kashyap et al. 2016; Nouwens et al. 2000; Peng et al. 2005; Pieper et al. 2006; Rai et al. 2016; Sharma et al. 2015). Several polymerase chain reaction (PCR) based technologies such as repetitive PCR, amplified fragment length polymorphism (AFLP), random amplification of polymorphic DNA (RAPD), and multiplex PCR have been improved (Kashyap et al. 2016; Rai et al. 2015, 2016; Srivastava et al. 2014). The NGS (next generation sequencing) and multi-omics (genomics, metagenomics, transcriptomics, and proteomics) technologies allow much deeper insight into the structure of plant-associated microbial communities and its interaction with the ecosystem which support and often extends the current body of knowledge (Berg et al. 2015; Jansson and Baker 2016). In addition to this, these tools also reveal the functional dynamics and plant–microbe interaction as well as new PGP (plant growth promoting) traits.

In brief this chapter includes several approaches for studying the plant microbiome, the past and present tools for the identification of the microbiome and their advancement. In addition to these we have discussed the different microbes present in different plant parts. This chapter gives an overview of the application of the microbes in agriculture and the allied sectors. Figure 1.1 describes the pictorial representation of the structure of the plant microbiome, its biotic (plant microbe interaction) and abiotic factors which govern the structure and composition, its study using classical and modern tools and its application in the field of agriculture and other allied sectors.

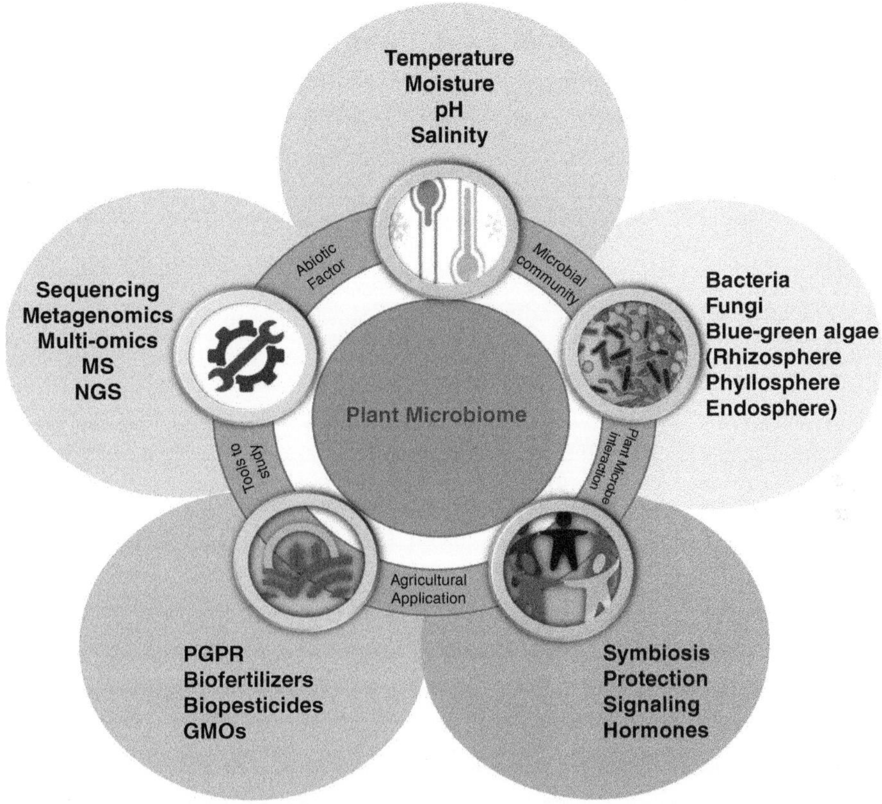

Figure 1.1 Plant microbiome structure, function, application, and their study using modern tools. The abiotic factors affect the microbiome which includes mainly bacteria, fungi, blue-green algae and several other microbes. They interact with the plant environment by means of different methods. Because of the interaction, they develop some characteristics which benefit each other. There are several tools to study all these components. The figure depicts the basics of plant microbiome components and their study. MS: mass-spectroscopy; NGS: next generation sequencing; GMO: genetically modified organism; PGPR: plant growth-promoting rhizobacteria.

1.2 Plant Microbiome

Microbes evolved on Earth approximately 3.5 billion years ago and eventually occupied every habitable environment in the planet's biosphere (Margulis 1981). Although microorganisms are known to be responsible for key functions on Earth, such as nutrient and biogeochemical cycling, and determining the health and disease state of the planet's plant and animal inhabitants, more than millions of microbes thought to exist have yet to be discovered. The plant microbiome includes only a fraction of the existing microbiome, which can be further divided on the basis of the localization and association to the plant part (Sanchez-Canizares et al. 2017). On the basis of localization, the microbiome may be divided into rhizospheric, phyllospheric, and endospheric. Figure 1.2 explains the components of the plant microbiome, and their interaction.

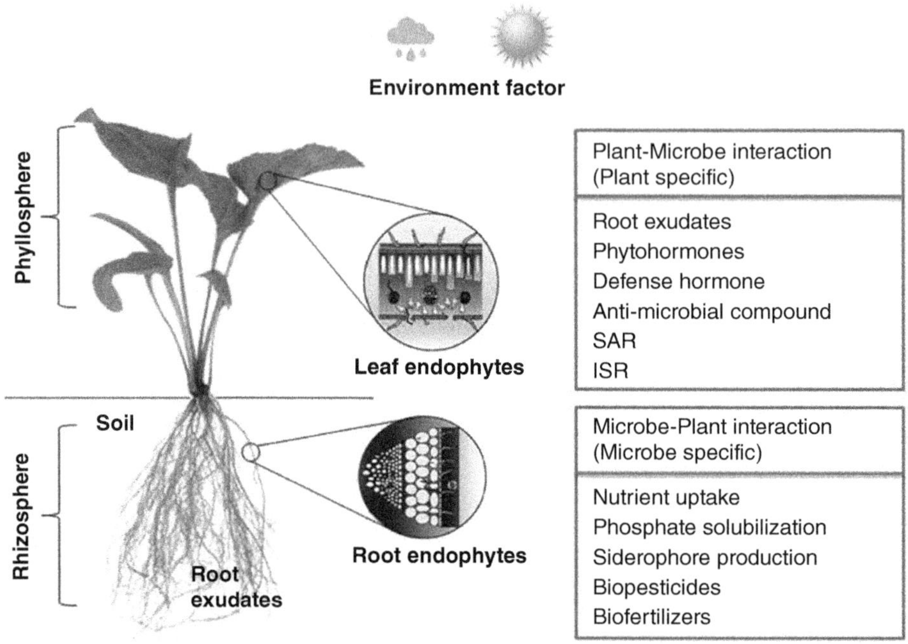

Figure 1.2 Components of the plant microbiome and its interrelations. SAR: systemic acquired resistance; ISR: Induced systemic resistance

1.2.1 Rhizospheric Microbiome

The rhizospheric microbiome is generally influenced by the deposition of mucilage secreted by root exudates and sloughed cells. The root exudates contain a number of organic acids, sugar, amino acids, fatty acids, vitamins, growth factors, hormones, and secondary metabolites. These compounds particularly decide the fate of the rhizospheric microbiome. The rhizospheric microbiome is of particular interest because of the plant growth promoting characteristics (Cai et al. 2017; Goswami et al. 2019; Guo et al. 2019; Pereira et al. 2019; Sharma et al. 2019; Singh et al. 2014; Solanki et al. 2012; Srivastava et al. 2013). They are generally described as plant growth promoting rhizobacteria (PGPR) and they act through a variety of mechanisms. They efficiently colonize the rhizosphere and stimulate plant growth through direct or indirect mechanisms (Lugtenberg and Kamilova 2009; Solanki et al. 2014, 2015). They possess general plant growth promoting properties such as mineralization and solubilization of phosphate, and siderephore formation, and they increase the bioavailability of inorganic phosphorus and iron (Cai et al. 2017; Guo et al. 2019; Hegazi et al. 2019; Pereira et al. 2019; Singh et al. 2019a). Because of these properties they are of interest to use as inoculants for plant growth promotion (Lugtenberg and Kamilova 2009). The rhizobacteria include the N_2 fixing free living *Azotobacter*, symbiont *Rhizobium* spp., *Bacillus* sp., *Pseudomonas* sp., etc.

1.2.2 Phyllospheric Microbiome

The phyllosphere is basically the aerial part of the plant which is poor in nutrients. There are a number of abiotic factors which influence the phyllospheric microbiome such as fluctuation in the temperature, moisture, radiation, wind, and precipitation; because of this it is much more dynamic compared with the rhizospheric microbiome (Kembel et al. 2014; Lindow 1996; Thapa and Prasanna 2018). Leaves also secrete some organic acids, sugar, and phytohormones through the stomata, hairs and veins which attract microbe for colonization on the leaf surface. The leaf surface colonizes $10^7/cm^2$ microbes (Lindow and Brandl 2003). The use of both traditional, culturing-based taxonomy and modern tools has illustrated that the diversity among bacterial members is mainly restricted to *Actinobacteria, Proteobacteria, Firmicutes, Bacteroidetes,* and less frequently, cyanobacteria; oomycetous communities are common inhabitants, besides fungi. The phyllospheric and rhizospheric microbes are also known as epiphytes.

1.2.3 Endospheric Microbiome

The microbiome inhabiting intra or inter-cellular plant tissues is defined as the endophytic microbiome and the microorganisms are known as endophytes. Some of the endophytes dedicate all their life cycle or part of the life cycle to the host plant. The term endophyte refers to a complex set of interactions, including the interactions between the different microbes that comprise the endophyte community and the host-plant defense mechanism to prevent these fungal species from becoming pathogens (Xu et al. 2016). They have the ability to colonize the inside of the tissue of different plant parts like healthy leaves, petioles, stems, twigs, bark, roots, fruit, flowers, and seeds without causing any harmful effect or infection in their host plant (Varma et al. 2012; Wassermann et al. 2019; White et al. 2019). Endophytic fungi belong to many different ecological and phylogenetic groups in a highly diverse environment. Endophytic microbes mostly belong to the *Ralstonia, Burkholderia, Pseudomonas, Staphylococcus, Mesorhizobium, Propionibacterium, Dyella, Bacillus,* and Ascomycetes; the rest are very rare (Kushwaha et al. 2019a, 2019b; Rosa et al. 2010). Some endophytes are listed in Table 1.1 along with their isolated plant parts. Apparently, more than one million different endophytic fungal strains inhabiting about 300,000 various plant species have been reported already (Wassermann et al. 2019; White et al. 2019). This amount of diversity is because of the presence of several endophytic fungi in an individual plant species, which can colonize the inside of the plant tissue. To mediate the plant–endophyte interaction, they produce a number of bioactive metabolites or secondary metabolites (Xu et al. 2016). These secondary metabolites have several applications in medicine, agriculture, and industry. In addition, these fungal endophytic metabolites have many roles in plant growth promotion by enhancing phosphate solubilization, siderophore production and other PGP traits, and they help in enhancing the production of the host plant (Taghavi et al. 2009; Varma et al. 2012; Wassermann et al. 2019; White et al. 2019). Many bioactive metabolites are originated from microbial organisms; fungi are the core important groups of eukaryotic organisms that have the wide capacity to produce numerous metabolites. Several bioactive compounds, including antifungal and antibacterial agents, have been isolated from fungi (Suryanarayanan et al. 2009).

Table 1.1 Some endophytic fungi isolated from different parts of the plants.

Host plants				
Common name	**Scientific names**	**Isolation parts**	**Frequent taxonomic groups**	**References**
Orange	*Citrus* spp.	Leaves and seeds	*Colletotrichum gloeosporioides*, *Guignardia citricarpa* and *Cladosporium* sp.	Zhang et al. (2013); Jalgaonwala et al. (2011)
Rice	*Oryza sativa* L.	Leaves, seeds, and roots	*Chaetomium globosum*, *Penicillium chrysogenum*, *Fusarium oxysporum* and *Cladosporium cladosporioides*	Wang et al. (2016); Walitang et al. (2017)
Sugarcane	*Saccharum* spp.	Leaves	Ascomycota phylum	Dong et al. (2018)
Wheat	*Triticum aestivum* L.	Leaves, stems, glumes, and grains	*Alternaria alternata*, *Cladosporium herbarum*, *Epicoccum nigrum*, *Cryptococcus* sp., *Rhodotorula rubra*, *Penicillium* sp., and *Fusarium graminearum*	Pagé et al. (2019), Conn and Franco (2004); Jalgaonwala et al. (2011)
Banana	*Musa acuminata Colla*	Leaves	*Xylaria* sp., *Colletotrichum musae* and *Cordana musae*	Gamez et al. (2019); Jalgaonwala et al. (2011)
Common bean	*Phaseolus vulgaris* L.	Leaves	*Colletotrichum*, *Hannaella*, *Cochliobolus*, and *Phomopsis*.	López-López et al. (2010)
Cowpea	*Vigna unguiculata* (L.) Walp.	Root nodules and seeds	*Aspergillus* spp., *Penicillium* spp. and *Fusarium* spp.	Leite et al. (2017)
Maize	*Zea mays* L.	Roots, leaves, and stems	*Alternaria alternata* and *Aureobasidium pullulans* var. *melanigerum*	Abedinzadeh et al. (2019); Jalgaonwala et al. (2011)

1.2.4 The Effect of Abiotic and Biotic Factors on the Plant Microbiome

The microbial composition associated with the plant may vary depending on several factors (Figure 1.2). The abiotic factors like temperature, moisture, soil pH, fertility, salinity, exudes, soil structure and soil organic matter affect the below ground microbiota (Fierer 2017). Whereas the environmental conditions, climate, pathogens, and human practices affect both the below and above ground microbiota (Hardoim et al. 2015; Hartmann et al. 2009). The plant species and genotype as well as the rhizodeposition and exudates also affect the recruitment of the microbes. It has been seen in many findings that the different plant microbiomes differ from one another, even though they are in the same soil conditions (Aleklett et al. 2015; Chaparro et al. 2014; Samad et al. 2017).

Beside these abiotic factors, the interaction between plants and microbes also influences the microbiome. This plant–microbe interaction is highly complex and dynamic. The plant defense system also affects the composition of the plant microbiome. It has been reported that a mutation in *Arabidopsis thaliana* deficient in the systemic acquired resistance (SAR) shows a difference in rhizospheric bacterial community as compared to the wild type plant (Doornbos et al. 2011; Hein et al. 2008). The salicylic acid (SA) mediated defense reduces the phyllospheric microbiome while the plant deficient in jasmonate mediated defense shows an increase in epiphytic diversity (Kniskern et al. 2007). Several other examples also show that the plant defense system also influence the microbe community, and SAR is responsible in controlling the bacterial population (Cheng et al. 2017; Egamberdieva et al. 2017). Some chemicals are produced by plant-like flavonoids and trigger diverse responses in rhizobia, and strigolactones induce hyphal branching in mycorrhizal fungi, etc. Plants also produce some antimicrobial compounds like phenolics, terpenoids, and alkaloids which also control the microbial population.

1.3 Approaches to Studying the Plant Microbiome

1.3.1 Classical Approaches

There are several approaches for studying the plant microbiome. The classical approaches based on culture isolation and purification of the microbes from the microbiome on different nutrient media and growth conditions depend on the target organism. Then these purified microbes are identified biochemically or phenotypically. This method has many drawbacks because only culturable microbes can be identified by this method and only 1% of the microbial population are culturable (Newcombe et al. 2018; Sarhan et al. 2019). The requirement of the pure culture misses the majority of the microbial population. To overcome this problem, several PCR-based sequencing technologies have been developed. For prokaryotic bacteria, 16S ribosomal gene sequencing is used, whereas for eukaryotic fungi the ITS and intraspecific ITS sequencing is utilized for identification (Chen et al. 2000; Ciardo et al. 2006; Nocker et al. 2007). To reduce the complexity in identification, several other targets have been chosen for the PCR amplification, for example *rpoB, gyrA, gyrB, sod*A, and *hsp* gene sequencing are utilized to differentiate the closely related species like *Mycobacterium chelonae* and *Mycobacterium abscessus*. The ITS regions also have some limitations, and because of that, the two variable domains D1, D2 near the 5′ end of the 28S ribosomal RNA genes, elongation factor α (e.g. *Fusarium* sp.), and β-tabulin (e.g. *Phaeoacremonium* sp.) are being used as alternative gene targets (Kashyap et al. 2017a; Stielow et al. 2015; Dagar et al. 2011).

1.3.2 Modern Approaches

Fingerprinting technology is also widely used in microbiome identification. Repetitive PCR, RAPD, and multiplex PCR are widely used methods (Cocconcelli et al. 1995; Kashyap et al. 2016; Lin et al. 1996; Rai et al. 2016). These methods utilize PCR along with the specific set of primers. The change in the sequence or the pattern can be utilized in identification. The restriction fragment length polymorphism (RFLP) is another method that utilizes restriction enzymes. These methods are used to take advantage of the DNA polymorphism in the related organism, on the basis of which the organism can be differentiated in the mixed samples (Cocconcelli et al. 1995; Lin et al. 1996; Versalovic et al. 1994). Microarray-based technology has also been developed for microbe identification which has several probes on the surface of the chip immobilized on the silica plate (Liu et al. 2001; Sharma et al. 2017; Wilson et al. 2002). Depending on the probe, it may be a protein or DNA microarray. Protein microarray utilizes specific antibody (protein) against the microbe while DNA microarray utilizes the specific marker DNA as a probe (Liu et al. 2001; Wilson et al. 2002).

Protein profiling is one of the recent methods for identification (Emerson et al. 2008). Microbes can be identified on the basis of the migration pattern of the protein on the gel which can be compared with the reference. The 2DE (two dimensional gel electrophoresis) is the fusion of the iso-electric focusing (IEF) and SDS-PAGE (sodium dodecyl sulfate-polyacrylamide gel electrophoresis), where proteins are first separated on the basis of the iso-electric point (charge) then on the basis of molecular weight. The 2DE map of different organisms can be stored and a database can be prepared which can be used as a reference (Malmstrom et al. 2002). The major drawback of this method is that it is a labor extensive and time consuming and requires ample amount of the protein. Proteome profiles of organisms can be compared with the existing databases and on the basis of the comparison the microorganism can be identified. This method is greatly enhanced by the introduction of MALDI-TOF-MS. Using this method the proteome of many microbes has been made available (Nouwens et al. 2000; Peng et al. 2005; Pieper et al. 2006).

Recently, metagenomics development has made the study of the microbiome an easy task. This has been possible because of the development of NGS technology and powerful databases. Metagenomics is the genome analysis of the population of microorganisms present in a microbiome. This doesn't include the culturing of the microbes and purification. The genomic DNA can be isolated directly from the rhizosphere or the plant part and a library can be made which can further be utilized for sequencing. Similarly, the transcriptomics and proteomics have also proven useful in microbiome studies (Handelsman 2005; Jiao et al. 2019; Loman et al. 2012; Petrosino et al. 2009; Stahl and Lundeberg 2012).

1.3.3 Plant Microbiome Studies: Limitations and Future Directions

Development of the single molecule sequencing and the MS (mass spectroscopy) technologies over the last two decades have made it possible to study microbiomes of a variety of complex ecosystems. In the beginning of the year 2000, Banfield and co-workers used the combination of the sequencing with MS approaches to study microbiomes of low microbial complexity (Ram et al. 2005). Subsequently, microbiomes of varying diversity and complexity have been investigated using either advance sequencing or MS or both (Wilmes et al. 2015). Because of the regularly improving technologies, the expectation of greater insights into the microbial communities of the previously studied communities has increased.

Knowledge of the molecular function of the microbiome is still a big challenge (Sergaki et al. 2018). The biggest obstacle is bioinformatic and computational analysis. Several other challenges also exist, such as the extraction of the biomolecules from the highly diverse and complex samples; assembly of the complete genome rather than sequence fragments directly from the complex ecosystem; high throughput MS technology for metaproteomics and metabolomics; the requirement of high speed computational systems for *de novo* assembly of large metagenomes and metatranscriptomes; sufficient storage; and the development of algorithms and mathematical tools for data interpretation to provide meaningful biological insights. We expect that positive initiatives will enable new technologies in the future that do not currently exist, such as high resolution ion mobility separation of peptides and metabolites; high speed and high storage systems; and improved databases and algorithms (Jansson and Baker 2016). These technologies with computational improvements will be vital for deciphering the role of microbes in the natural habitat and determining the role of the complex interplay between the members of microbial communities and ecosystems.

1.4 Microbiome and Agriculture in Past and Current Scenarios

In the last few decades, microbiome research has customized the insight on the complexity of microbial community structure. Presently, microbiome research is at the initial stages of starting to understand the set-up of microbial complex communities, the inter-genera and intra-generic dependencies, and extended to the biotic as well as abiotic factors (del Pilar Martínez-Diz et al. 2019; Rodriguez et al. 2019; Sanchez-Canizares et al. 2017). The enhancing demand for alternative experimental protocols, as well as the development of new tools has given a new dimension to the comprehension of the dynamics that exit within the microbiomes and their interaction with host organisms (Goodrich et al. 2017).

There have been numerous attempts to understand the complexity of plant microbiomes and soil (Cui et al. 2019; Goswami et al. 2019; Lundberg et al. 2012; Rodriguez et al. 2019), prompting new innovations for the next green revolution with sustainable crop production (Jez et al. 2016). In the last decade, a number of reports have been published (Figure 1.3) which show extensive research in the rhizospheric microbiome as well as the endospheric

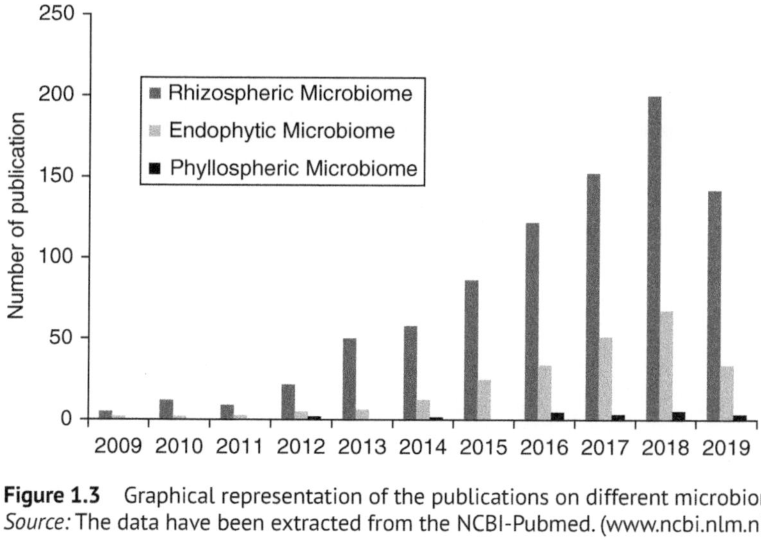

Figure 1.3 Graphical representation of the publications on different microbiomes in the last decade. *Source:* The data have been extracted from the NCBI-Pubmed. (www.ncbi.nlm.nih.gov/pubmed).

microbiome, but very little information about the phyllospheric microbiome. Comprehension of the extended potential of microbiomes, demands the innovation of new analytical strategies to gain higher crop production through the higher application of microbial communities with maximum efficiency (Bashiardes et al. 2018). The significance of maintaining a diverse microbiome and well-balanced, diversity in the rhizophere is important in crop production. Microbiome applications, however, have been implemented on farms with the aim of improving crop production via maintenance of soil health, fertility and nutrient availability to the plant through various solubilization mechanisms (Ab Rahman et al. 2018). Pertaining to this, the prime challenge is the delivery of laboratory generated innovations to the field and further, working out the structure of the rhizospheric microbiome (Schlaeppi and Bulgarelli 2015; Sergaki et al. 2018). It essentially involves correlating the dynamics of the microbial community with functioning of the microbiome (Sanchez-Canizares et al. 2017).

Plant disease control is important for the production of biomaterials, consumable resources and food. According to estimates, the most important international issue in recent times is that global food production must be increased by 70 percent until 2050, to meet the growing concern for global food security (Ingram 2011; Keinan and Clark 2012; Valdes 2019). Presently, the current food production system is responsible for the loss of 60 percent of global terrestrial biodiversity, along with 25 percent of global greenhouse gas emissions (Ab Rahman et al. 2018). The expanding world population demands an efficient management system and control of diseases in crop production, as crop protection plays a vital role in protecting crop productivity against competition from pathogens (Oerke and Dehne 2004).

Beneficial biocontrol of microbiomes is one of the few options that shows the ability to control disease, and can provide benefits by competing with pathogens or by directly antagonizing plant pathogens by release of antimicrobial chemical compounds (Chen et al. 2019; Choudhary et al. 2007; Kepler et al. 2017; Mansfield 2000). The localized infection with phyto-pathogens can lead to SAR, but the biopriming of beneficial non-pathogenic rhizobacteria to plants for induced systemic resistance (ISR) against pathogens is also practiced (Cheng et al. 2017; Egamberdieva et al. 2017). Further, biopriming offers a superior facilitation to plants that are capable of responding more quickly and strongly to pathogen attack. For instance, organic particles primed with *Gliocladium virens* (KA 2301) and *Trichoderma harzianum* (KA 159.2), were found efficient in inhibiting *Phytophthora cinnamomi* in avocado roots when used as surface mulch (Costa et al. 2000; Singh et al. 2019b). Similarly, disease severity and stem lesion length of *Phytophthora capsici* root and crown rot of bell pepper was noticeably reduced and further, the total microbial population and biocontrol activity was increased when delivering soil was mixed with compost holding chitosan, crab shell waste along with citrus pulp and cane molasses (del Carmen Orozco-Mosqueda et al. 2018; Kim et al. 1997).

1.4.1 Plant Growth Promoting Activities by Microbiome

The microbiome is capable of various functional traits like nitrogen fixation; solubilization; and mineralization of zinc (Zn), iron (Fe), potassium (K), phosphorus (P); and production of siderophores and hydrocyanic acid (HCN); and hormone production tagged microbiomes with the entitled plant growth promoting microorganisms (Khan et al. 2018; Meena et al. 2017; Singh et al. 2016). Reports pertaining to microbial metabolic and multi-functional traits have covered the protocols for developing microbiome inoculants for plant growth promotion (PGP), biological control of phyto-pests and pathogens, biodegradation of contaminated soil and degradation of agricultural wastes (Kumar et al. 2018; Singh et al. 2017). Hence, the soils with a high abundance of organisms like N-fixers, P-solubilizers, phytohormone producers and bioremediators are considered high quality soils, and support crop health. The ecofriendly microbiome strain application in agriculture minimized the use of chemical fertilizers. The application of microbiomes, either individually or in consortium, are rich in their specific functional traits for making soils fertile, productive, and sustainable (Bhardwaj et al. 2014; Hunter 2016). There are several successful instances of microbiome application in various crops. Application of plant growth promoting biofertilizer in paddy fields in Vietnam resulted in a significant reduction in usage of synthetic fertilizers by 52 percent (Nguyen et al. 2017). This reduction in fertilizers is not gained within one or two years, but the application of biofertilizers were continuously applied to the paddy fields for over 15 years, and co-inoculation of fertilizers and biofertilizers improved sweet potato yields in Uganda.

Inoculation of arbuscular mycorrhiza in sweet potato cultivar NASPOT 11 with NPK fertilizers enhanced production between 12.8 and 20.1 ton ha^{-1} compared to previous production of 4.5 ton ha^{-1} (Mukhongo et al. 2017). From numerous reports, it has been concluded that inoculation of host-specific microbiomes especially endophytic and rhizospheric microflora, promotes growth and results in a higher yield (Thokchom et al. 2017).

Foeniculum vulgare shows a significant improvement in growth and essential oil content, when treated with bio-fertilizers (El-Azim et al. 2017). The results have revealed and supported that microbial input not only increases plant growth, but is also capable of bringing qualitative and quantitative changes. Application of microbial inoculations having plant growth promoting traits in agriculture have a significant value and do not add greenhouse gas in the environment (Sabir et al. 2012; Sharma and Sharma 2017; Singh et al. 2019b). An experiment was conducted with the objective of assessing the impact of bio-fertilizers on growth of paddy fields and greenhouse gas emission in alluvial soils of Indonesia. The results showed that emission of greenhouse gases such as nitrous oxide (N_2O), methane (CH_4), and carbon dioxide (CO_2) was significantly condensed with inoculation of biofertilizer in the paddy plot and further, it was concluded that biofertilizers improved the paddy production. Simultaneously, chemical fertilizer input was also reduced with 75 percent of the recommended dose, along with a decrease of greenhouse gas emissions (Hadi and Nur 2017).

The application of *Trichoderma*-based bio-fertilizer increased plant growth and productivity up to 12.9 percent. Further, mineral content and antioxidant level of ascorbic acid, β-carotene and lycopene in plants and fruits was increased. Asl (2017) has reported that using nitrogen fixing and P-solubilizing biofertilizers in sesame improved growth promotion, harvest index and nutrient use efficiency. The discussed instances corroborate that the application of specific microbial formulation functions have a high degree of ability to elevate the growth promotion and nutritional health of crop plants.

1.4.2 Biopesticide Potential of the Microbiome

Among the available options, biological control of plant pests and phyto-pathogens appears to be the best options for low cost, eco-friendly, and sustainable management approaches for protecting crops. Nowadays, microbial biocontrol is accepted as an important tool for the control of diseases in plants, and this method is sustainable in agriculture practices (Al-Ani and Albaayit 2018; Azcón-Aguilar and Barea 1997; Mona et al. 2017). A large number of biological controls are reported as options, and a deep comprehension of the complex interaction between plants, the environment, and pathogens is essential for future exploration (Mirzaee et al. 2015).

In terms of plant pathology, biocontrol has been accepted as the co-relationship of various environmental factors used to weaken the non-favorable effects of factors detrimental to the growth of organisms, and increase the growth of useful organisms like crops, beneficial insects, and microorganisms (Al-Ani and Albaayit 2018; Pal and Gardener 2006). The use of natural products and synthetic compounds extracted from plant or modified organisms or gene products are included under biological control (Levy et al. 2018; Pal and Gardener 2006). The prime goal of performing research on biocontrol is to minimize the dependence of agrochemical application and reduce the risks to human health and the environment (Berg 2009). Biocontrol through microbes was developed in the last 30–40 years, when plant pathology research attracted more interest in the use of beneficial microorganisms for the management of plant diseases (Berg 2009; Nihorimbere et al. 2011). The various modes of interactions between the populations are categorized as mutualism, neutralism, protocooperation, predation, commensalism, competition, parasitism, and amensalism (Compant et al. 2005; Kiely et al. 2006). All these biological controlling mechanism/interactions between plants and microbes takes place naturally at the microscopic level (Berg 2009). During the course of the plant life cycle, plants are susceptible to the various environmental challenges of biotic (fungi, oomycetes, nematodes, bacteria and viruses etc.) and abiotic stresses (drought, cold, salinity, flooding etc.). Counter to the pathogenic attack, plants have developed a wide range of mechanisms to counter attack and ward off attackers (Ponce de León and Montesano 2013). Broadly, plant protection mechanisms can be divided into passive defenses; non-host resistance (NHR); physical and chemical barriers (PCB); rapid active defenses (RAD); and delayed active defenses (DAD).

RAD includes alteration in membrane function, initial oxidative damage, cell wall reinforcement, hypersensitive response (HR) and finally in programmed cell death (PCD) (Pieterse et al. 2014), while DAD involves pathogen containment and wound repair, pathogenesis-related (PR) gene expression and SAR (Reichling 2018). Plant protecting signaling molecules includes salicylic acid (SA), and it is accepted as a necessary molecule for defense against biotrophic pathogens and SAR, while jasmonic acid (JA) and ethylene (ET) are involved in defense against necrotrophic pathogens or in beneficial plant microbe interactions. Stimulation and activation of JA and ET are triggers for action through biopriming and ISR (Agrios 2005). The SAR and ISR are higher states of alertness in the plant that have the potential for a quick and stronger response against pathogen attack.

1.4.3 Correlation of Microbiomes and Genetic Engineering as GMO

The World Health Organization (WHO) defined genetic modification as biological practice that manipulates the genetic material of all types of living organisms through the use of rDNA technology (Desquilbet and Bullock 2009). Similarly, the Food and Agriculture Organization (FAO) and the European Commission elaborates genetically modified organisms (GMO) as a product "which does not exist naturally" as happens through the mating and/or natural recombination. The food industry defines GM foods as foods produced from genetically modified microbes, which are responsible for the increment of nutritional values, alter the taste and texture, increases yields, shelf-life or reduces maturation time (Paarlberg 2010). Some of the GM foods that are in reach of common people are soy, milk, cheeses, yogurt, food additives, amino acids, vitamins, wine, and valuable nutrition foods.

Microorganisms not only add to the quality of foods, but also enhance the bioavailability of nutrients, improve food texture, give aroma to the product, and possibly deliver certain inhibitory compounds that resist food deterioration and resulting in the capacity and security of the food product (Tamang et al. 2016). The potential of microorganisms to grow on a wide range of raw substrates and ferment them into desirable products with biological activity and enhanced bio-functionality of the food products, ensures the food security. Different biological activities of microbes used in the fermented foods include probiotic properties, antimicrobial properties, antioxidant effects, peptide production, polyglutamic acid, degradation of anti-nutritive compounds and fibrinolytic activity (Lucht 2015). The functional properties of microorganisms in different fermented food products have been considered to impart health benefits to consumers.

Bacillus thuringiensis (Bt) bacterium produces unique crystal proteins encoded by *cry* genes that have insecticidal effects against different insect species, but are not harmful to mammals, birds, or fish (Wu et al. 2008). The crystal proteins form a complex to receptors on the intestinal lining of susceptible insects, causing cell lysis. Because of these unique features, *Bt* has been used as a safe alternative to chemical pesticides for several decades, and a successful example is genetically modified Bt cotton (Huang et al. 2009; Kathage and Qaim 2012).

Crown gall disease in a wide range of broad leaf plants is caused by *Agrobacterium tumefaciens*, transferring part of its DNA (T-DNA) into the plant cell through tumor-inducing (Ti) plasmids (Hoekema et al. 1983). Once the T–DNA is integrated into the plant host's chromosome, the genes on Ti-DNA are expressed, and the consequence is overexpression of plant growth hormones and opines (Carlier et al. 2003). Overexpression of plant growth hormones causes uncontrolled growth, and opines serve as nutrients for the bacterium. *Agrobacterium radiobacter* K84 produces a bacteriocin, agrocin 84, to which pathogenic *A. tumefaciens* strains are susceptible (Vicedo et al. 1993) and K84 competes for the nutrients on which the pathogen survive. Hence, *A. radiobacter* K84 became the first commercial biological control agent against crown gall disease, and is currently also exploited as a tool for genetic manipulation in plants (Gleba et al. 2013).

1.4.4 Microbiome Directed Mitigation in Abiotic Stresses

Interactions of microbes with plants are important for adaptation and survival of both living entities in abiotic environments. Microbe-mediated induction of abiotic stress responses are achieved by induced systemic tolerance (IST). In the past few decades, the role of microorganisms for alleviation of abiotic stress in plants has attracted great attention

(Kashyap et al. 2018; Souza et al. 2015; de Zelicourt et al. 2013). Microbial metabolism and genetic capabilities have been extended to alleviate abiotic stress in plants (Gopalakrishnan et al. 2015) and some of the compounds or exudates performing various beneficial functions are listed in Table 1.2. The rhizospheric microbes of different genera such as *Pseudomonas* (Sorty et al. 2016), *Azotobacter* (Sahoo et al. 2014), *Azospirillum* (Omar et al. 2009), *Pantoea, Rhizobium, Bacillus, Enterobacter* (Nadeem et al. 2007; Sorty et al. 2016; Tiwari et al. 2011), *Bradyrhizobium* (Panlada et al. 2013), *Methylobacterium* (Meena et al. 2012), *Burkholderia* (Oliveira et al. 2009), *Trichoderma* (Ahmad et al. 2015) and cyanobacteria (Singh et al. 2014, 2019a) have the potential for plant growth promotion and are capable of mitigation of different kinds of abiotic stress. Pandey et al. (2016) have reported the potential of *T. harzianum* on stress mitigation in paddy genotypes through aquaporin, dehydrin, and malondialdehyde gene upregulation, along with support of important physiological parameters. Desiccation endurance and resilience governed by rhizobacteria control changes in phytohormone levels, defense-related enzymes and proteins, antioxidants and epoxypolysaccharides, which are known for microbe-derived plant responses. Such mechanisms enable plants to withstand abiotic stresses efficiently (Kaushal and Wani 2016).

Research pertaining to plant stress and mitigation of plant stress through the application of microorganisms includes selection, screening, and standardization of delivery mechanisms of stress tolerant microorganisms, and some related reports have been tabulated in Table 1.3. Hence, use of microbes with stress relieving properties for agriculture productivity in stress prone regions (Ahmad et al. (2015) have reported that *T. harzianum* significantly

Table 1.2 Bioactive signaling for beneficial plant–microbe interactions.

Exudate/compound	Beneficial microbe attracted	Function/benefit to plant	References
Flavonoids	*Rhizobia*	N-fixation to legumes and other plants when free-living; leads to increased yields	Fierer et al. (2007); Cooper (2007); Poole et al. (2018);
Jasmonic acid	*Bacillus thuringiensis, Bacillus cereus, Planococcaceae,* *Paenibacillus amylolyticus*	Antifungal, antibacterial; antioomycete; insecticidal; assist in defense against root pathogens and herbivorous insects; reduced yield losses	Carvalhais et al. (2013); Xie et al. (2018)
Arabinose, glucose, fructose, ribose, inositol Strigolactone	Fungus and bacteria	P- solubilization; leads to increased yields, defense; leads to increased yields	Carvalhais et al. (2011); Reischke et al. (2014)
Malate, succinate, fumarate	*Mycorrhizae, Burkholderia, Frateuria, Sphingomonas,* *Legionella, Talaromyces, Peziza Pseudomonas fluorescens*	P- solubilization, water supply, defense; leads to increased yields and resilience against abiotic and biotic stresses. Suppresses pathogens; reduced yield losses from pathogen attacks	Oku et al. (2014); Zhao et al. (2018)

Table 1.3 Microbial mitigation of abiotic stress tolerance in plants.

Abiotic stress	Microbe inoculation	Plant	Tolerance strategy	References
Salt	*Pseudomonas simiae*	*Glycine max*	4-nitroguaiacol and quinoline promote soybean seed germination	Vaishnav et al. (2016)
Salt	Root-associated plant growth-promoting rhizobacteria	*Oryza sativa*	Expression of salt stress-related *RAB18* plant gene	Jha et al. (2014)
Salt	Cyanobacteria and cyanobacterial extracts	*Oryza sativa, Triticum aestivum, Zea mays, Gossypium hirsutum*	Phytohormones as elicitor molecule	Singh (2014)
Salt	*Pseudomonas koreensis* strain AK-1	*Glycine max* L. Merrill	Reduction in Na^+ level and increase in K^+ level	Kasotia et al. (2015)
Osmotic stress	*Bacillus megaterium*	*Zea mays*	High hydraulic conductance, increased root expression of two ZmPIP isoforms	Marulanda et al. (2010)
Salinity	*Azospirillum brasilense* and *Pantoea dispersa*	*Capsicum annuum*	High stomatal conductance and photosynthesis	del Amor and Cuadra-Crespo (2012)
Salinity	*Pseudomonas putida* Rs-198	*Gossypium hirsutum*	Prevented salinity-induced abscisic acid (ABA) accumulation in seedlings	Yao et al. (2011)
Drought	*Burkholderia phytofirmans* *Enterobacter* sp. FD17	*Zea mays*	Increased photosynthesis, root and shoot biomass under drought conditions	Naveed et al. (2014)
Drought	*Bacillus thuringiensis* AZP2	*Triticum aestivum*	Production of volatile organic compounds	Timmusk et al. (2014)
Drought	*Bacillus licheformis* strain K11	*Capsicum annum*	Stress related genes and proteins	Lim and Kim (2013)

increased the oil content in NaCl affected *Brassica juncea*, as *T. harzianum* application had enhanced the uptake of required nutrients, increased accumulation of osmolytes and antioxidants, simultaneously decreasing Na^+ uptake and upregulating enzyme monodehydroascorbate reductase. Further, it was confirmed by mutational investigation that *Trichoderma* improves salinity stress by secreting ACC-deaminase (Brotman et al. 2013).

Pseudomonas sp. and *Acinetobacter* sp. in barley and oats were found to increase production of indole acetic acid and ACC deaminase in the salt affected rhizosphere (Chang et al. 2014). Palaniyandi et al. (2014) reported mitigation of salt stress by *Streptomyces* sp. strain PGPA39 and a significantly enhanced growth promotion in tomato plants. *Burkholderia phytofirmans* strain relieved drought stress in maize (Naveed et al. 2014), wheat (Naveed et al. 2014) and salt stress in *Arabidopsis* (Egamberdieva et al. 2017; Pinedo et al. 2015).

The rhizosphere constitutes a microenvironment near the root region where an average count of microbes is comparatively higher than the rest of the soil, and it is obvious that plant roots exposed to a diversity of nutrients, minerals, and metabolites would fare better than in a low level of rhizospheric microbes (Al-Ani and Albaayit 2018; Mona et al. 2017). The root exudates secreted by plants is an important factor for microbial colonization within the rhizosphere. Microbial chemotactic displacement toward the root exudates play a vital role in attracting microbial communities for root colonization in the rhizospheric region.

1.5 Conclusion

The plant microbiome can be considered as an extension of the host genome. The microbiome associated with the different parts of the plant is different in its composition, structure, and function. The rhizospheric microbiome is different from that of the phyllospheric and the endospheric microbiomes. The rhizospheric microbiomes of the plant may vary even when they are grown in the same soil. It may vary from the species to species as each plant species secretes different chemoattractants and repellants, hormones, alkaloids, terpens and several secondary metabolites which decide the microbiota. It depends on the interaction of the plant and the microbes. Not only the rhizospheric but the phyllospheric and endospheric microbiomes are also very specific and vary from species to species. Besides the biotic factors, abiotic factors also decide the structure of the microbiome which includes several factors such as pH, temperature, salinity of soil, moisture, climate, precipitation, etc. The study of the plant microbiome has increased knowledge about plant–microbe interactions and how shifts in biodiversity impact functions in the context of ecosystem services (plant productivity, biogeochemical cycles, etc.). The study of the plant microbiome increases understanding of the regulation of such complex communication pathways within the plant microbiome. They possess many plant growth promoting properties which have proven their usefulness in agriculture and use as inoculants their utility other.

Conflicting Interests

The authors declare that they have no conflicting interests.

Acknowledgments

We thank the Indian Council of Agricultural Research (ICAR) for financial support.

References

Ab Rahman, S.F.S., Singh, E., Pieterse, C.M., and Schenk, P.M. (2018). Emerging microbial biocontrol strategies for plant pathogens. *Plant Science* 267: 102–111.

Abedinzadeh, M., Etesami, H., and Alikhani, H.A. (2019). Characterization of rhizosphere and endophytic bacteria from roots of maize (*Zea mays* L.) plant irrigated with wastewater with biotechnological potential in agriculture. *Biotechnology Reports* 21: e00305.

Adekambi, T. and Drancourt, M. (2004). Dissection of phylogenetic relationships among 19 rapidly growing Mycobacterium species by 16S rRNA, hsp65, sodA, recA and rpoB gene sequencing. *International Journal of Systematic and Evolutionary Microbiology* 54 (6): 2095–2105.

Adesemoye, A.O., Torbert, H.A., and Kloepper, J.W. (2009). Plant growth-promoting rhizobacteria allow reduced application rates of chemical fertilizers. *Microbial Ecology* 58 (4): 921–929.

Agrios, G.N. (2005). *Plant Pathology*, 5e. United States of America: Department of Plant Pathology. University of Florida.

Ahmad, P., Hashem, A., Abd-Allah, E.F. et al. (2015). Role of *Trichoderma harzianum* in mitigating NaCl stress in Indian mustard (*Brassica juncea* L) through antioxidative defense system. *Frontiers in Plant Science* 6: 868.

Al-Ani, L.K.T. and Albaayit, S.F.A. (2018). Antagonistic of some Trichoderma against *Fusarium oxysporum* sp. f. *cubense* tropical race 4 (FocTR4). *The Eurasia Proceedings of Science, Technology, Engineering & Mathematics* 2: 35–38.

Aleklett, K., Leff, J.W., Fierer, N., and Hart, M. (2015). Wild plant species growing closely connected in a subalpine meadow host distinct root-associated bacterial communities. *Peer J* 3: e804.

del Amor, F.M. and Cuadra-Crespo, P. (2012). Plant growth-promoting bacteria as a tool to improve salinity tolerance in sweet pepper. *Functional Plant Biology* 39 (1): 82–90.

Andrews, J.H. (1992). Biological control in the phyllosphere. *Annual Review of Phytopathology* 30 (1): 603–635.

Asl, A.N. (2017). Effects of nitrogen and phosphate biofertilizers on morphological and agronomic characteristics of sesame (*Sesamum indicum* L.). *Open Journal of Ecology* 7 (02): 101.

Azcón-Aguilar, C. and Barea, J.M. (1997). Arbuscular mycorrhizas and biological control of soil-borne plant pathogens–an overview of the mechanisms involved. *Mycorrhiza* 6 (6): 457–464.

Bashiardes, S., Godneva, A., Elinav, E., and Segal, E. (2018). Towards utilization of the human genome and microbiome for personalized nutrition. *Current Opinion in Biotechnology* 51: 57–63.

Berg, G. (2009). Plant–microbe interactions promoting plant growth and health: perspectives for controlled use of microorganisms in agriculture. *Applied Microbiology and Biotechnology* 84 (1): 11–18.

Berg, G., Rybakova, D., Grube, M., and Köberl, M. (2015). The plant microbiome explored: implications for experimental botany. *Journal of Experimental Botany* 67 (4): 995–1002.

Berg, G., Köberl, M., Rybakova, D. et al. (2017). Plant microbial diversity is suggested as the key to future biocontrol and health trends. *FEMS Microbiology Ecology* 93 (5): fix050. https://doi.org/10.1093/femsec/fix050.

Bhardwaj, D., Ansari, M.W., Sahoo, R.K., and Tuteja, N. (2014). Biofertilizers function as key player in sustainable agriculture by improving soil fertility, plant tolerance and crop productivity. *Microbial Cell Factories* 13 (1): 66.

Bloemberg, G.V. and Lugtenberg, B.J. (2001). Molecular basis of plant growth promotion and biocontrol by rhizobacteria. *Current Opinion in Plant Biology* 4 (4): 343–350.

Brader, G., Compant, S., Vescio, K. et al. (2017). Ecology and genomic insights into plant-pathogenic and plant-nonpathogenic endophytes. *Annual Review of Phytopathology* 55: 61–83.

Brotman, Y., Landau, U., Cuadros-Inostroza, A. et al. (2013). *Trichoderma*-plant root colonization: escaping early plant defense responses and activation of the antioxidant machinery for saline stress tolerance. *PLoS Pathogens* 9 (3): e1003221.

Cai, F., Pang, G., Miao, Y. et al. (2017). The nutrient preference of plants influences their rhizosphere microbiome. *Applied Soil Ecology* 110: 146–150.

Carlier, A., Uroz, S., Smadja, B. et al. (2003). The Ti plasmid of *Agrobacterium tumefaciens* harbors an attM-paralogous gene, aiiB, also encoding N-acyl homoserine lactonase activity. *Applied and Environmental Microbiology* 69 (8): 4989–4993.

del Carmen Orozco-Mosqueda, M., del Carmen Rocha-Granados, M., Glick, B.R., and Santoyo, G. (2018). Microbiome engineering to improve biocontrol and plant growth-promoting mechanisms. *Microbiological Research* 208: 25–31.

Carvalhais, L.C., Dennis, P.G., Fedoseyenko, D. et al. (2011). Root exudation of sugars, amino acids, and organic acids by maize as affected by nitrogen, phosphorus, potassium, and iron deficiency. *Journal of Plant Nutrition and Soil Science* 174 (1): 3–11.

Carvalhais, L.C., Dennis, P.G., Badri, D.V. et al. (2013). Activation of the jasmonic acid plant defence pathway alters the composition of rhizosphere bacterial communities. *PLoS One* 8 (2): e56457.

Chang, P., Gerhardt, K.E., Huang, X.D. et al. (2014). Plant growth-promoting bacteria facilitate the growth of barley and oats in salt-impacted soil: implications for phytoremediation of saline soils. *International Journal of Phytoremediation* 16 (11): 1133–1147.

Chaparro, J.M., Badri, D.V., and Vivanco, J.M. (2014). Rhizosphere microbiome assemblage is affected by plant development. *International Society for Microbial Ecology Journal* 8 (4): 790.

Chen, Y.C., Eisner, J.D., Kattar, M.M. et al. (2000). Identification of medically important yeasts using PCR-based detection of DNA sequence polymorphisms in the internal transcribed spacer 2 region of the rRNA genes. *Journal of Clinical Microbiology* 38 (6): 2302–2310.

Chen, Q.L., Cui, H.L., Su, J.Q. et al. (2019). Antibiotic resistomes in plant microbiomes. *Trends in Plant Science* 24 (6): 530–541.

Cheng, X., van de Mortel, J.E., Etalo, D.W. et al. (2017). Genome-wide analysis of bacterial determinants of plant growth promotion and induced systemic resistance by *Pseudomonas fluorescens*. *Environmental Microbiology* 19 (11): 4638–4656. https://doi.org/10.1111/1462-2920.13927.

Choudhary, D.K., Prakash, A., and Johri, B.N. (2007). Induced systemic resistance (ISR) in plants: mechanism of action. *Indian Journal of Microbiology* 47 (4): 289–297.

Ciardo, D.E., Schär, G., Böttger, E.C. et al. (2006). Internal transcribed spacer sequencing versus biochemical profiling for identification of medically important yeasts. *Journal of Clinical Microbiology* 44 (1): 77–84.

Cocconcelli, P.S., Porro, D., Galandini, S., and Senini, L. (1995). Development of RAPD protocol for typing of strains of lactic acid bacteria and enterococci. *Letters in Applied Microbiology* 21 (6): 376–379.

Compant, S., Duffy, B., Nowak, J. et al. (2005). Use of plant growth-promoting bacteria for biocontrol of plant diseases: principles, mechanisms of action, and future prospects. *Applied and Environmental Microbiology* 71 (9): 4951–4959.

Compant, S., Samad, A., Faist, H., and Sessitsch, A. (2019). A review on the plant microbiome: ecology, functions and emerging trends in microbial application. *Journal of Advanced Research* 19: 29–37.

Conn, V.M. and Franco, C.M. (2004). Analysis of the endophytic actinobacterial population in the roots of wheat (*Triticum aestivum* L.) by terminal restriction fragment length polymorphism and sequencing of 16S rRNA clones. *Applied and Environmental Microbiology* 70 (3): 1787–1794.

Cooper, J.E. (2007). Early interactions between legumes and rhizobia: disclosing complexity in a molecular dialogue. *Journal of Applied Microbiology* 103 (5): 1355–1365.

Costa, J.L.D.S., Menge, J.A., and Casale, W.L. (2000). Biological control of *Phytophthora* root rot of avocado with microorganisms grown in organic mulches. *Brazilian Journal of Microbiology* 31 (4): 239–246.

Cui, E., Fan, X., Li, Z. et al. (2019). Variations in soil and plant-microbiome composition with different quality irrigation waters and biochar supplementation. *Applied Soil Ecology* 142: 99–109.

Dagar, S.S., Kumar, S., Mudgil, P. et al. (2011). D1/D2 domain of large-subunit ribosomal DNA for differentiation of *Orpinomyces* spp. *Applied and Environmental Microbiology* 77 (18): 6722–6725.

Desquilbet, M. and Bullock, D.S. (2009). Who pays the costs of non-GMO segregation and identity preservation? *American Journal of Agricultural Economics* 91 (3): 656–672.

Dong, M., Yang, Z., Cheng, G. et al. (2018). Diversity of the bacterial microbiome in the roots of four *Saccharum* species: *S. spontaneum, S. robustum, S. barberi,* and *S. officinarum. Frontiers in Microbiology* 9: 267.

Doornbos, R.F., Geraats, B.P., Kuramae, E.E. et al. (2011). Effects of jasmonic acid, ethylene, and salicylic acid signaling on the rhizosphere bacterial community of *Arabidopsis thaliana. Molecular Plant-Microbe Interactions* 24 (4): 395–407.

Egamberdieva, D., Reckling, M., and Wirth, S. (2017). Biochar-based *Bradyrhizobium* inoculum improves growth of lupin (*Lupinus angustifolius* L.) under drought stress. *European Journal of Soil Biology* 78: 38–42.

El-Azim, A., Rania, M.R., and Badawy, M.Y.M. (2017). Effect of bio-fertilization and different licorice extracts on growth and productivity of *Foeniculum vulgare*, Mill. plant. *Middle East Journal of Agriculture* 6 (1): 1–12.

Emerson, D., Agulto, L., Liu, H., and Liu, L. (2008). Identifying and characterizing bacteria in an era of genomics and proteomics. *Bioscience* 58 (10): 925–936.

Feng, H., Zhang, N., Du, W. et al. (2018). Identification of chemotaxis 1 compounds in root exudates and their sensing chemoreceptors in plant-growth-promoting rhizobacteria *Bacillus amyloliquefaciens* SQR9. *Molecular Plant-Microbe Interactions* 31 (10): 995–1005.

Fierer, N. (2017). Embracing the unknown: disentangling the complexities of the soil microbiome. *Nature Reviews Microbiology* 15 (10): 579.

Fierer, N., Bradford, M.A., and Jackson, R.B. (2007). Toward an ecological classification of soil bacteria. *Ecology* 88 (6): 1354–1364.

Gamez, R.M., Rodríguez, F., Vidal, N.M. et al. (2019). Banana (*Musa acuminata*) transcriptome profiling in response to rhizobacteria: *Bacillus amyloliquefaciens* Bs006 and *Pseudomonas fluorescens* Ps006. *BMC Genomics* 20 (1): 378.

Gleba, Y.Y., Tusé, D., and Giritch, A. (2013). Plant viral vectors for delivery by agrobacterium. In: *Plant Viral Vectors*, Current Topics in Microbiology and Immunology, vol. 375 (eds. K. Palmer and Y. Gleba), 155–192. Berlin, Heidelberg: Springer.

Goodrich, J.K., Davenport, E.R., Clark, A.G., and Ley, R.E. (2017). The relationship between the human genome and microbiome comes into view. *Annual Review of Genetics* 51: 413–433.

Gopalakrishnan, S., Sathya, A., Vijayabharathi, R. et al. (2015). Plant growth promoting rhizobia: challenges and opportunities. *3 Biotech* 5 (4): 355–377.

Goswami, S.K., Kashyap, P.L., and Awasthi, S. (2019). Deciphering rhizosphere microbiome for the development of novel bacterial consortium and its evaluation for salt stress management in solanaceous crops in India. *Indian Phytopathology* 72 (3), 479–488. https://doi.org/10.1007/s42360-019-00174-1.

Guo, J.H., Zhang, L., Wang, D. et al. (2019). Consortium of plant growth-promoting Rhizobacteria strains suppresses sweet pepper disease by altering the Rhizosphere microbiota. *Frontiers in Microbiology* 10: 1668.

Hadi, A. and Nur, H.S. (2017). The use of biofertilizers increased plant growth with no trade-off effect on GreenHouse gas emission. *Journal of Wetlands Environmental Management* 5 (1): 18–23.

Handelsman, J. (2005). Metagenomics: application of genomics to uncultured microorganisms. *Microbiology and Molecular Biology Reviews* 69 (1): 195–195.

Hardoim, P.R., Van Overbeek, L.S., Berg, G. et al. (2015). The hidden world within plants: ecological and evolutionary considerations for defining functioning of microbial endophytes. *Microbiology and Molecular Biology Reviews* 79 (3): 293–320.

Hartmann, A., Schmid, M., Van Tuinen, D., and Berg, G. (2009). Plant-driven selection of microbes. *Plant and Soil* 321 (1–2): 235–257.

Hegazi, N., Hartmann, A., and Ruppel, S. (2019). The plant microbiome: exploration of plant-microbe interactions for improving agricultural productivity. *Journal of Biotechnology* 235: 171–180.

Hein, J.W., Wolfe, G.V., and Blee, K.A. (2008). Comparison of rhizosphere bacterial communities in *Arabidopsis thaliana* mutants for systemic acquired resistance. *Microbial Ecology* 55 (2): 333–343.

Hoekema, A., Hirsch, P.R., Hooykaas, P.J., and Schilperoort, R.A. (1983). A binary plant vector strategy based on separation of vir-and T-region of the *Agrobacterium tumefaciens* Ti-plasmid. *Nature* 303 (5913): 179.

Huang, D.W., Sherman, B.T., and Lempicki, R.A. (2009). Systematic and integrative analysis of large gene lists using DAVID bioinformatics resources. *Nature Protocols* 4 (1): 44.

Hunter, P. (2016). Plant microbiomes and sustainable agriculture. *EMBO Reports* 17 (12): 1696–1699.

Ingram, J. (2011). A food systems approach to researching food security and its interactions with global environmental change. *Food Security* 3 (4): 417–431.

Jalgaonwala, R.E., Mohite, B.V., and Mahajan, R.T. (2011). A review: natural products from plant associated endophytic fungi. *Journal of Microbiology and Biotechnology Research* 1 (2): 21–32.

Jansson, J.K. and Baker, E.S. (2016). A multi-omic future for microbiome studies. *Nature Microbiology* 1 (5): 10–1038.

Jesumirhewe, C., Ogunlowo, P.O., Olley, M. et al. (2016). Accuracy of conventional identification methods used for Enterobacteriaceae isolates in three Nigerian hospitals. *PeerJ* 4: e2511.

Jez, J.M., Lee, S.G., and Sherp, A.M. (2016). The next green movement: plant biology for the environment and sustainability. *Science* 353 (6305): 1241–1244.

Jha, Y., Sablok, G., Subbarao, N. et al. (2014). Bacterial-induced expression of RAB18 protein in *Orzya sativa* salinity stress and insights into molecular interaction with GTP ligand. *Journal of Molecular Recognition* 27 (9): 521–527.

Jiao, S., Chen, W., and Wei, G. (2019). Resilience and assemblage of soil microbiome in response to chemical contamination combined with plant growth. *Applied and Environmental Microbiology* 85 (6): e02523–e02518.

Kanianska, R. (2016). Agriculture and its impact on land-use, environment, and ecosystem services. In: *Landscape Ecology-The Influences of Land Use and Anthropogenic Impacts of Landscape Creation*, 1–26. http://dx.doi.org/10.5772/63719.

Kashyap, P.L., Rai, S., Kumar, S., and Srivastava, A.K. (2016). Genetic diversity, mating types and phylogenetic analysis of Indian races of *Fusarium oxysporum* f. sp. *ciceris* from chickpea. *Archives of Phytopathology and Plant Protection* 49 (19–20): 533–553.

Kashyap, P.L., Rai, P., Kumar, S. et al. (2017a). DNA barcoding for diagnosis and monitoring of fungal plant pathogens. In: *Molecular Markers in Mycology* (eds. B.P. Singh and V.K. Gupta), 87–122. Springer.

Kashyap, P.L., Rai, P., Srivastava, A.K., and Kumar, S. (2017b). *Trichoderma* for climate resilient agriculture. *World Journal of Microbiology and Biotechnology* 33 (8): 155.

Kashyap, P.L., Srivastava, A.K., Tiwari, S.P., and Kumar, S. (2018). *Microbes for Climate Resilient Agriculture*, 376. Hoboken: Wiley.

Kasotia, A., Varma, A., and Choudhary, D.K. (2015). Pseudomonas-mediated mitigation of salt stress and growth promotion in *Glycine max*. *Agricultural Research* 4 (1): 31–41.

Kathage, J. and Qaim, M. (2012). Economic impacts and impact dynamics of Bt (*Bacillus thuringiensis*) cotton in India. *Proceedings of the National Academy of Sciences of the United States of America* 109 (29): 11652–11656.

Kaushal, M. and Wani, S.P. (2016). Plant-growth-promoting rhizobacteria: drought stress alleviators to ameliorate crop production in drylands. *Annals of Microbiology* 66 (1): 35–42.

Keinan, A. and Clark, A.G. (2012). Recent explosive human population growth has resulted in an excess of rare genetic variants. *Science* 336 (6082): 740–743.

Kembel, S.W., O'Connor, T.K., Arnold, H.K. et al. (2014). Relationships between phyllosphere bacterial communities and plant functional traits in a neotropical forest. *Proceedings of the National Academy of Sciences of the United States of America* 111 (38): 13715–13720.

Kent, A.D. and Triplett, E.W. (2002). Microbial communities and their interactions in soil and rhizosphere ecosystems. *Annual Reviews in Microbiology* 56 (1): 211–236.

Kepler, R.M., Maul, J.E., and Rehner, S.A. (2017). Managing the plant microbiome for biocontrol fungi: examples from Hypocreales. *Current Opinion in Microbiology* 37: 48–53.

Khan, N., Bano, A., and Zandi, P. (2018). Effects of exogenously applied plant growth regulators in combination with PGPR on the physiology and root growth of chickpea (*Cicer arietinum*) and their role in drought tolerance. *Journal of Plant Interactions* 13 (1): 239–247.

Kiely, P.D., Haynes, J.M., Higgins, C.H. et al. (2006). Exploiting new systems-based strategies to elucidate plant-bacterial interactions in the rhizosphere. *Microbial Ecology* 51 (3): 257–266.

Kim, K.D., Nemec, S., and Musson, G. (1997). Control of *Phytophthora* root and crown rot of bell pepper with composts and soil amendments in the greenhouse. *Applied Soil Ecology* 5 (2): 169–179.

Kniskern, J.M., Traw, M.B., and Bergelson, J. (2007). Salicylic acid and jasmonic acid signaling defense pathways reduce natural bacterial diversity on *Arabidopsis thaliana*. *Molecular Plant-Microbe Interactions* 20 (12): 1512–1522.

Kumar, A., Imran, M., Kumar, S.P.J. et al. (2018). Assessment of soil urease activity and its kinetic parameters in chilli rhizosphere under the influence of pesticides. *Biopesticides Internationale* 14 (1): 47–57.

Kushwaha, P., Kashyap, P.L., Kuppusamy, P. et al. (2019a). Functional characterization of endophytic bacilli from pearl millet (*Pennisetum glaucum*) and their possible role in multiple stress tolerance. *Plant Biosystems* https://doi.org/10.1080/11263504.2019.1651773.

Kushwaha, P., Kashyap, P.L., Srivastava, A.K., and Tiwari, R.K. (2019b). Plant growth promoting and antifungal activity in endophytic *Bacillus* strains from pearl millet (*Pennisetum glaucum*). *Brazilian Journal of Microbiology* https://doi.org/10.1007/s42770-019-00172-5.

Leite, J., Fischer, D., Rouws, L.F. et al. (2017). Cowpea nodules harbor non-rhizobial bacterial communities that are shaped by soil type rather than plant genotype. *Frontiers in Plant Science* 7: 2064.

Lemaire, G., Franzluebbers, A., de Faccio Carvalho, P.C., and Dedieu, B. (2014). Integrated crop–livestock systems: strategies to achieve synergy between agricultural production and environmental quality. *Agriculture, Ecosystems and Environment* 190: 4–8.

Lemanceau, P., Blouin, M., Muller, D., and Moënne-Loccoz, Y. (2017). Let the core microbiota be functional. *Trends in Plant Science* 22 (7): 583–595.

Levy, A., Conway, J.M., Dangl, J.L., and Woyke, T. (2018). Elucidating bacterial gene functions in the plant microbiome. *Cell Host & Microbe* 24 (4): 475–485.

Lim, J.H. and Kim, S.D. (2013). Induction of drought stress resistance by multi-functional PGPR *Bacillus licheniformis* K11 in pepper. *The Plant Pathology Journal* 29 (2): 201–208.

Lin, J.J., Kuo, J., and Ma, J. (1996). A PCR-based DNA fingerprinting technique: AFLP for molecular typing of bacteria. *Nucleic Acids Research* 24 (18): 3649–3650.

Lindow, S.E. (1996). Role of immigration and other processes in determining epiphytic bacterial populations. In: *Aerial Plant Surface Microbiology* (ed. C.E. Morris), 155–168. Boston, MA: Springer.

Lindow, S.E. and Brandl, M.T. (2003). Microbiology of the phyllosphere. *Applied and Environmental Microbiology* 69 (4): 1875–1883.

Liu, W.T., Mirzabekov, A.D., and Stahl, D.A. (2001). Optimization of an oligonucleotide microchip for microbial identification studies: a non-equilibrium dissociation approach. *Environmental Microbiology* 3 (10): 619–629.

Loman, N.J., Constantinidou, C., Chan, J.Z. et al. (2012). High-throughput bacterial genome sequencing: an embarrassment of choice, a world of opportunity. *Nature Reviews Microbiology* 10 (9): 599.

López-López, A., Rogel, M.A., Ormeno-Orrillo, E. et al. (2010). *Phaseolus vulgaris* seed-borne endophytic community with novel bacterial species such as *Rhizobium endophyticum* sp. nov. *Systematic and Applied Microbiology* 33 (6): 322–327.

Lucht, J. (2015). Public acceptance of plant biotechnology and GM crops. *Viruses* 7 (8): 4254–4281.

Lugtenberg, B. and Kamilova, F. (2009). Plant-growth-promoting rhizobacteria. *Annual Review of Microbiology* 63: 541–556.

Lundberg, D.S., Lebeis, S.L., Paredes, S.H. et al. (2012). Defining the core *Arabidopsis thaliana* root microbiome. *Nature* 488 (7409): 86.

Malmstrom, L., Malmström, J., Marko-Varga, G., and Westergren-Thorsson, G. (2002). Proteomic 2DE database for spot selection, automated annotation, and data analysis. *Journal of Proteome Research* 1 (2): 135–138.

Mansfield, J.W. (2000). Antimicrobial compounds and resistance. In: *Mechanisms of Resistance to Plant Diseases* (eds. A.J. Slusarenko et al.), 325–337. Dordrecht: Springer.

Marenya, P.P. and Barrett, C.B. (2009). Soil quality and fertilizer use rates among smallholder farmers in western Kenya. *Agricultural Economics* 40 (5): 561–572.

Margulis, L. (1981). *Symbiosis in Cell Evolution: Life and Its Environment on the Early Earth*, 419. San Francisco: W.H. Freeman.

Marrone, P.G. (2019). Pesticidal natural products–status and future potential. *Pest Management Science* 75 (9): 2325–2340.

Marulanda, A., Azcón, R., Chaumont, F. et al. (2010). Regulation of plasma membrane aquaporins by inoculation with a *Bacillus megaterium* strain in maize (*Zea mays* L.) plants under unstressed and salt-stressed conditions. *Planta* 232 (2): 533–543.

Meena, K.K., Kumar, M., Kalyuzhnaya, M.G. et al. (2012). Epiphytic pink-pigmented methylotrophic bacteria enhance germination and seedling growth of wheat (*Triticum aestivum*) by producing phytohormone. *Antonie Van Leeuwenhoek* 101 (4): 777–786.

Meena, K.K., Sorty, A.M., Bitla, U.M. et al. (2017). Abiotic stress responses and microbe-mediated mitigation in plants: the omics strategies. *Frontiers in Plant Science* 8: 172.

Mendes, R., Garbeva, P., and Raaijmakers, J.M. (2013). The rhizosphere microbiome: significance of plant beneficial, plant pathogenic, and human pathogenic microorganisms. *FEMS Microbiology Reviews* 37 (5): 634–663.

Mirzaee, H., Shuey, L., and Schenk, P.M. (2015). Transcriptomics of plants interacting with pathogens and beneficial microbes. In: *Genomics, Proteomics and Metabolomics in Nutraceuticals and Functional Foods* (eds. D. Bagchi, B. Manashi and S. Anand), 525–536. USA: Wiley.

Mitter, B., Pfaffenbichler, N., and Sessitsch, A. (2016). Plant–microbe partnerships in 2020. *Microbial Biotechnology* 9 (5): 635–640.

Mona, S.A., Hashem, A., Abd_Allah, E.F. et al. (2017). Increased resistance of drought by *Trichoderma harzianum* fungal treatment correlates with increased secondary metabolites and proline content. *Journal of Integrative Agriculture* 16 (8): 1751–1757.

Mukhongo, R.W., Tumuhairwe, J.B., Ebanyat, P. et al. (2017). Combined application of biofertilizers and inorganic nutrients improves sweet potato yields. *Frontiers in Plant Science* 8: 219.

Müller, C.A., Obermeier, M.M., and Berg, G. (2016). Bioprospecting plant-associated microbiomes. *Journal of Biotechnology* 235: 171–180.

Nadeem, S.M., Zahir, Z.A., Naveed, M., and Arshad, M. (2007). Preliminary investigations on inducing salt tolerance in maize through inoculation with rhizobacteria containing ACC deaminase activity. *Canadian Journal of Microbiology* 53 (10): 1141–1149.

Naveed, M., Hussain, M.B., Zahir, Z.A. et al. (2014). Drought stress amelioration in wheat through inoculation with Burkholderia phytofirmans strain PsJN. *Plant Growth Regulation* 73 (2): 121–131.

Newcombe, G., Harding, A., Ridout, M., and Busby, P. (2018). A hypothetical bottleneck in the plant microbiome. *Frontiers in Microbiology* 9: 1645.

Nguyen, T.H., Phan, T.C., Choudhury, A.T. et al. (2017). BioGro: a plant growth-promoting biofertilizer validated by 15 Years' research from laboratory selection to Rice Farmer's fields of the Mekong Delta. In: *Agro-Environmental Sustainability* (eds. J.S. Singh et al.), 237–254. Cham: Springer.

Nihorimbere, V., Ongena, M., Smargiassi, M., and Thonart, P. (2011). Beneficial effect of the rhizosphere microbial community for plant growth and health. *Biotechnologie, Agronomie, Société et Environnement* 15 (2): 327–337.

Nocker, A., Burr, M., and Camper, A.K. (2007). Genotypic microbial community profiling: a critical technical review. *Microbial Ecology* 54 (2): 276–289.

Nouwens, A.S., Cordwell, S.J., Larsen, M.R. et al. (2000). Complementing genomics with proteomics: the membrane subproteome of *Pseudomonas aeruginosa* PAO1. *Electrophoresis: An International Journal* 21 (17): 3797–3809.

Oerke, E.C. and Dehne, H.W. (2004). Safeguarding production – losses in major crops and the role of crop protection. *Crop Protection* 23 (4): 275–285.

Oku, S., Komatsu, A., Nakashimada, Y. et al. (2014). Identification of *Pseudomonas fluorescens* chemotaxis sensory proteins for malate, succinate, and fumarate, and their involvement in root colonization. *Microbes and Environments* 29 (4): 413–419.

Oliveira, C.A., Alves, V.M.C., Marriel, I.E. et al. (2009). Phosphate solubilizing microorganisms isolated from rhizosphere of maize cultivated in an oxisol of the Brazilian Cerrado Biome. *Soil Biology and Biochemistry* 41 (9): 1782–1787.

Omar, M.N.A., Osman, M.E.H., Kasim, W.A., and El-Daim, I.A. (2009). Improvement of salt tolerance mechanisms of barley cultivated under salt stress using *Azospirillum brasilense*. In: *Salinity and Water Stress* (eds. M. Ashraf et al.), 133–147. Dordrecht: Springer.

Paarlberg, R. (2010). GMO foods and crops: Africa's choice. *New Biotechnology* 27 (5): 609–613.

Pagé, A.P., Tremblay, J., Masson, L., and Greer, C.W. (2019). Nitrogen-and phosphorus-starved *Triticum aestivum* show distinct belowground microbiome profiles. *PLoS One* 14 (2): e0210538.

Pal, K.K. and Gardener, B.M. (2006). Biological control of plant pathogens. *Plant Health Instructor* 2: 1117–1142.

Palaniyandi, S.A., Damodharan, K., Yang, S.H., and Suh, J.W. (2014). *Streptomyces* sp. strain PGPA39 alleviates salt stress and promotes growth of 'Micro Tom' tomato plants. *Journal of Applied Microbiology* 117 (3): 766–773.

Pandey, V., Ansari, M.W., Tula, S. et al. (2016). Dose-dependent response of *Trichoderma harzianum* in improving drought tolerance in rice genotypes. *Planta* 243 (5): 1251–1264.

Panlada, T., Pongdet, P., Aphakorn, L. et al. (2013). Alleviation of the effect of environmental stresses using co-inoculation of mungbean by *Bradyrhizobium* and rhizobacteria containing stress-induced ACC deaminase enzyme. *Soil Science and Plant Nutrution* 59: 559–571.

Peng, X., Xu, C., Ren, H. et al. (2005). Proteomic analysis of the sarcosine-insoluble outer membrane fraction of *Pseudomonas aeruginosa* responding to ampicilin, kanamycin, and tetracycline resistance. *Journal of Proteome Research* 4 (6): 2257–2265.

Pereira, L.B., Andrade, G.S., Meneghin, S.P. et al. (2019). Prospecting plant growth-promoting bacteria isolated from the Rhizosphere of sugarcane under drought stress. *Current Microbiology* 76 (11): 1345–1354.

Petrosino, J.F., Highlander, S., Luna, R.A. et al. (2009). Metagenomic pyrosequencing and microbial identification. *Clinical Chemistry* 55 (5): 856–866.

Philippot, L., Hallin, S., Börjesson, G., and Baggs, E.M. (2009). Biochemical cycling in the rhizosphere having an impact on global change. *Plant and Soil* 321 (1–2): 61–81.

Pieper, R., Gatlin-Bunai, C.L., Mongodin, E.F. et al. (2006). Comparative proteomic analysis of *Staphylococcus aureus* strains with differences in resistance to the cell wall-targeting antibiotic vancomycin. *Proteomics* 6 (15): 4246–4258.

Pieterse, C.M., Zamioudis, C., Berendsen, R.L. et al. (2014). Induced systemic resistance by beneficial microbes. *Annual Review of Phytopathology* 52: 347–375.

del Pilar Martínez-Diz, M., Andrés-Sodupe, M., Bujanda, R. et al. (2019). Soil-plant compartments affect fungal microbiome diversity and composition in grapevine. *Fungal Ecology* 41: 234–244.

Pinedo, I., Ledger, T., Greve, M., and Poupin, M.J. (2015). *Burkholderia phytofirmans* PsJN induces long-term metabolic and transcriptional changes involved in *Arabidopsis thaliana* salt tolerance. *Frontiers in Plant Science* 6: 466.

Ponce de León, I. and Montesano, M. (2013). Activation of defense mechanisms against pathogens in mosses and flowering plants. *International Journal of Molecular Sciences* 14 (2): 3178–3200.

Poole, P., Ramachandran, V., and Terpolilli, J. (2018). Rhizobia: from saprophytes to endosymbionts. *Nature Reviews Microbiology* 16 (5): 291.

Rai, P., Sharma, A., Saxena, P. et al. (2015). Comparison of molecular and phenetic typing methods to assess diversity of selected members of the genus *Bacillus*. *Microbiology* 84 (2): 265–275.

Rai, S., Kashyap, P.L., Kumar, S. et al. (2016). Identification, characterization and phylogenetic analysis of antifungal Trichoderma from tomato rhizosphere. *Springerplus* 5: 1939.

Ram, R.J., VerBerkmoes, N.C., Thelen, M.P. et al. (2005). Community proteomics of a natural microbial biofilm. *Science* 308 (5730): 1915–1920.

Reichling, J. (2018). Plant-microbe interactions and secondary metabolites with antiviral, antibacterial and antifungal properties. In: *Functions and Biotechnology of Plant Secondary Metabolites*, Annual Plant Reviews Book Series, vol. 39 (eds. M. Dickinson and J. Beynon), 214–347. Wiley Blackwell https://doi.org/10.1002/9781119312994.apr0420.

Reinhold-Hurek, B., Bünger, W., Burbano, C.S. et al. (2015). Roots shaping their microbiome: global hotspots for microbial activity. *Annual Review of Phytopathology* 53: 403–424.

Reischke, S., Rousk, J., and Bååth, E. (2014). The effects of glucose loading rates on bacterial and fungal growth in soil. *Soil Biology and Biochemistry* 70: 88–95.

Rodriguez, P.A., Rothballer, M., Chowdhury, S.P. et al. (2019). Systems biology of plant microbiome interactions. *Molecular Plant* 12 (6): 804–821.

Rosa, L.H., Almeida Vieira, M.D.L., Santiago, I.F., and Rosa, C.A. (2010). Endophytic fungi community associated with the dicotyledonous plant *Colobanthus quitensis* (Kunth) Bartl. (Caryophyllaceae) in Antarctica. *FEMS Microbiology Ecology* 73 (1): 178–189.

Sabir, A., Yazici, M.A., Kara, Z., and Sahin, F. (2012). Growth and mineral acquisition response of grapevine rootstocks (*Vitis* spp.) to inoculation with different strains of plant growth-promoting rhizobacteria (PGPR). *Journal of the Science of Food and Agriculture* 92 (10): 2148–2153.

Sahoo, R.K., Ansari, M.W., Dangar, T.K. et al. (2014). Phenotypic and molecular characterisation of efficient nitrogen-fixing *Azotobacter* strains from rice fields for crop improvement. *Protoplasma* 251 (3): 511–523.

Samad, A., Trognitz, F., Compant, S. et al. (2017). Shared and host-specific microbiome diversity and functioning of grapevine and accompanying weed plants. *Environmental Microbiology* 19 (4): 1407–1424.

Sanchez-Canizares, C., Jorrin, B., Poole, P.S., and Tkacz, A. (2017). Understanding the holobiont: the interdependence of plants and their microbiome. *Current Opinion in Microbiology* 38: 188–196.

Sarhan, M.S., Hamza, M.A., Youssef, H.H. et al. (2019). Culturomics of the plant prokaryotic microbiome and the dawn of plant-based culture media–a review. *Journal of Advanced Research* 19: 15–27.

Schlaeppi, K. and Bulgarelli, D. (2015). The plant microbiome at work. *Molecular Plant-Microbe Interactions* 28 (3): 212–217.

Sergaki, C., Lagunas, B., Lidbury, I. et al. (2018). Challenges and approaches in microbiome research: from fundamental to applied. *Frontiers in Plant Science* 9: 1205.

Sharma, I.P. and Sharma, A.K. (2017). Physiological and biochemical changes in tomato cultivar PT-3 with dual inoculation of mycorrhiza and PGPR against root-knot nematode. *Symbiosis* 71 (3): 175–183.

Sharma, A., Singh, P., Kumar, S. et al. (2015). Deciphering diversity of salt-tolerant bacilli from saline soils of eastern Indo-Gangetic plains of India. *Geomicrobiology Journal* 32 (2): 170–180.

Sharma, S., Rai, P., Rai, S. et al. (2017). Genomic revolution in crop disease diagnosis: a review. In: *Plants and Microbes in an Ever-Changing Environment* (ed. S.S. Singh), 257–293. Hauppauge: Nova Science Publishers.

Sharma, A., Kashyap, P.L., Srivastava, A.K. et al. (2019). Isolation and characterization of halotolerant bacilli from chickpea (*Cicer arietinum* L.) rhizosphere for plant growth promotion and biocontrol traits. *European Journal of Plant Pathology* 153 (3): 787–800.

Singh, R. (2014a). Microorganism as a tool of bioremediation technology for cleaning environment: a review. *Proceedings of the International Academy of Ecology and Environmental Sciences* 4 (1): 1–6.

Singh, S. (2014b). A review on possible elicitor molecules of cyanobacteria: their role in improving plant growth and providing tolerance against biotic or abiotic stress. *Journal of Applied Microbiology* 117 (5): 1221–1244.

Singh, B.K., Bardgett, R.D., Smith, P., and Reay, D.S. (2010). Microorganisms and climate change: terrestrial feedbacks and mitigation options. *Nature Reviews Microbiology* 8 (11): 779–790.

Singh, R.K., Kumar, D.P., Singh, P. et al. (2014). Multifarious plant growth promoting characteristics of chickpea rhizosphere associated Bacilli help to suppress soil-borne pathogens. *Plant Growth Regulation* 73 (1): 91–101.

Singh, J.S., Kumar, A., Rai, A.N., and Singh, D.P. (2016). Cyanobacteria: a precious bio-resource in agriculture, ecosystem, and environmental sustainability. *Frontiers in Microbiology* 7: 529.

Singh, M., Deokaran, M.J.S., and Bhatt, B.P. (2017). Effect of integrated nutrient management on production potential and quality of summer mungbean (*Vigna radiata* L.). *Journal of Krishi Vigyan* 5: 39–45.

Singh, D., Raina, T.K., Kumar, A. et al. (2019a). Plant microbiome: a reservoir of novel genes and metabolites. *Plant Gene* 18: 100177.

Singh, S., Tripathi, A., Maji, D. et al. (2019b). Evaluating the potential of combined inoculation of *Trichoderma harzianum* and *Brevibacterium halotolerans* for increased growth and oil yield in *Mentha arvensis* under greenhouse and field conditions. *Industrial Crops and Products* 131: 173–181.

Solanki, M.K., Kumar, S., Pandey, A.K. et al. (2012). Diversity and antagonistic potential of *Bacillus* spp. associated to the rhizosphere of tomato for the management of *Rhizoctonia solani*. *Biocontrol Science and Technology* 22 (2): 203–217.

Solanki, M.K., Singh, R.K., Srivastava, S. et al. (2014). Isolation and characterization of siderophore producing antagonistic rhizobacteria against *Rhizoctonia solani*. *Journal of Basic Microbiology* 54 (6): 585–597.

Solanki, M.K., Singh, R.K., Srivastava, S. et al. (2015). Characterization of antagonistic-potential of two *Bacillus* strains and their biocontrol activity against *Rhizoctonia solani* in tomato. *Journal of Basic Microbiology* 55 (1): 82–90.

Sorty, A.M., Meena, K.K., Choudhary, K. et al. (2016). Effect of plant growth promoting bacteria associated with halophytic weed (*Psoralea corylifolia* L.) on germination and seedling growth of wheat under saline conditions. *Applied Biochemistry and Biotechnology* 180 (5): 872–882.

Souza, R.D., Ambrosini, A., and Passaglia, L.M. (2015). Plant growth-promoting bacteria as inoculants in agricultural soils. *Genetics and Molecular Biology* 38 (4): 401–419.

Spinler, J.K., Karri, V., and Hirschi, K.D. (2019). Planting the microbiome. *Trends in Microbiology* 27 (2): 90–93.

Srivastava, R., Tripathi, B.M., Singh, R.K. et al. (2013). Profiling of plant growth promoting bacteria associated with Jaunpuri giant raddish rhizosphere. *International Journal of Agriculture, Environment and Biotechnology* 6: 187–196.

Srivastava, A.K., Singh, P., Singh, R.K. et al. (2014). Identification and characterization of ethanol utilizing fungal flora of oil refinery contaminated soil. *World Journal of Microbiology and Biotechnology* 30 (2): 705–714.

Stahl, P.L. and Lundeberg, J. (2012). Toward the single-hour high-quality genome. *Annual Review of Biochemistry* 81: 359–378.

Stielow, J.B., Levesque, C.A., Seifert, K.A. et al. (2015). One fungus, which genes? Development and assessment of universal primers for potential secondary fungal DNA barcodes. *Persoonia* 35: 242–263.

Suryanarayanan, T.S., Thirunavukkarasu, N., Govindarajulu, M.B. et al. (2009). Fungal endophytes and bioprospecting. *Fungal Biology Reviews* 23 (1–2): 9–19.

Taghavi, S., Garafola, C., Monchy, S. et al. (2009). Genome survey and characterization of endophytic bacteria exhibiting a beneficial effect on growth and development of poplar trees. *Applied and Environmental Microbiology* 75 (3): 748–757.

Tamang, J.P., Watanabe, K., and Holzapfel, W.H. (2016). Diversity of microorganisms in global fermented foods and beverages. *Frontiers in Microbiology* 7: 377.

Templeton, S.R. and Scherr, S.J. (1999). Effects of demographic and related microeconomic change on land quality in hills and mountains of developing countries. *World Development* 27 (6): 903–918.

Thapa, S. and Prasanna, R. (2018). Prospecting the characteristics and significance of the phyllosphere microbiome. *Annals of Microbiology* 68 (5): 229–245.

Thokchom, E., Thakuria, D., Kalita, M.C. et al. (2017). Root colonization by host-specific rhizobacteria alters indigenous root endophyte and rhizosphere soil bacterial communities and promotes the growth of mandarin orange. *European Journal of Soil Biology* 79: 48–56.

Timmusk, S., El-Daim, I.A.A., Copolovici, L. et al. (2014). Drought-tolerance of wheat improved by rhizosphere bacteria from harsh environments: enhanced biomass production and reduced emissions of stress volatiles. *PLoS One* 9 (5): e96086.

Tiwari, S., Singh, P., Tiwari, R. et al. (2011). Salt-tolerant rhizobacteria-mediated induced tolerance in wheat (*Triticum aestivum*) and chemical diversity in rhizosphere enhance plant growth. *Biology and Fertility of Soils* 47 (8): 907–911.

Uroz, S., Courty, P.E., and Oger, P. (2019). Plant symbionts are engineers of the plant-associated microbiome. *Trends in Plant Science* 24 (10): 905–916.

Vaishnav, A., Kumari, S., Jain, S. et al. (2016). PGPR-mediated expression of salt tolerance gene in soybean through volatiles under sodium nitroprusside. *Journal of Basic Microbiology* 56 (11): 1274–1288.

Valdes, A. (2019). *Food Security for Developing Countries*, 374. Routledge.

Varma, A., Bakshi, M., Lou, B. et al. (2012). *Piriformospora indica*: a novel plant growth-promoting mycorrhizal fungus. *Agricultural Research* 1 (2): 117–131.

Versalovic, J., Schneider, M., De Bruijn, F.J., and Lupski, J.R. (1994). Genomic fingerprinting of bacteria using repetitive sequence-based polymerase chain reaction. *Methods in Molecular and Cellular Biology* 5 (1): 25–40.

Vicedo, B., Peñalver, R., Asins, M.J., and López, M.M. (1993). Biological control of *Agrobacterium tumefaciens*, colonization, and pAgK84 transfer with *Agrobacterium radiobacter* K84 and the Tra-mutant strain K1026. *Applied and Environmental Microbiology* 59 (1): 309–315.

Vorholt, J.A. (2012). Microbial life in the phyllosphere. *Nature Reviews Microbiology* 10 (12): 828–840.

Vorholt, J.A., Vogel, C., Carlström, C.I., and Mueller, D.B. (2017). Establishing causality: opportunities of synthetic communities for plant microbiome research. *Cell Host & Microbe* 22 (2): 142–155.

Walitang, D.I., Kim, K., Madhaiyan, M. et al. (2017). Characterizing endophytic competence and plant growth promotion of bacterial endophytes inhabiting the seed endosphere of rice. *BMC Microbiology* 17 (1): 209.

Wang, W., Zhai, Y., Cao, L. et al. (2016). Endophytic bacterial and fungal microbiota in sprouts, roots and stems of rice (*Oryza sativa* L.). *Microbiological Research* 188: 1–8.

Wassermann, B., Cernava, T., Müller, H. et al. (2019). Seeds of native alpine plants host unique microbial communities embedded in cross-kingdom networks. *Microbiome* 7 (1): 108.

White, J.F., Kingsley, K.L., Zhang, Q. et al. (2019). Endophytic microbes and their potential applications in crop management. *Pest Management Science* 75 (10): 2558–2565.

Wilmes, P., Heintz-Buschart, A., and Bond, P.L. (2015). A decade of metaproteomics: where we stand and what the future holds. *Proteomics* 15 (20): 3409–3417.

Wilson, K.H., Wilson, W.J., Radosevich, J.L. et al. (2002). High-density microarray of small-subunit ribosomal DNA probes. *Applied and Environmental Microbiology* 68 (5): 2535–2541.

Wu, K.M., Lu, Y.H., Feng, H.Q. et al. (2008). Suppression of cotton bollworm in multiple crops in China in areas with Bt toxin–containing cotton. *Science* 321 (5896): 1676–1678.

Xie, S., Jiang, H., Ding, T. et al. (2018). *Bacillus amyloliquefaciens* FZB42 represses plant miR846 to induce systemic resistance via a jasmonic acid-dependent signalling pathway. *Molecular Plant Pathology* 19 (7): 1612–1623.

Xu, X., He, Q., Chen, C., and Zhang, C. (2016). Differential communications between fungi and host plants revealed by secretome analysis of phylogenetically related endophytic and pathogenic fungi. *PLoS One* 11 (9): e0163368.

Yao, D., Zhang, X., Zhao, X. et al. (2011). Transcriptome analysis reveals salt-stress-regulated biological processes and key pathways in roots of cotton (*Gossypium hirsutum* L.). *Genomics* 98 (1): 47–55.

de Zelicourt, A., Al-Yousif, M., and Hirt, H. (2013). Rhizosphere microbes as essential partners for plant stress tolerance. *Molecular Plant* 6 (2): 242–245.

Zhang, M., Powell, C.A., Benyon, L.S. et al. (2013). Deciphering the bacterial microbiome of citrus plants in response to '*Candidatus liberibacter asiaticus*' – infection and antibiotic treatments. *PLoS One* 8 (11): e76331.

Zhao, Y., Luo, L., Xu, J. et al. (2018). Malate transported from chloroplast to mitochondrion triggers production of ROS and PCD in *Arabidopsis thaliana*. *Cell Research* 28 (4): 448.

2

The Plant Microbiome in Agricultural Sustainability

From Microbe to Microbiome

Jose Pedro Fonseca, Yuan Wang, and Kirankumar S. Mysore

The Noble Research Institute, Ardmore, OK, USA

2.1 Introduction

Agricultural sustainability can be understood as the means to obtain productive performance of a cropping system over time without hampering the future potential of natural resources such as soil health (Barea 2015). While the green revolution of the twentieth century was key for increasing crop yields and feeding a growing world population through breeding programs, chemical fertilizers and pesticides have had a negative impact on soil health (Mauchline and Malone 2017). A new green revolution in the twenty-first century is needed not only to maintain increased crop production and yield but also to improve agricultural sustainability with the use of microbial biotechnology (Schlaeppi and Bulgarelli 2015). This could help to meet the food requirement for a rapidly increasing global population while addressing possible environmental and food contamination issues caused by overuse of chemical fertilizers or pesticides. To solve these problems, the concepts of sustainable agricultural production and low-input agriculture have been put forward in recent years. A critical component of both concepts is exploitation of the potential of beneficial soil microorganisms to better serve agricultural production.

Present soil management strategies depend heavily on inorganic chemical fertilizers that can have undesirable effects on the environment and human health. One way to circumvent this problem is to employ more sustainable methods, such as using beneficial microbes as biofertilizers (Bhardwaj et al. 2014). The use of plant growth-promoting rhizobacteria (PGPR), beneficial fungi (e.g. mycorrhiza) and others has led to improvement of many plant physiological processes, including nutrient uptake, growth and tolerance to abiotic and biotic stresses (Antoniou et al. 2017; Bhardwaj et al. 2014; Bulgarelli et al. 2015; Dudenhöffer et al. 2016; Haney et al. 2015; Kashyap et al. 2017; Pandey et al. 2012; Pieterse et al. 2016; Santhanam et al. 2015; Vacheron et al. 2013).

But in order to harness the full potential of the plant microbiome for application in agriculture, we will first need to understand its interactions with the plant genotype and the environment's role in shaping microbiome diversity and function. Knowing "who" is there

The Plant Microbiome in Sustainable Agriculture, First Edition. Edited by Alok Kumar Srivastava, Prem Lal Kashyap and Madhumita Srivastava.

(taxonomy and also what they are doing (function) is a necessary first step in determining how plants sense and recruit microbes to help them grow, reproduce, and cope with a myriad of abiotic and biotic stresses.

Plant microbiome composition has been shown to be a dynamic entity that quickly responds to many genetic, chemical, and environmental factors, such as host genotype and agricultural practices including fertilizer or pesticide application (Bulgarelli et al. 2012; Castrillo et al. 2017; Lebeis et al. 2015; Trognitz et al. 2016).

Advancing agro-sustainability is a key to increasing our current understanding of the mechanisms controlling such plant–microbe interactions in order to increase agriculture output/yields and improve soil health by using beneficial microbes. Indeed, Busby et al. (2017) have defined a set of research priorities for optimizing microbiome usage in sustainable agriculture and increased crop productivity: (i) the development of model host microbiomes across different plant species including fruits, vegetables, grasses, and others; (ii) the identification of a "core microbiome" for different plant species; (iii) the design of synthetic microbial communities that can be beneficial to the host; and (iv) determining the functional mechanisms controlling plant-microbiome interactions.

2.2 Beneficial Microbes

Plants and microorganisms have co-evolved for millions of years. Plant-associated microbes are so important to their hosts that it has been suggested that a plant and all its associated endophytic and rhizospheric microbes be conceptualized as a "holobiome" (Guerrero et al. 2013). From the perspective of agricultural production, how soil microorganisms influence crop performance and productivity has been a question for decades. Soil, being one of the most complex ecosystems, is the habitat of various microorganisms, including bacteria, fungi, oomycetes, phages and protists, with bacteria and fungi being the most abundant. Beneficial soil microbes enhance plant health, fitness, and productivity, while soil-borne pathogenic microbes cause plant diseases.

The "probiotic" soil microorganisms can benefit their plant hosts in a variety of ways, including through growth promotion, nutrient acquisition and stress alleviation. In the soil bacterial communities, PGPR competitively colonize plant roots and stimulate plant growth by a wide variety of mechanisms, such as enhanced nutrient uptake, activation of systemic immunity and suppression of soil-borne pathogens (Pieterse et al. 2016). The common mechanism of PGPR-mediated growth promotion involves secretion of signaling molecules and compounds that influence the host hormone signaling pathway and/or improve root architecture (Figure 2.1) (Vacheron et al. 2013). On the other hand, soil microorganisms facilitating plant nutrition have been well-studied in root-nodulating rhizobia and arbuscular mycorrhizal fungi (AMF). Rhizobia and AMF form a nutritional symbiosis with their hosts to facilitate plant survival and growth under unfavorable nutrient conditions. Rhizobia specifically infect roots of plants belonging to the Leguminosae family. In response to rhizobia infection, the plants form new organs called root nodules, in which rhizobia reside and transform atmospheric nitrogen gas (N_2) into a plant-available nitrogen source(-NH_4+) (Oldroyd et al. 2011). In contrast, AMF symbiosis is fairly ubiquitous throughout the plant kingdom (Remy et al. 1994). AMF invade root cortical cells but also extend into the soil

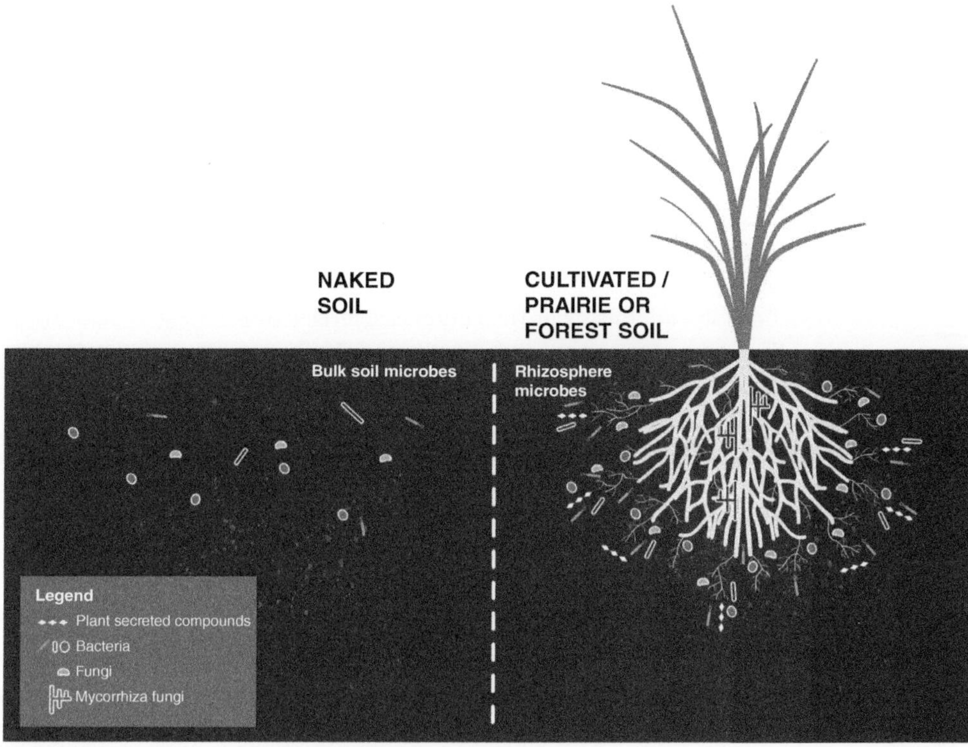

Figure 2.1 Schematic diagram showing host plant root secretions that influence the rhizosphere microbiota. The microbiota located at the rhizosphere interacts with root secreted compounds like carbon, sugar, and mucilages which can promote growth and selection of several microbes and also increase organic matter through rhizo-deposition where microbes can thrive. Microbes in turn can help plants with several physiological processes such as biotic and abiotic stress amelioration and growth. Bulk soil microbes in contrast rely exclusively on soil type properties as a driver of microbiota changes.

matrix, thus functionally acting as extensions of the plant root system. The AMF symbionts acquire phosphorus (P) originating from sparingly soluble inorganic and organic P-compounds and translocate P and other micronutrients to their plant hosts in exchange for photosynthetically derived carbon (White and Hammond 2008).

A study in the legume alfalfa (*Medicago sativa*) reported that pesticide treatment can inhibit and delay recruitment of rhizobial bacteria to host plant roots, causing a reduction in efficiency of nitrogen-fixing bacteria (Fox et al. 2007). The authors performed enzymatic and symbiotic nitrogen fixing (SNF) inoculation assays and determined fewer root nodules were produced with lower rates of nitrogenase activity, thus resulting in reduction of overall plant yield at the time of harvest for pesticide-treated plants (Fox et al. 2007). This result strongly indicates that sustainable agricultural practices may be negatively impacted by pesticides, which have the ability to disrupt SNF and decrease plant yield, directly impacting soil health. Currently there is a lack of studies showing the effect of pesticides in the overall plant microbiome community structure, but some studies already point to significant changes, at least in SNF bacterial populations and agricultural sustainability (Fox et al. 2007).

One way to circumvent negative effects of pesticide use is to use microbial biotechnology such as biofertilizers, biopesticides, bioherbicides, and bioinsecticides.

Despite the fact that quite a few beneficial soil microbes have been well-characterized in laboratory conditions, the overwhelming majority of interactions between plants and the billions of root-associated microbes in field conditions remain unknown.

2.3 Rhizosphere Microbiome

Traditionally, beneficial soil microbes were usually characterized by isolating and culturing microbes with different media and growth conditions. However, culturable microbes account for only about 1 per cent of the microbial communities in an environment (Vartoukian et al. 2010) or in plant-associated habitats, which indicates that the vast majority of microbial diversity was missed in such culture-based approaches. In the early twenty-first century, with the rise of molecular biology, culture-independent molecular markers such as the 16S ribosomal RNA (rRNA) genes were widely adopted in environmental microbial community studies (Vartoukian et al. 2010).

Recent advances in next-generation sequencing allowed the study of not only individual genomes but also the collection of genomes (metagenomes) present in a given environment (Bai et al. 2015; Barea 2015; Bulgarelli et al. 2013; Fonseca et al. 2018; Reinhold-Hurek et al. 2015). The rise of metagenomics allowed researchers to capture the complexity and diversity of the microbial community inhabiting agricultural landscapes such as soils, plant roots and plant tissues. Recent advances in metagenomics allowed researchers to characterize for the first time the composition ("who is there") and function ("what are they doing") of rhizosphere microbiomes in the model plant *Arabidopsis*, and several crops including rice, maize, wheat, barley, soybean, and pearl millet were investigated (Table 2.1). These studies revealed that soil microbiomes inhabiting different compartments – rhizosphere, rhizoplane, endosphere, and unplanted bulk soil – have distinct community compositions (Edwards et al. 2015).

In microbiome studies, the rhizosphere microbiome is of particular interest. The rhizosphere, the soil that is closely attached (a few millimeters) from roots, is continuously influenced by plant roots through the rhizodeposition of exudates, mucilages, and sloughed cells (Bais et al. 2006).

The emerging picture on the effect of plant root activity in the rhizosphere microbiota indicates a dynamic environment where several biological processes and exchanges take place between root secreted compounds and rhizosphere dwelling microbes (Figure 2.1) which are absent in bulk soil microbiota causing shifts in microbial population. One of the main processes responsible for these shifts in microbial abundance and diversity observed at the rhizosphere is known as rhizosphere priming. Rhizosphere priming refers to the process where an increase in the amount of nutrients including carbon and P by decomposition can change the soil organic matter (SOM) usually by rhizodeposition (Dijkstra et al. 2013). Rhizosphere priming has been speculated to increase the release of nitrogen through decomposition of SOM (Dijkstra et al. 2013) providing in this way a thriving environment for plants and microbes to inhabit.

Signaling molecules or secreted compounds derived from plant roots or soil microorganisms are also enriched in the rhizosphere, indicating active communications between them

Table 2.1 Effect of the rhizosphere microbiome in crops.

Crop	Rhizosphere associated Microorganism or compound	References
Banana	*Bacillus*	Xue et al. (2015)
Barley	Comamonadaceae, Flavobacteriaceae, and Rhizobiaceae	Bulgarelli et al. (2015)
Barley	*Pseudomonas* (fluorescent pseudomonads)	Dudenhöffer et al. (2016)
Canola	Rhizobacteria	Lay et al. (2018)
Citrus	Nitrogen, carbon cycling, and resistance-associated bacteria	Trivedi et al. (2011)
Citrus	*Bradyrhizobium* and *Burkholderia*	Zhang et al. (2017)
Maize	*Proteobacteria, Bacteroidetes,* and *Actinobacteria*	Li et al. (2014)
Rice	*Proteobacteria, Chloroflexi,* and *Acidobacteria*	Edwards et al. (2015)
Soybean	*Actinobacteria, Acidobacteria,* and *Chloroflexi*	Mendes et al. (2014)
Soybean	*Bacillus, Bradyrhizobium,* and *Rhizobium*	Sugiyama et al. (2014)
Soybean/Tomato	Phenolic compounds	Elhady et al. (2018)
Switchgrass	Lipid	Liang et al. (2016)
Tobacco	*Pseudomonas, Bacillus*	Santhanam et al. (2015)
Tomato	*Proteobacteria, Actinobacteria,* and *Bacillus*	Antoniou et al. (2017)
Watermelon	*Bacillus, Lysobacter,* and *Pseudomonas*	Zhao et al. (2018)
Wheat	*Bacteroidetes* and *Proteobacteria*	Mavrodi et al. (2018)
Solanaceous crops (potato, tomato, chili, and brinjal)	*Brevibacillus fluminis, Brevibacillus agri, Bacillus paralicheniformis, and Microbacterium arborescens*	Goswami et al. (2019)

(Venturi and Keel 2016) (Figure 2.1). Thus, specific bacterial and fungal communities are selected from background soil populations by a variety of biotic, chemical, and physical drivers in the rhizosphere. The rhizosphere microbiota can influence the growth, development, nutrition, and immunity of its plant hosts. In this sense, rhizosphere microbiomes have also been referred to as a "second genome" of their plant hosts (Berendsen et al. 2012). In addition to the direct influences on the plants, rhizosphere microbiomes enhance soil ecosystem services by moderating soil nutrient cycles, increasing soil fertility and promoting diversified soil communities.

A number of recent studies demonstrated the importance of the rhizosphere from several crop cultivars for plant growth and defense responses as well as shifts observed in microbiome community diversity between crop rhizosphere and bulk soil (Table 2.1).

Analysis of the microbiota from soybean rhizosphere and bulk soil from field conditions revealed a higher metabolic capacity, increase in proteobacteria and growth promoting rhizobacteria such as *Bacillus*, *Bradyrhizobium*, and *Rhizobium* for rhizosphere compared to bulk soil (Sugiyama et al. 2014).

The idea that the microbiome from rhizosphere can play a role in plant fitness was demonstrated in switch grass where the authors showed the rhizosphere had an impact in increasing microbial biomass at the same time it depleted necromass in agriculture soils (Liang et al. 2016). The authors found an increase in lipid markers for AMF and Gram-negative bacteria in rhizosphere compared to bulk soil samples but found no significant differences using 16S rRNA sequencing (Liang et al. 2016).

Another study aimed to characterize the canola core root microbiome composition from roots and rhizosphere to the core root microbiome of wheat and pea from the Canadian prairie biome (Lay et al. 2018). The authors found significant difference in microbiome composition between canola (*Brassica napus*) and the other two crops (wheat and pea) and found a positive correlation between potential PGPR microbes (e.g. *Amycolatopsis* sp., *Serratia proteamaculans*, *Pedobacter* sp.) and canola yield (Lay et al. 2018).

A recent study contrasting rhizosphere and bulk soil microbiomes from a pristine forest ecosystem in the Amazon forest reported significant shifts in microbial abundance between rhizosphere and bulk soil samples (Fonseca et al. 2018). The authors found that several bacterial and fungal phyla were highly enriched in the rhizosphere when compared with bulk soil samples, while archaea showed the opposite trend, with specific phyla being more abundant in bulk soil samples. The study also captured a significant enrichment in the rhizosphere samples of several beneficial bacteria and fungi taxa that have a role in basic metabolic functions such as nitrogen and carbon cycling as well as decomposition of organic matter in forest ecosystems (e.g. *Catenulispora acidiphila* and *Postiaplacenta*) compared with bulk soil, suggesting that the rhizosphere has the potential to select beneficial bacteria (Fonseca et al. 2018). Another study also demonstrated a reduction of microbes, especially methanotrophs, in bulk soil compared with pristine forest ecosystems which are under more influence from root systems than bulk soil from pasture (Meyer et al. 2017).

2.4 Effect of Host Genotype on Microbiome

The finding that plants are able not only to recruit certain microbes but also shape the microbial communities in the root rhizosphere and endophytic compartment revealed the potential regulatory role of microbes in plant physiological processes, which was ignored until recently (Bais et al. 2006; Lundberg et al. 2012; Philippot et al. 2013). Several studies using *Arabidopsis* as a model have shown different underground compartments in the root system, notably the rhizosphere and endophytic root compartments, where microbes can interact with plants (Bulgarelli et al. 2015; Lebeis et al. 2015; Lundberg et al. 2012).

A large study using over 600 *Arabidopsis* plants provided key evidence on the effect of the host genotype on microbial community structure across different root compartments such as the endophytic compartment and the rhizosphere (Lundberg et al. 2012). The authors showed that rhizosphere and soil bacterial communities tend to be more similar when compared with endophytic root communities (Lundberg et al. 2012). Another work showed the effect of a key defense phytohormone salicylic acid on the modulation of the microbiome population structure (Lebeis et al. 2015), further demonstrating the role of the host genotype in shaping the root microbiome. The same study also demonstrated that the plant genotype modulated bacterial taxa at the phylum level for the endophytic root compartment. Different *Arabidopsis* plant immune mutants, such as *cpr5*, *pad4,* and *npr1*, displayed altered relative abundance levels of *Actinobacteria*, *Proteobacteria,* and *Firmicutes* (Lebeis et al. 2015).

A recent study also confirmed the effect of host plants in shaping the root-associated microbiome (Bulgarelli et al. 2015). In this study, the authors found evidence for positive selection for certain protein families involved with host–pathogen interaction in the barley rhizosphere microbiome such as the type 3 secretion system from the bacterial phytopathogen *Pseudomonas syringae,* but also microbe-associated molecular patterns (MAMPs) from the plant immune system (e.g. regulatory proteins containing a GGDEF domain involved in bacterial signaling) (Bulgarelli et al. 2015).

The above examples illustrate how the host genotype can shape microbial population diversity and abundance and also plant immune signaling (e.g. effect of defense-related phytohormones such as SA). The emerging picture is that the plant microbiome is extremely dynamic and adaptable to help plants to cope with a plethora of biotic and abiotic stress responses. Indeed, a recent work showed that plants can modulate their microbiome in response to adaptive phosphate starvation, which they undergo in order to manage low phosphate (Pi) availability from soil (Castrillo et al. 2017). The abundance of several bacterial taxa present in root endophytic communities varied across different plant phosphate-starvation response (PSR) mutants, reinforcing the role of genes involved in stress responses as key drivers of microbiome changes besides soil composition.

In addition, it has been shown that the rhizosphere microbiome was “drawn” from the bulk soil microbiome and endophytic microbiome that is highly selected by host genotypes (Bai et al. 2015; Bulgarelli et al. 2012; Lundberg et al. 2012; Mendes et al. 2014; Ofek-Lalzar et al. 2014; Schlaeppi et al. 2014). Moreover, soil type and environmental parameters are major factors shaping the composition of the rhizosphere microbial communities (Bulgarelli et al. 2012; Lundberg et al. 2012; Wagner et al. 2016). Additionally, agricultural practices such as irrigation and land use intensity affect root-associated microbiome as well (Estendorfer et al. 2017; Mavrodi et al. 2018).

Two models were proposed to explain the observed root compartment microbiota differentiation: the two-step selection model (Bulgarelli et al. 2013) and the three-step enrichment model (Reinhold-Hurek et al. 2015). In these models, the bulk soil provides a highly diverse microbial reservoir, while the effect of the host genotype, which gradually increases toward the roots, selects specialized communities into the rhizosphere, rhizoplane, and endosphere. Thus, the microbial degree of specialization increases toward the roots. In bulk soil, the assembly of microbial communities depends on the soil type, environmental factors and the vegetation history. Toward the roots, the gradients of rhizodeposits imposed by roots affect the community composition in the rhizosphere. Later, a more specialized community is selected and enriched by the strong effect of host genotypes on the rhizoplane and lastly, the degree of microbial selection and specialization by host plant further increases at the endophytic compartment.

Although the soil type is identified as the major factor shaping the composition of the rhizosphere microbial communities, under identical environmental conditions and soils, the plant genotype can also be considered one of the main drivers of microbial diversity in endophytic root communities. Plant roots can act as filters for its own microbiome (Reinhold-Hurek and Hurek 2011). Consistent with the selection model of microbiome differentiations, a few studies have portrayed the notion that the plant genotype can determine the endophytic microbiome. For example, different *Arabidopsis* accessions grown in laboratory conditions exhibited distinct microbial community compositions (Micallef et al. 2009).

In the other hand, plant genetic variations impose moderate but significant effects on the structure and function of the microbial communities in the rhizosphere. Variations in rhizosphere microbial diversity showed a significant correlation with different maize inbred lines across different fields (Peiffer et al. 2013). Recently, a follow-up study identified the specific microbes that reproducibly associated with plant genotype (Walters et al. 2018). The samples were collected from 27 maize inbred lines grown in three fields at multiple time points, in combination with partial replication five years later. Bacteria profiling of over 4000 samples revealed a collection of 143 operational taxonomic units (OTUs), the relative abundances of which are regulated by plant genotype. Although the root microbiome is largely affected by soil type, plant growth and weather events, specific taxa responded to differences in host genotype. If these taxa have beneficial functions, they may serve as candidate traits for selective breeding in the next-generation agriculture.

It has been proposed that the effect of plant genotype may be bigger in natural ecosystems with long-term co-evolution of root and rhizosphere microbiota (Philippot et al. 2013). Notably, a recent study showed that the host genotype-mediated effects on microbiome composition were translated into differential support of beneficial *Pseudomonas* growth, which in turn has a great effect on plant growth and health (Haney et al. 2015). Other evidence comes from significant differences between the composition of the rhizosphere microbiomes of domesticated barley and its wild ancestors (Bulgarelli et al. 2015), which suggests that the genetic makeup of host genomes affects the composition of the rhizosphere community structure. Taken together, this evidence indicates that host genotype plays a key role in influencing the microbial community structure in the rhizosphere and the endosphere.

In addition, interactions between plants and the soil microbiome are subject to the regulation of plant developmental stages. Rhizosphere microbiomes of a common annual grass, *Avena fatua*, undergo a compositional succession, and the successional pattern is remarkably consistent across multiple growing seasons (Shi et al. 2015). Changes in rhizosphere microbial communities throughout plant development have also been observed in maize and *Arabidopsis* (Chaparro et al. 2014; Li et al. 2014). Soil bacteria that respond to the presence of grassroots display phylogenetic clustering, suggesting that rhizosphere competency is phylogenetically conserved and an evolved trait (Shi et al. 2015). Thus, the strong and consistent successional patterns of rhizosphere microbes may represent a plant fitness trait. Furthermore, *Arabidopsis* roots at distinct developmental stages are capable of rapidly selecting their own microbiomes as demonstrated in a laboratory experimental system (Yuan et al. 2015). The mechanisms underlying the assembly and succession of rhizosphere microbial communities are just beginning to be explored.

2.5 Microbial Biotechnology in Agricultural Sustainability: The Role of Microbiomes in Plant Defense

Plants are susceptible to opportunistic microbial pathogens that can cause disease. These include fungi, bacteria, oomycete, viruses, and insects etc.. Crop loss due to disease caused by plant pathogens is a common threat to agricultural productivity. The impact of this on the global economy is a loss of about 200 billion US dollars (USD) per year (Oerke 2006; Strange and Scott 2005). Pathogens are therefore a major threat to global food security and agricultural sustainability.

Until recently, plant-pathogen interaction was regarded mainly as a battle between the plant host immune system and the pathogen's ability to evade the host defenses. However, recent findings (Antoniou et al. 2017; Castrillo et al. 2017; Lebeis et al. 2015; Xue et al. 2015) of an active role of the root microbiota in plant biotic stresses introduced a third element to plant-microbe interactions: the microbiome.

Microbes can help plants to fight disease in different plant compartments and organs but notably in areas where pathogens can gain access or invade, such as leaves (e.g. stomata) and roots. There are broad applications for microbes in sustainable agriculture, mainly coming from the application of beneficial microbes to improve crop tolerance to stresses or improve growth instead of using chemical fertilizers and pesticides.

Plants can recruit beneficial microbes, which in turn can help plants to resist pathogen infection. There are many ways in which these so-called beneficial microbes can help the plant to fight against infection: (i) Beneficial microbes can outnumber and outgrow pathogenic microbes. (ii) Beneficial microbes can secrete toxins that inhibit and kill pathogenic microbes. (iii) Signaling pathways like hormone biosynthesis by the plant host can recruit and select certain microbes that help the plant to fight against infection.

The close relationship between root endophytes and plant species can often aid plants in the battle against pathogen infection. Root endophytes can secrete antifungal metabolites and antibiotics that help in suppressing pathogen virulence (Backman and Sikora 2008; Kushwaha et al. 2019). Several fungal genera have been reported to antagonize pathogen infections. These include species of *Trichoderma* (Kashyap et al. 2017; Narisawa et al. 2002), *Aureobasidium* (Wachowska and Głowacka 2014), *Fusarium* (Narisawa et al. 2002) and *Candida* (de Capdeville et al. 2002) etc.

2.5.1 Disease Suppressive Soils

The constant growth of crops in agricultural soils can have two contrasting effects: It can cause pathogens to accumulate in the soil (Santhanam et al. 2015), or it can give rise to soils that are resistant to plant pathogens with the accumulation of biocontrol agents (Raaijmakers and Mazzola 2016) in a phenomenon known as disease-suppressive soils.

Disease-suppressive properties of microorganisms from suppressive composts were demonstrated recently in tomato plants grown in soils containing suppressive compost and sterile compost without microorganisms (Antoniou et al. 2017) and subjected to two different strains of *Fusarium oxysporum*. The authors found that disease intensity was reduced in suppressive soil compared with sterile soil for both *F. oxysporum* strains and found shifts in phylum composition in the rhizosphere such as *Proteobacteria* and *Actinobacteria* compared to bulk soil.

The causal agent of Panama disease in banana species (*Musa* spp.), the soil-borne fungi *F. oxysporum* f. sp. *cubense* (FOC), has decimated the banana cultivar Gros Michel (Xue et al. 2015). New variants of FOC have appeared, threatening banana production worldwide in other banana cultivars such as Cavendish. By using a metagenomics approach (16S rRNA gene sequencing) in diseased soils and disease-free soils (also known as disease-suppressive soils), researchers were able to find a bacterial species, *Bacillus amyloliquefaciens* NJN-6, enriched in disease-suppressive soils (Xue et al. 2015). The use of this species as a biocontrol in heavily infected banana production areas decreased disease by almost 70%, resulting in a higher yield (Xue et al. 2015). Not only was Panama disease almost completely controlled but it was done in a sustainable way without the need to use heavy chemicals or fungicides, avoiding environmental contamination. The use of bacteria enriched in disease-suppressive soils was also faster than traditional breeding methods and genetic engineering, allowing a quick recovery of soil health by halting FOC proliferation.

Perhaps one of the most drastic examples of how the use of the microbiota to confer resistance to host plants against pathogens without the necessity of generating transgenic plants or breeding programs was shown by Santhanam et al. (2015). The authors demonstrated how the use of native, root-associated bacteria helped attenuate sudden wilt disease in a tobacco relative, *Nicotiana attenuata*, from a diseased field in Utah that likely emerged because of continuous cropping. In this work, the authors studied diseased and healthy roots from *N. attenuata* plants. They found the phytopathogens *F. oxysporum* and *Alternaria* spp. enriched in field diseased roots. In addition, treatment of seeds *in vitro* with six native bacteria (*Arthrobacter nitroguajacolicus* E46, *Bacillus cereus* CN2, *Bacillus megaterium* B55, *Bacillus mojavensis* K1, *Pseudomonas azotoformans* A70, and *Pseudomonas frederiksbergensis* A176) isolated from healthy plants significantly reduced *N. attenuata* seedling mortality to wilting disease (Santhanam et al. 2015). The authors found that the use of a consortium of native bacteria compared with a single bacterial species provided improved disease control.

A recent study showed that manipulation of associated rhizosphere microbial communities through application of a bio-organic fertilizer (BOF) suppressed *Fusarium* wilt disease and improved watermelon quality (Zhao et al. 2018). BOFs, which combine functional microbes with a suitable substrate, are widely recognized as a promising approach for suppressing soil-borne pathogens and promoting plant growth. The BOF used in this study is highly complex in microbial composition. Application of this BOF induced remarkable changes in the rhizosphere microbial communities with altered bacterial and fungal diversity. Interestingly, the BOF induced changes in rhizosphere microbial communities were correlated with the reduced *Fusarium* wilt disease and improved watermelon quality.

2.5.2 Rhizosphere Microbiome Changes Caused by Pathogen Infection

Plant-associated microbes living in the rhizosphere or at the root endophytic compartment can aid and complement plant defense responses, especially the first layer of defense known as pathogen-associated molecular pattern-triggered immunity (PAMP-triggered immunity [PTI]), which is the first stage of recognition of pathogen-associated microbial patterns (PAMPs). Barley plant's root microbial communities can suppress *Fusarium graminearum* infection by the recruitment of fluorescent pseudomonads, which are correlated to disease suppression and secretion of antifungal compounds (Dudenhöffer et al. 2016). The rhizosphere microbiome from different crops (soybean, maize, and tomato) were inoculated in soybean and tomato plants prior to addition of disease-related nematodes; the authors used bulk soil inoculant and un-inoculated plants as control, and showed that addition of rhizosphere microbiome significantly reduced nematode invasion (Elhady et al. 2018). The authors also found a small increase in soybean and tomato biomass after transplantation with maize and soybean rhizosphere microbiome but no specific microbial phylum or species was directly associated with this increase (Elhady et al. 2018).

Citrus trees infected with the bacterium *Candidatus liberibacter* (Huanglongbing, HLB), the causal agent of citrus greening, displayed a shift in microbial community composition compared with non-infected trees, affecting genes involved in carbon degradation and nitrogen cycling (Trivedi et al. 2012). HLB was also shown to impair the citrus roots' microbiome rhizosphere-to-rhizoplane enrichment process by reducing the abundance of transcripts involved in plant–microbe interactions (Zhang et al. 2017).

The bacterial endophyte *Pseudomonas aeruginosa* has been shown to induce systemic resistance in cucumber plants infected with the pathogen *Sclerotium rolfsii* and subjected to high salinity (Pandey et al. 2012). The endophyte reduced seedling mortality significantly (60%); at the same time it also increased biomass accumulation upon biotic and abiotic stresses by 7 and 18 per cent, respectively (Pandey et al. 2012).

Another way plants can defend against pathogen infection is by secreting root exudates, which can alter the soil microbiota composition. Recently, it has been shown that the presence of exudates from healthy and *Ralstonia solanacearum*-infected tomato plants caused shifts in soil microbiome composition. This change can be attributed to differences in phenolic compounds present in higher amounts in the infected exudates, such as caffeic acid. Since plants invest a considerable share of carbon into root exudates in the rhizosphere, it would not be too far-fetched to imagine a microbiota regulatory role for plant exudation mechanisms.

2.6 Future Prospects of Microbiome-based Plant Breeding

With the demand to feed an increasing world population and concern about the overuse of agrochemicals, development of environment-friendly, low-input, and sustainable agricultural production has become a necessity. Exploiting beneficial soil microbes to facilitate crop production is one of the keys for sustainable agricultural production. However, the domestication and breeding of modern crop cultivars aimed at maximized yield with the use of chemical fertilizers and pesticides have affected the root-associated microbiome and compromised certain beneficial soil microorganisms (Bulgarelli et al. 2015). With a better understanding of soil microbiomes, optimizing plant–microbiome interactions will be a potential breeding strategy within the context of sustainable agricultural production.

Current advances in soil microbiomes have already shed light on the potential of the future microbiome-based breeding strategy. High reproducibility of selected soil microbiomes in altering plant functions (Panke-Buisse et al. 2015) suggests that microbiomes can be selected to modify plant traits. This also suggests the feasibility of applications of soil microbiomes in agricultural production. In addition, the efficiency of plants in selecting for beneficial microbiomes may be an important trait that can be potentially exploited for plant breeding purposes.

To design the microbiome-based breeding strategy, we first need in-depth analysis of rhizosphere soil microbiota in crop plants, including spatial and temporal changes of soil microbiota. Further, understanding how plant productivity and fitness traits are synergistically controlled by plant genomes and plant-associated microbiomes will contribute to the design of future crops to achieve sustainable agriculture. In the past few years, plant microbiome advances provided important insights into processes underlying the structure and assembly of the microbial community. There are still numerous undiscovered "black boxes" and knowledge gaps, including characterization of rhizosphere signaling compounds, mechanisms involved in the molecular dialog between plant roots and the rhizosphere microbiome, and identification of key plant genes controlling recruitment of beneficial microbes. All of these are important aspects to take into account for crop breeding.

2.7 Conclusion

Plants are sessile organisms and therefore must secrete an arsenal of chemicals and molecules to adapt to their surrounding environment and cope with a plethora of biotic and abiotic stresses. We described in this chapter many examples of how plants can modulate their root microbiome (mostly in the rhizosphere and endophytic compartment), causing shifts in microbial abundance and diversity, especially under biological stresses. The findings discussed in this chapter may help researchers, producers, and farmers to use microbes not only as biocontrol agents against pathogens, but also as biofertilizers to improve plant growth under stress conditions such as nutrient deficiency and drought, and ultimately help to maintain more sustainable agriculture.

The use of microbes to boost agriculture holds great potential for future generations. The number of identified plant growth-promoting microbes (PGPM) has increased steadily over the last years, but the mechanistic base governing these host plant–microbe beneficial interactions is still largely unknown. Precisely how host plants can recruit microbes to cope with several plant physiological processes, such as growth, abiotic stresses and pathogen resistance, needs to be described in more detail. More specifically, what signaling pathways and genes are involved in the process? What is the effect of host genotype in driving shifts in microbial populations upon physiological cues such as growth and stress responses? Many studies point to specific compartmentalization of the shifts in microbial composition in response to different conditions faced by plants mainly in the rhizosphere and root endophytic compartment.

Another challenge is to understand different sources of variation impacting plant-modulated shifts in the microbiome, whether environmental, genetic, or chemical. Many environmental factors, such as soil composition, pH, depth, and temperature, can contribute to the observed changes in microbial composition besides the host genotype (Wagner et al. 2016).

One of the key goals in sustainable agriculture is therefore to harness the microbiota particularly associated with a beneficial phenotype and to maintain it in field conditions. In order to do this, studies with successfully tested inoculants that are able not only to gain access to host plants but also to persist in field conditions will be required.

References

Antoniou, A., Tsolakidou, M.D., Stringlis, I.A., and Pantelides, I.S. (2017). Rhizosphere microbiome recruited from a suppressive compost improves plant fitness and increases protection against vascular wilt pathogens of tomato. *Frontiers in Plant Science* 8: 2022.

Backman, P.A. and Sikora, R.A. (2008). Endophytes: an emerging tool for biological control. *Biological Control* 46 (1): 1–3.

Bai, Y., Müller, D.B., Srinivas, G. et al. (2015). Functional overlap of the *Arabidopsis* leaf and root microbiota. *Nature* 528 (7582): 364–369.

Bais, H.P., Weir, T.L., Perry, L.G. et al. (2006). The role of root exudates in rhizosphere interactions with plants and other organisms. *Annual Review of Plant Biology* 57: 233–266.

Barea, J.M. (2015). Future challenges and perspectives for applying microbial biotechnology in sustainable agriculture based on a better understanding of plant-microbiome interactions. *Journal of Soil Science and Plant Nutrition* 15 (2): 261–282.

Berendsen, R.L., Pieterse, C.M., and Bakker, P.A. (2012). The rhizosphere microbiome and plant health. *Trends in Plant Science* 17 (8): 478–486.

Bhardwaj, D., Ansari, M.W., Sahoo, R.K., and Tuteja, N. (2014). Biofertilizers function as key player in sustainable agriculture by improving soil fertility, plant tolerance and crop productivity. *Microbial Cell Factories* 13 (1): 66–76.

Bulgarelli, D., Rott, M., Schlaeppi, K. et al. (2012). Revealing structure and assembly cues for *Arabidopsis* root-inhabiting bacterial microbiota. *Nature* 488 (7409): 91–95.

Bulgarelli, D., Schlaeppi, K., Spaepen, S. et al. (2013). Structure and functions of the bacterial microbiota of plants. *Annual Review of Plant Biology* 64 (1): 807–838.

Bulgarelli, D., Garrido-Oter, R., Münch, P.C. et al. (2015). Structure and function of the bacterial root microbiota in wild and domesticated barley. *Cell Host & Microbe* 17 (3): 392–403.

Busby, P.E., Soman, C., Wagner, M.R. et al. (2017). Research priorities for harnessing plant microbiomes in sustainable agriculture. *PLoS Biology* 15 (3): e2001793.

de Capdeville, G., Wilson, C.L., Beer, S.V., and Aist, J.R. (2002). Alternative disease control agents induce resistance to blue mold in harvested 'red delicious' apple fruit. *Phytopathology* 92 (8): 900–908.

Castrillo, G., Teixeira, P.J.P.L., Paredes, S.H. et al. (2017). Root microbiota drive direct integration of phosphate stress and immunity. *Nature* 543 (7646): 513–518.

Chaparro, J.M., Badri, D.V., and Vivanco, J.M. (2014). Rhizosphere microbiome assemblage is affected by plant development. *The ISME Journal* 8 (4): 790–803.

Dijkstra, F.A., Carrillo, Y., Pendall, E., and Morgan, J.A. (2013). Rhizosphere priming: a nutrient perspective. *Frontiers in Microbiology* 4: 216.

Dudenhöffer, J.H., Scheu, S., and Jousset, A. (2016). Systemic enrichment of antifungal traits in the rhizosphere microbiome after pathogen attack. *Journal of Ecology* 104 (6): 1566–1575.

Edwards, J., Johnson, C., Santos-Medellín, C. et al. (2015). Structure, variation, and assembly of the root-associated microbiomes of rice. *Proceedings of the National Academy of Sciences of the United States of America* 112 (8): E911–E920.

Elhady, A., Adds, S., Hallmann, J., and Heuer, H. (2018). Rhizosphere microbiomes modulated by pre-crop assisted plants in defence against plant-parasitic nematodes. *Frontiers in Microbiology* 9: 1133.

Estendorfer, J., Stempfhuber, B., Haury, P. et al. (2017). The influence of land use intensity on the plant-associated microbiome of *Dactylis glomerata* L. *Frontiers in Plant Science* 8: 930.

Fonseca, J.P., Hoffmann, L., Cabral, B.C.A. et al. (2018). Contrasting the microbiomes from forest rhizosphere and deeper bulk soil from an Amazon rainforest reserve. *Gene* 642: 389–397.

Fox, J.E., Gulledge, J., Engelhaupt, E. et al. (2007). Pesticides reduce symbiotic efficiency of nitrogen-fixing rhizobia and host plants. *Proceedings of the National Academy of Sciences of the United States of America* 104 (24): 10282–10287.

Goswami, S.K., Kashyap, P.L., and Awasthi, S. (2019). Deciphering rhizosphere microbiome for the development of novel bacterial consortium and its evaluation for salt stress management in solanaceous crops in India. *Indian Phytopathology* https://doi.org/10.1007/s42360-019-00174-1.

Guerrero, R., Margulis, L., and Berlanga, M. (2013). Symbiogenesis: the holobiont as a unit of evolution. *International Microbiology* 16 (3): 133–143.

Haney, C.H., Samuel, B.S., Bush, J., and Ausubel, F.M. (2015). Associations with rhizosphere bacteria can confer an adaptive advantage to plants. *Nature Plants* 1 (6): 15051.

Kashyap, P.L., Rai, P., Srivastava, A.K., and Kumar, S. (2017). *Trichoderma* for climate resilient agriculture. *World Journal of Microbiology and Biotechnology* 33 (8): 155.

Kushwaha, P., Kashyap, P.L., Kuppusamy, P. et al. (2019). Functional characterization of endophytic bacilli from pearl millet (*Pennisetum glaucum*) and their possible role in multiple stress tolerance. *Plant Biosystems* https://doi.org/10.1080/11263504.2019.1651773.

Lay, C.Y., Bell, T.H., Hamel, C. et al. (2018). Canola root–associated microbiomes in the Canadian prairies. *Frontiers in Microbiology* 9: 1188.

Lebeis, S.L., Paredes, S.H., Lundberg, D.S. et al. (2015). Salicylic acid modulates colonization of the root microbiome by specific bacterial taxa. *Science* 349 (6250): 860–864.

Li, X., Rui, J., Mao, Y. et al. (2014). Dynamics of the bacterial community structure in the rhizosphere of a maize cultivar. *Soil Biology and Biochemistry* 68: 392–401.

Liang, C., Jesus, E.D.C., Duncan, D.S. et al. (2016). Switchgrass rhizospheres stimulate microbial biomass but deplete microbial necromass in agricultural soils of the upper Midwest, USA. *Soil Biology and Biochemistry* 94: 173–180.

Lundberg, D.S., Lebeis, S.L., Paredes, S.H. et al. (2012). Defining the core *Arabidopsis thaliana* root microbiome. *Nature* 488 (7409): 86–90.

Mauchline, T.H. and Malone, J.G. (2017). Life in earth–the root microbiome to the rescue? *Current Opinion in Microbiology* 37: 23–28.

Mavrodi, D.V., Mavrodi, O.V., Elbourne, L.D. et al. (2018). Long-term irrigation affects the dynamics and activity of the wheat rhizosphere microbiome. *Frontiers in Plant Science* 9: 345.

Mendes, L.W., Kuramae, E.E., Navarrete, A.A. et al. (2014). Taxonomical and functional microbial community selection in soybean rhizosphere. *The ISME Journal* 8 (8): 1577–1587.

Meyer, K.M., Klein, A.M., Rodrigues, J.L. et al. (2017). Conversion of Amazon rainforest to agriculture alters community traits of methane-cycling organisms. *Molecular Ecology* 26 (6): 1547–1556.

Micallef, S.A., Shiaris, M.P., and Colón-Carmona, A. (2009). Influence of *Arabidopsis thaliana* accessions on rhizobacterial communities and natural variation in root exudates. *Journal of Experimental Botany* 60 (6): 1729–1742.

Narisawa, K., Kawamata, H., Currah, R.S., and Hashiba, T. (2002). Suppression of *Verticillium* wilt in eggplant by some fungal root endophytes. *European Journal of Plant Pathology* 108 (2): 103–109.

Oerke, E.C. (2006). Crop losses to pests. *The Journal of Agricultural Science* 144 (1): 31–43.
Ofek-Lalzar, M., Sela, N., Goldman-Voronov, M. et al. (2014). Niche and host-associated functional signatures of the root surface microbiome. *Nature Communications* 5: 4950.
Oldroyd, G.E., Murray, J.D., Poole, P.S., and Downie, J.A. (2011). The rules of engagement in the legume-rhizobial symbiosis. *Annual Review of Genetics* 45: 119–144.
Pandey, P.K., Yadav, S.K., Singh, A. et al. (2012). Cross-species alleviation of biotic and abiotic stresses by the endophyte *Pseudomonas aeruginosa* PW09. *Journal of Phytopathology* 160 (10): 532–539.
Panke-Buisse, K., Poole, A.C., Goodrich, J.K. et al. (2015). Selection on soil microbiomes reveals reproducible impacts on plant function. *The ISME Journal* 9 (4): 980–989.
Peiffer, J.A., Spor, A., Koren, O. et al. (2013). Diversity and heritability of the maize rhizosphere microbiome under field conditions. *Proceedings of the National Academy of Sciences of the United States of America* 110 (16): 6548–6553.
Philippot, L., Raaijmakers, J.M., Lemanceau, P., and Van Der Putten, W.H. (2013). Going back to the roots: the microbial ecology of the rhizosphere. *Nature Reviews Microbiology* 11 (11): 789.
Pieterse, C.M., de Jonge, R., and Berendsen, R.L. (2016). The soil-borne supremacy. *Trends in Plant Science* 21 (3): 171–173.
Raaijmakers, J.M. and Mazzola, M. (2016). Soil immune responses. *Science* 352 (6292): 1392–1393.
Reinhold-Hurek, B. and Hurek, T. (2011). Living inside plants: bacterial endophytes. *Current Opinion in Plant Biology* 14 (4): 435–443.
Reinhold-Hurek, B., Bünger, W., Burbano, C.S. et al. (2015). Roots shaping their microbiome: global hotspots for microbial activity. *Annual Review of Phytopathology* 53: 403–424.
Remy, W., Taylor, T.N., Hass, H., and Kerp, H. (1994). Four hundred-million-year-old vesicular arbuscular mycorrhizae. *Proceedings of the National Academy of Sciences of the United States of America* 91 (25): 11841–11843.
Santhanam, R., Weinhold, A., Goldberg, J. et al. (2015). Native root-associated bacteria rescue a plant from a sudden-wilt disease that emerged during continuous cropping. *Proceedings of the National Academy of Sciences of the United States of America* 112 (36): E5013–E5020.
Schlaeppi, K. and Bulgarelli, D. (2015). The plant microbiome at work. *Molecular Plant-Microbe Interactions* 28 (3): 212–217.
Schlaeppi, K., Dombrowski, N., Oter, R.G. et al. (2014). Quantitative divergence of the bacterial root microbiota in *Arabidopsis thaliana* relatives. *Proceedings of the National Academy of Sciences of the United States of America* 111 (2): 585–592.
Shi, S., Nuccio, E., Herman, D.J. et al. (2015). Successional trajectories of rhizosphere bacterial communities over consecutive seasons. *MBio* 6 (4): e00746–e00715.
Strange, R.N. and Scott, P.R. (2005). Plant disease: a threat to global food security. *Annual Review of Phytopathology* 43: 83–116.
Sugiyama, A., Ueda, Y., Zushi, T. et al. (2014). Changes in the bacterial community of soybean rhizospheres during growth in the field. *PLoS One* 9 (6): e100709.
Trivedi, P., Spann, T., and Wang, N. (2011). Isolation and characterization of beneficial bacteria associated with citrus roots in Florida. *Microbial Ecology* 62 (2): 324–336.
Trognitz, F., Hackl, E., Widhalm, S., and Sessitsch, A. (2016). The role of plant–microbiome interactions in weed establishment and control. *FEMS Microbiology Ecology* 92 (10): 1–15.
Vacheron, J., Desbrosses, G., Bouffaud, M.L. et al. (2013). Plant growth-promoting rhizobacteria and root system functioning. *Frontiers in Plant Science* 4: 356.

Vartoukian, S.R., Palmer, R.M., and Wade, W.G. (2010). Strategies for culture of "unculturable" bacteria. *FEMS Microbiology Letters* 309 (1): 1–7.

Venturi, V. and Keel, C. (2016). Signaling in the rhizosphere. *Trends in Plant Science* 21 (3): 187–198.

Wachowska, U. and Głowacka, K. (2014). Antagonistic interactions between *Aureobasidium pullulans* and *Fusarium culmorum*, a fungal pathogen of winter wheat. *BioControl* 59 (5): 635–645.

Wagner, M.R., Lundberg, D.S., Tijana, G. et al. (2016). Host genotype and age shape the leaf and root microbiomes of a wild perennial plant. *Nature Communications* 7: 12151.

Walters, W.A., Jin, Z., Youngblut, N. et al. (2018). Large-scale replicated field study of maize rhizosphere identifies heritable microbes. *Proceedings of the National Academy of Sciences of the United States of America* 115 (28): 7368–7373.

White, P.J. and Hammond, J.P. (2008). *The Ecophysiology of Plant-Phosphorus Interactions*, vol. 7. Springer Science & Business Media. https://doi.org/10.1007/978-1-4020-8435-5_1.

Xue, C., Penton, C.R., Shen, Z. et al. (2015). Manipulating the banana rhizosphere microbiome for biological control of Panama disease. *Scientific Reports* 5: 11124.

Yuan, J., Chaparro, J.M., Manter, D.K. et al. (2015). Roots from distinct plant developmental stages are capable of rapidly selecting their own microbiome without the influence of environmental and soil edaphic factors. *Soil Biology and Biochemistry* 89: 206–209.

Zhang, Y., Xu, J., and Riera, N. (2017). Huanglongbing impairs the rhizosphere-to-rhizoplane enrichment process of the citrus root-associated microbiome. *Microbiome* 5 (97): 1–17.

Zhao, J., Liu, J., Liang, H. et al. (2018). Manipulation of the rhizosphere microbial community through application of a new bio-organic fertilizer improves watermelon quality and health. *PLoS One* 13 (2): e0192967.

3

Seed Microbiome and Its Implication in Plant Growth Promotion and Health

Padmavathi Tallapragada[1] *and Usha Seshachla*[2]

[1] *Department of Microbiology, School of Sciences, Jain University, Bangalore, Karnataka, India*
[2] *Department of Environmental Sciences, Sri Krishna Degree College, Bangalore, Karnataka, India*

3.1 Introduction

One of the most important challenges of the present times is to provide adequate food for the vast population of the world. Food production in modern times has varied limitations in the form of severe climatic changes and is affected by several factors. According to the estimates of the Food and Agriculture Organization, there will be an enhanced requirement of another billion tons of cereal every year to quench the hunger of the world human population (FAO et al. 2018). One possible method is to use developmental programs in the agricultural field designed to bring "lab to farm" concepts which include "the success of the captive-breeding and field program" (Ronald 2014). Since the existing methodology to improve plant growth is inadequate, newer techniques are being tried out continuously in the field of agricultural research. One recent technique comprises the simultaneous cultivation of plants and microbes in their natural habitat, which promotes plant growth and sustainability even under stressful conditions (Jones 2013). The vast array of fungal organisms related to plant populations (Berendsen et al. 2012) which range to several thousands of species imparts the plants with a second set of genomes with a great beneficial capacity; this is the genome of the microbes (Berg et al. 2014; Turner et al. 2013).

The plant microbiome encompasses several microorganisms residing on the various tissues of the plant body imparting beneficial aspects to the plant and the soil in total. The plant microbiome comprises miscellaneous groups of microbial populations colonizing the inner and outer tissues of the plants, all tissue layers reachable by the microbes and complementing various plant functions that promote the growth of healthy plants (Mueller and Sachs 2015).

Wipps (1992) has defined the microbiome as "A characteristic microbial community occupying a reasonably well defined habitat which has distinct physico-chemical properties."

The Plant Microbiome in Sustainable Agriculture, First Edition. Edited by Alok Kumar Srivastava, Prem Lal Kashyap and Madhumita Srivastava.

The endophytic microbes inhabit the tissues of the plants without causing any deleterious effects to them, but rather form a mutualistic relationship with the plants, thus enhancing the plant growth and metabolism. Therefore, a seed microbiome can be defined as an eternal set of endophytic microorganisms present in the seed of a plant that play a key role in plant growth and development. They may be considered as plant symbionts necessarily present within the seed.

3.2 Seed Structure

Angiospermic seeds develop from sexually fertilized zygotes resulting in an embryo. The embryo is enveloped by food storage and protective tissues (Bewley et al. 2006). Different plant species produce varied seeds differing in their structural and anatomical complexity and in size ranging from 0.003 mg to 20 kg in orchids to double coconuts. In spite of the wide range of differences, all seeds share the same basic tissue morphology (Figure 3.1). They comprise three stages of development.

a) Tissue development within the embryo, histodifferentiation
b) Cell growth and enlargement
c) Increase in the dry weight by accumulation of dry matter, which ends as the seeds mature (Bewley et al. 2006)

The seed is comprised of tissues that represent three different generations of plants.

1) The sporophyte that develops into the seed, or ovule
2) The gametophyte that results in the formation of ova inside the ovule
3) Thirdly, the new sporophytic embryo

The other tissues present in the seed include the endosperm that imparts nutrition to the growing embryo, and the protective seed coat, the scutellum. The cotyledon includes the coleoptiles, plumule, radical, and the coleorhizae (Figure 3.2).

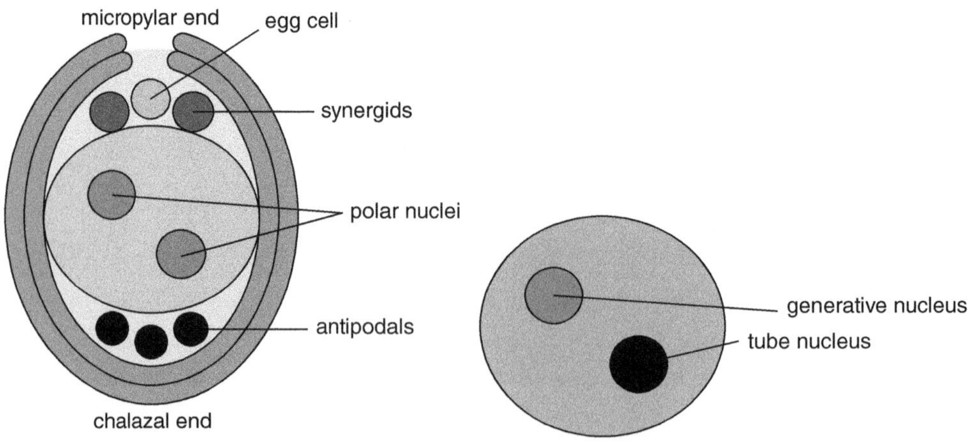

Figure 3.1 Basic tissue morphology of a plant seed. (*Source:* Courtesy: www.thefullwiki.org/Ovule).

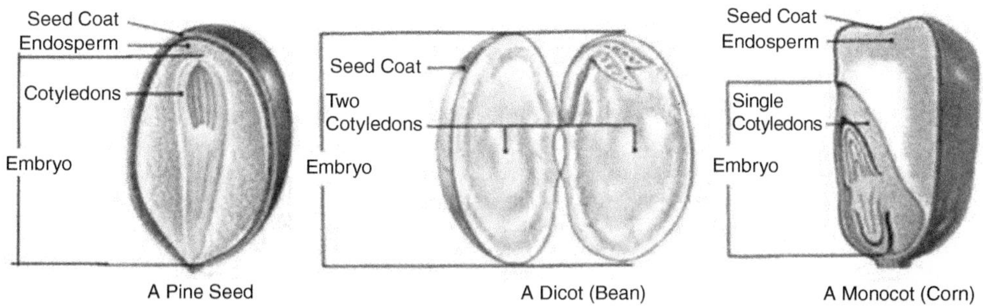

Figure 3.2 Seed cotyledon and its different parts. (*Source:* Courtesy: www.robinsonlibrary.com/science/botany/anatomy/graphics/seeds.gif.)

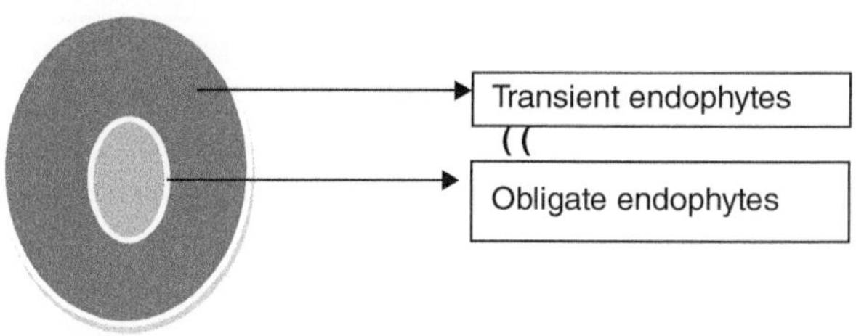

Figure 3.3 Endophytic region of the seed.

3.3 Endophytes of the Seed

The regions of the seed that are colonized by microorganisms are termed endophytic regions, and microbes, as endophytes. The endophytic region of the seed can be classified into two regions namely, the transient region harboring the microbes that stay for a short period, and the core region colonized mainly by the obligate endophytes that are permanent residents of the seed residing over many generations (Figure 3.3).

3.4 Seed Microbiome

The microbiome of the plant can be grouped into different zones based on the site of microbial colonization, namely, rhizosphere, endosphere, phyllosphere, and the endophytic microbial flora. The main source of microbes for all these microbial regions is the soil, and has been documented in studies on the microbiome of *Arabidopsis*, maize, rice, grapevine, cannabis, and cucurbits (Edwards et al. 2005; Winston et al. 2014). The microbiome of plants comprise of endophytic zones harboring varied microbial populations of archaea, endobacteria, and fungal organisms (Hardoim et al. 2015). The significance of these endophytic microbiomes with respect to plant health has been acknowledged in recent years. Their role in enhancing plant growth and their interactions with other microbial communities in the soil and in plant tissues has been documented and indicating the long-term beneficial impact of these microbes for plant health and growth enhancement. The rhizospheric soil microbes also play a key role in deciding the microbiome of the seeds (Philippot et al. 2013). The rhizosperic microorganisms enter plant root tissues thus colonizing the internal tissues of plant root and establish themselves as endophytes. The endophytes are nevertheless found in all other tissues of the plant, however, the seed endophytes form the microbiome that is the most important for the growth of plants even before they come in contact with soil microbes. As the microbes present in the tissues of the seed are the first microbes that a germinating seed encounters, their role is much greater than the role of rhizospheric microbes. However, not many studies have been done on this juncture of the seed microbiome (Barret et al. 2015; Klaedtke et al. 2016).

In order to have a better understanding, a look at the journey of the seed through its course to emerge as a mature plant needs to be visualized as shown in Figure 3.4.

The seed passes through the different phases of growth and development *viz.*, propagation, survival, and hibernation. As it passes through the different phases the seed is gradually colonized by different sets of microorganisms present endophytically within the tissues.

The dry seeds imbibe enough water to start the germination and culminate in the formation of the radical tip which emerges out of the seed coat. The process of imbibition is a tri-phasic stage which begins at a faster rate marking the initial phase. This leads to the plateau phase or phase II (Bewley et al. 2013). The final phase shows three stages, firstly, recommencement of uptake of water by the seeds, splitting of the endosperm, and formation of the radical protrusion by the process of elongation of the axes of embryo. This completes the process of germination of the seeds (Weitbrecht et al. 2011).

The microbial colonization of the seed occurs during the different phases of growth in the various regions of the seed. The core microbiomes colonize during the phases of activation, growth and reproductive stages of the seed, leading to competition among different microbial communities. The microorganisms spread to the other internal tissues of the plant through the process of internalization, and disseminate through vertical and horizontal transmissions. Systemic colonization is established in the plant tissues when the

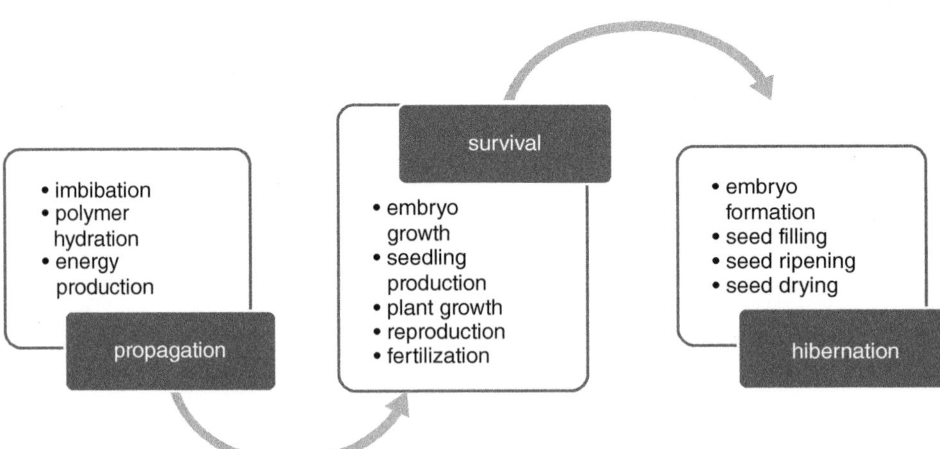

Figure 3.4 Different phases of growth and development of seed.

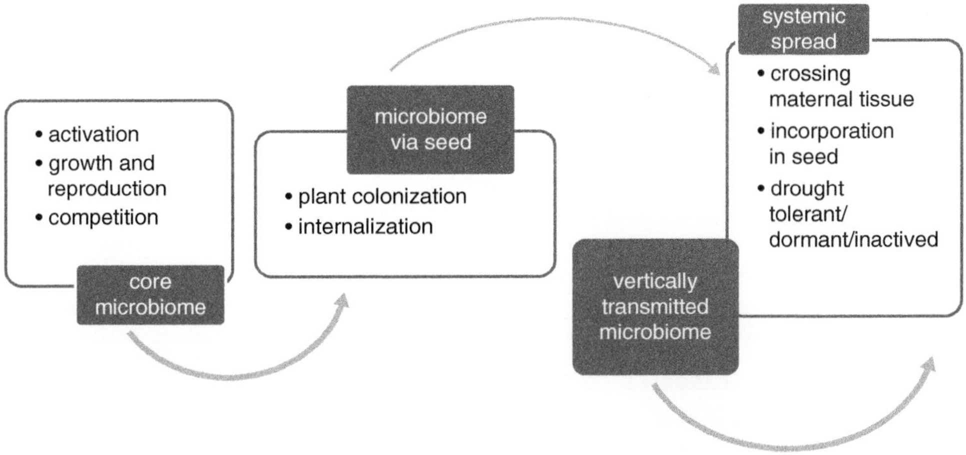

Figure 3.5 Assortment of microbial species inhabiting the tissues of the seed.

seed microbiome penetrates the maternal tissues of the plant and spreads through the other tissues. During the process of seed formation, the microbiome of the seed becomes established by a reversed transmission of microbes from the plant tissues to the seed tissues (Mitter et al. 2017).

The array of microorganisms detected inside the seed of a plant is usually few in number and type, as detected by various workers (Truyens et al. 2015). The assortment of microbial species inhabiting the tissues of the seed differs with the plant species thus establishing a unique characteristic for each plant (Figure 3.5).

The microbiome typically found in seeds consists of a limited range of microbial species (Truyens et al. 2015). It appears to have evolved by co-selection with the plant species, providing important traits for plant, affecting its health and survival capacity (Hardoim et al. 2012). Studies available have indicated that the genetic makeup of the microbes is in correlation with the genes coded by the plant genetic machinery. This establishes the fact that the inheritance of the plant characters is largely influenced by the combined effect of the genes of both the microbial population and that of the plants (Delaux et al. 2014).

3.5 Spermosphere

The region of soil present immediately around the seeds where the soil particles, microbioflora, and the germinating seeds interact with each other is termed the spermosphere. The term spermosphere is strictly limited to the time of germination of the seeds. It represents a very small region which is formed at the time of germination and later culminates to form the rhizosphere around the growing plant. Even though spermosphere is a temporary zone around the germinating seed, it plays a crucial role in that it harbors an exclusive set of microorganisms whose metabolic activities result in a unique blend of metabolites which serve as nutrients for the growing plant and also impact several key physiological functions of the plant (Figure 3.6).

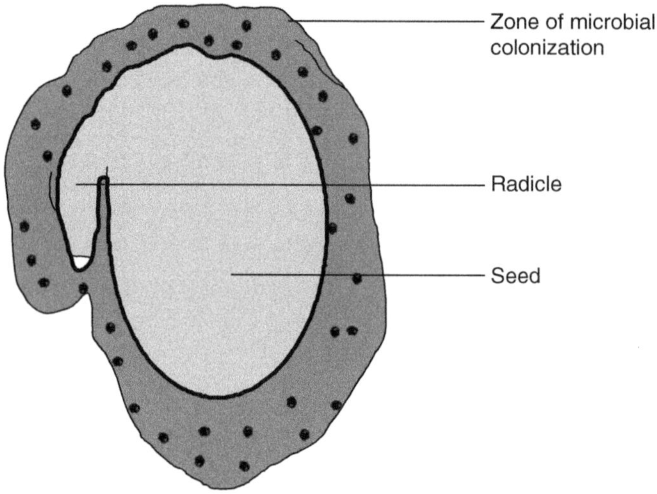

Figure 3.6 Spermosphere of a seeds.

The microbial activities have a long-term beneficial effect on the plant metabolism and thus are of utmost importance in the soil (Schiltz et al. 2015).

The spermosphere is affected by two main factors namely: (i) Seed exudates secreted by the germinating seed into the surrounding soil. These can comprise biochemical compounds that are either inhibitive in nature or may act as growth promoters. (ii) The kind and numbers of the microbial flora that are present in the vicinity of the seed.

An early use of the term spermosphere was by Verona in 1958. Further, a definition of spermosphere was given by Nelson in 2004 who stated it as "The zone of microbial interaction around the seed that is under the influence of seed carbon deposition."

The microbial population that colonizes the surface of the seeds and the surrounding region of the soil is of utmost importance to the plants since it marks the beginning of the mutual association between the plants, soil microbes and the pathogenic microorganisms present in the region which may lead to either positive or negative impacts on the growth of the plants, their development and their vigor (Singh et al. 2011). Even though the spermosphere is a temporary zone formed in the soil, the effects are seen for a longer time on the plants and play a key role in the formation of the rhizosphere of the plant. Although the role of spermosphere in the formation of the rhizosphere was known to several investigators, studies on the topic of initial interactions of soil microbes, seed microorganisms and their culmination into rhizospheric microflora are lacking.

The widest lacunae in the study of the spermosphere are the vast array of biomolecules excreted by the seeds. These biomolecules are excreted in minuscule amounts and remain in the spermosphere for an extremely short duration of time and impart a highly complex nature to the soil environment. The microbial population associated with this developing spermosphere is equally exclusive forming a unique association with the germinating seeds. Figure 3.7 shows an example of such a complex association between the seed embryo and certain soil fungi like *Fusarium culmorum* and *Epichloe typhina*, and species of *Clavibacter*, and *Verticillium* in storage tissues, and the bacteria, *Bacillus* and pathogenic *Alternaria* spp. in the seed coat tissues (Figure 3.7).

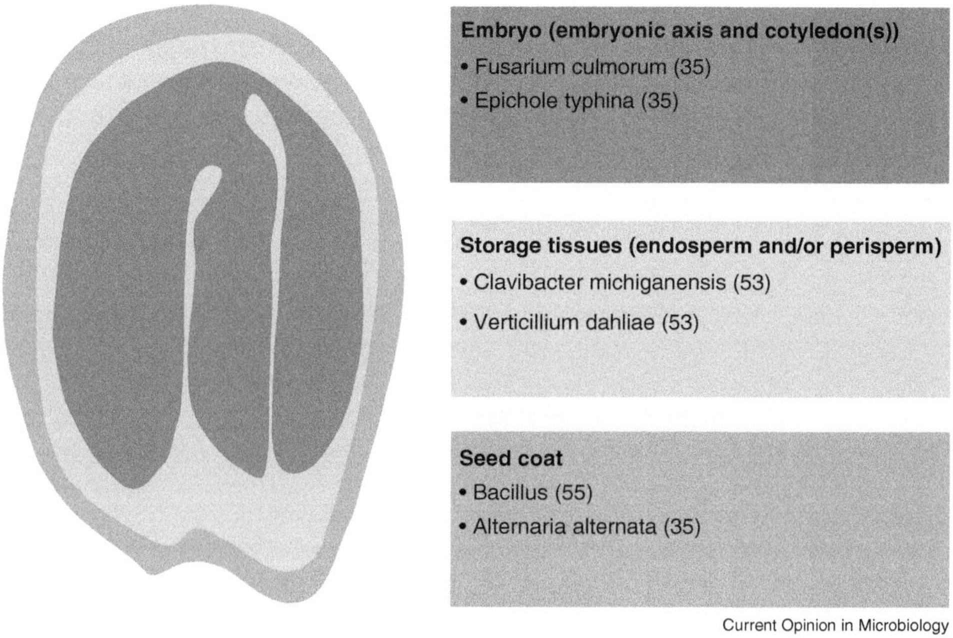

Figure 3.7 Microbes linked the developing spermosphere of a seed.

3.6 Interactions of the Spermosphere with Microorganisms

3.6.1 Mutualism and Symbiosis Among the Plant Microbiome

It has long been evident that there is a close association between plants and endophytic microbes. Studies reveal that endophytic microbes establish a strong mutual interaction within the tissues of the plants while not harming the plants adversely. The metabolic and physiological byproducts produced by plants are in turn used up by the endophytic microorganisms in the seed and the biomolecules produced by the endophytes are beneficial to the plants, thus establishing a mutually beneficial interaction between the two (Reinhold-Hurek and Hurek 2011). The plants endow the microbes with shelter and nutrients while the endophytes provide the plants with elements like nitrogen (N) and phosphorus (P). The plants also protect the endophytes from the antagonistic effect of other microbes, from external stress and from premature senescence (Ali et al. 2014).

The term spermosphere had been coined for the first time by Slykhuis (1947). Slykhuis defined the spermosphere as the region comprising a unique set of microbes and its importance, connected to the microbial activities in soil which were helpful in controlling metabolic processes of the seed and the control of seedling pathogens (Saritha and Sreeramulu 2013).

In certain conditions, the endophytes are known to produce secondary metabolites having medicinal properties, endowing the plants with the same (Alvin et al. 2014). Thus, they have been confused to be produced by plants (Chimwamurombe et al. 2016).

Detailed studies on plant–microbiome associations have revealed that there is a need for a comprehensive study of fungal endophytes in symbiosis with plants. Many fungal species are known to form close symbiotic relationship with plants that have strong environmental implications. These associations are exclusive in that they are more discreet than the arbuscular-mycorrhizal (AM) fungal associations (Vujanovic and Germida 2017).

3.6.2 Environmental Factors Affecting Mutualism

The environmental factors that directly impact the kind and numbers of microbial associations with plants include: climate, plant genomic make-up and stage of development of the plant (Hardoim et al. 2015). The selection of the set of microbial endophytes is entirely an attribute of the individual plant which has a direct effect on the characteristics of the plants, such as competitiveness, plant health, and yield (Berg et al. 2014). The unique set of microbial biome in the plant seeds encompasses the plants with special benefits which are reflected in terms of plant health, nature of metabolites formed and an edge in competition among other microbes. The plants have the capacity to select specific microbes from the vast variety of soil microflora to establish a mutually beneficial association, though the underlying principles involved in such a selection remain obscure.

Studies have been conducted on the seed microbiome of plants like grasses and the fungus *Neotyphodium* and its sexual telomorph, the clavicipitaceous, type 1, or "*epichloe*" endophytes, that are a unique set of endophytes found with a few species of grasses like tall fescue (*Festuca arundinacea*), meadow fescue (*Festuca pratensis*) or perennial ryegrass (*Lolium perenne*). The endophyte *Neotyphodium endophytes* has been found to be associated with several species of the grasses and their sexual forms (Leuchtmann 1997). Such microbiomes are known to be transmitted vertically by seeds on to the aerial parts of the plants such as the leaves and stem, thus resulting in a better dispersal of plant seeds (Sanchez Marquez et al. 2012). In such plants when the fungal endobiont happens to be lost, the corresponding characteristic feature in the plant is also lost, as shown by studies conducted by Saikkonen et al. (2004). Though these fungi were earlier considered as hazardous to animal lives, in the present day, they have been known to produce important beneficial effects such as production of chemicals that deter insect pests in agriculture (Trognitz et al. 2016).

The bacterial endophytes are known to gain entry into the seeds through surface colonization. This is a characteristic feature enabling the endophytes to ensure a close association with the plant, their survival and their mutualistic behavior with the plant (Truyens et al. 2015). The endophytic bacteria are vertically transmitted as seen in case of rice in studies by Hardoim et al. (2012). They observed that rice seeds contained vast microbiota in their seed for generations.

Studies on the maize ancestor plant *Teosinte* and in other modern maize races have shown the presence of unique microbial taxa inside the "seed-borne endophyte communities" and are vertically transmitted (Johnston-Monje and Raizada 2011). The species which were associated belonged to *Clostridium* and *Paenibacillus*. Some of the species of seed microbiota also show interactions with the soil microflora, thus colonizing the soil around the seeds and the neighboring plants; this form a co-inhabitation of the microbes with more than one plant.

Van Overbreek et al. (2011) have rightly stated that "*the contribution of the seed microbiome to the overall plant–soil microbial communities might be significant in terms of potential effects of the respective microorganisms on soil functioning and crop growth.*"

The extent of the contribution by the seed endophytes in the colonization of the soil and setting up of the soil microflora and the nature and length of their interactions with plants and other microbes is yet to be studied in detail (Trognitz et al. 2016).

3.6.3 Interactions of Spermosphere with Rhizosphere Microorganisms

The diversity of microorganisms involved with the plants and seeds are as vast as the different variety of seeds and plants present on the earth. The association of the microbes with the seeds and the plant tissues is progressive with the various stages of development of the seed to a plant. The nature and kinds of microbes associating with the seeds and later with the plants differ at each stage of the development of the plant from the seed (Kanivets and Pishchur 2001). The varieties of microbes are seen associating with the seed from the early seed stage. They are known to be present during the dormant stage of the seed, right through the germination stage then later entering through the seedling stage into the plant tissues. The spermospheric interactions of the microbe–seed–plant can be either a mutualistic one or it can at times be a deleterious reaction that may lead to plant diseases.

Studies reveal that the numerous species of microbes in the microbiome of the seed impart the plants with an alternate genome set which is highly beneficial to the plant species (Berg et al. 2014; Turner et al. 2013). These endophytic microbes reside in the tissues in a latent phase without causing any harm to the tissues of the plant and are transformed into symbionts over the passage of time forming beneficial microbes of the rhizosphere. Efforts have been put forward by ecologists in North America on the use of these endophytic microbes to strengthen the plants and make use of them to promote plant health. Simultaneously, the European Union's Cooperation in Science and Technology coalition have strived to expand the usefulness of endophytes of seeds in the field of agriculture and biotechnology (Vujanovic and Germida 2017).

It has been noted from several studies that plants exhibited a phenomenon known as "mycotrophy," wherein plants injest microscopic fungal organisms; this has been considered as a primary phase during evolution in the migration of symbiotic plants toward land which might have taken place some 400 million years before (Singh et al. 2011). This concept resulted in considering fungi as endophytic organisms inside the plant body imparting beneficial characteristics like "competitiveness, drought resistance, water-stress resistance and resistance to seed predators" and colonizing a vast plethora of plant species (Rodriguez et al. 2009). There have been only few research studies available on this topic, yet these point to the fact that the fungi present inside the tissues of plants as endophytes can occupy several different niches and can infect a vast majority of plants.

3.7 Treatment of Seed with Microorganisms

The endophytes of the seed otherwise called "the seed microbiome" have been found to be transmitted from one generation to the next among the plant species and are known to impact processes like seed germination, physiological functions of the plant, and its survival. Efforts have been made to strengthen the seed microbiomes of important crop plants such as wheat, rice, etc. and these have been observed to affect breeding practices in plants in a highly beneficial manner. The seed microbiomes have been implicated in enhancing the yield of crops, thereby increasing the productivity of food-crops, animal-feed, and production of plant fibers. Thus, a novel method of regulating the seed microbiome of some important crops has been tried and tested by various investigators and an attempt has been made to obtain certain specific microbiomes in selected seeds which can impart desired characteristics to the future plant. Specific sets of endophytic organisms have been used to treat the seeds so that when the seeds grow into a plant, the desired characteristics are expressed by the plants. In one such study, the endophytic organism *Paraburkholderia phytofirmans* strain PsJN was introduced to the flowers of parent plants and it was noticed that this microorganism was seen to be included in the next generation of seeds. This kind of inheritance is known as vertical inheritance wherein microbes are passed on from parents to the next filial generation (Mitter et al. 2017).

Further, the PsJN organism was also used in the treatment of seeds of wheat plant and the results indicated the presence of the microbe in wheat seeds paving the way for a new methodology of enhancing plant yield and other factors using seed-microbiomes. The above example has shown that it is possible to alter the endophytic microflora of the seeds specifically by introducing desired microorganisms into the seeds in a more explicit way. Studies have indicated the high success rate of using the bacterium *P. phytofirmans* PsJN, since it has shown great potential as a plant growth promoting microbe in genetically different plants. Seeds introduced with the PsJN strain were reported to exhibit increased growth and greater resistance to stress in many varieties of crop plants including the vegetables (Mitter et al. 2017).

Another important fact to be noted is that the organisms in the seed microbiome show a close interaction with the host plants. This interaction has been known to occur at molecular level and imparts beneficial aspects by altering the gene expression in the host plants. Such mechanisms have been noted in case of the strain PsJN on various plants (Lara-Chavez et al. 2015; Pinedo et al. 2015; Poupin et al. 2016). The strain PsJN is endophytic inside the seeds and exhibits active metabolism. The host plant-genomes exhibited as high as 60 percent of their genome expressing for *P. phytofirmans* PsJN during endophytic colonization of seeds. Examples of such endophytic activity were seen in case of potato plant and in *Arabidopsis thaliana*. In the case of potato, the gene expression for drought stress was shown and in *Arabidopsis*, the gene expression for storage and transport of iron was shown (Sheibani-Tezerji et al. 2015; Zhao et al. 2016).

3.8 Experimental Manipulation

Investigations have been carried out to find out the effect of controlled introduction of microbes into the microbiomes of the seeds of different plant species. It has been noted that when particular strains of microorganisms were inoculated into gnotobiotic plant hosts, and were used in place of the original microbiomes of the plant body on an experimental basis, this resulted in various beneficial as well as deleterious effects. The microbial replacements have been carried out by altering metabolic or physiological conditions like pH, nutrient compositions, temperature, and other such conditions. Beneficial effects include enhancement of host plant growth and performance and higher tolerance levels to environmental stress, while negative impacts include damage due to the use of antibiotics, and microbiome culture being unsuitable for growth, or resulting in diseases (Mueller and Sachs 2015).

3.9 Synthetic Microbiomes

Synthetic microbiomes are the culture of microbes used as inoculants to create an artificial microbiome inside the seed or plants. Specific cultures of microbes comprising a group of microbial species are used here as inoculants which are introduced into the plant seeds or the tissues in order to enhance plant growth or as efficient probes of microbes in natural conditions. Such synthetic microbiomes are beneficial in controlling the microbiomes in the seeds and regulating their functions, as well as for investigating the positive as well as negative effects of cultures on the plant tissues and in evaluating the performance of the inoculants on the plant species (Schlaeppi and Bulgarelli 2015). These microbiomes extend into the tissues of the plant and promote establishment of microflora. They play a role in reversal of dysbiosis in the case of infections in humans such as those caused by *Clostridium* sp. In spite of the benefits, these microbiomes can comprise only such organisms that are culturable and easy to grow in a synthetic environment (Bodenhausen et al. 2014).

The number and kinds of culturable microbes in the synthetic microbiomes are exponentially dependent on the kinds of microflora present in the synthetic biome. The variety of microflora in the synthetic microbiomes cannot be controlled and hence all microflora must be tested in detail. The synthetic microbiome thus creates a "hyperspace" which denotes the zone of microbial colonization and greatly varies compared with the composition of a natural set up (Mueller and Sachs 2015).

3.10 Microbiome Engineering by Artificial Selection on Host

The methods of artificial selection of the microbiome can be classified as follows:

3.10.1 Host-Mediated Selection of Microbiome

The host plant for microbiome inoculation is selected with known microflora from the plant rhizosphere. The plants are selected with either high biomass of low biomass. Soil used is sterilized and starter culture from the soil microflora is introduced into the seed and plant tissues. Evaluation of plant characteristics is done at the end of each growth cycle. This method has been tried on *A. thaliana* with early flowering and late flowering varieties (Swenson et al. 2000).

3.10.2 Host-Mediated Selection with Submicrobiome

This is a modified version of the above mentioned method and involves the use of a filter to remove all larger cells retaining only the smaller bacteria and viruses in the plant microbiome. Sub-microbiome inoculation has been used with legume plants comprising nodules and involving rhizospheric bacteria.

3.10.3 Selection of Host Plant with Different Microflora

Selection of host plant from two opposite topographical conditions with varying microflora was carried out by Lau and Lennon (2012) on *Brassica rapa*. The plants were selected from two regions, one with wet soil and the other with dry soil. The microbial flora of the two regions differed. The best lot of seeds from plants of two soil types was inoculated with a soil mix containing the complete microflora of both soil types which included fungi, nematodes, bacteria, mites, and arthropods. The resultant plants from such seeds showed a greater tolerance to drought than the original plants.

3.10.4 Simultaneous Selection of the Host and Microbiome

This method involves a combination of both host selection as well as selection of the microbiome together comprising a two-way selection process. The method as of now has not been tried out with any plant species. In nature, though, it exists among insect–fungal symbionts which show the two-way selection method. The method is more complex as it needs manipulation of both host and microbiomes (Mueller et al. 2005).

Benefits of using artificial microbiomes consist of use of the culturable as well as non-culturable microflora that can be regulated and used for microbiome engineering. Such synthetic microbiomes can be altered according to the need of various parameters like growth enhancement, disease susceptibility, etc. The role played by microbes can be experimentally evaluated and interpreted for comparative studies using different kinds of treatments. Comparison of taxonomic and genetic basis of microbiome communities can be easily done.

Microbiome engineering using artificial selection of microbes results in increased growth, plant health, plant performance in terms of better yield, which may contain "biologically, medically, or economically important" characteristics. The methods of artificial selection involve applications for up to several generations and are required to achieve the desired characteristics in the host plant by regulating the microbiome to "'reproduce' via experimental passage to the next generation of hosts." In most experiments, the host plant chosen is maintained in a stable form while manipulating the microflora alone (Mueller and Sachs 2015).

3.11 Applications

3.11.1 Microbiome of Coffee Seeds

The microflora associated with coffee seeds is of utmost importance in the curing of coffee as they impart the flavor and aroma typical to the variety of seeds. The coffee seed processing methods and technology provide easy access to certain microbes from the soil and from the machinery used to gain an entry into the seeds, thus setting up a microbial community.

In view of a better understanding of the microflora involved, it requires one to look into the techniques and methodology of coffee seed processing.

The coffee plants are cultivated using selected seed varieties that are allowed to germinate in small planters under controlled growth conditions suitably in a greenhouse. The colonization of the seedlings by the endophytic microflora occurs during the germination stage with some species attached to the surface and some others in the soil. When the seedlings are transplanted to the field six months later, they become further colonized by the soil microflora and are acted upon by environmental factors (Vaughan et al. 2015). The microbiome of the coffee seeds depends upon the method of processing used, dry or wet processing, environmental factors, conditions of storage, and include microbes from the fruit, seed, handlers, water, and processing machinery. Major groups of organisms in the microbiome include fungi belonging to Ascomycota and Basidiomycota, with major organisms such as *Aspergillus niger* and *Penicillium citrinum*. The major groups of bacteria include Proteobacteria, Firmicutes, and Actinobacteria, with *Bacillus* being the predominant form along with *Pseudomonas* and *Serratia* spp.

3.11.2 Use of *Bacillus subtilis* Strain GB03 on Cotton Seeds

The impact of the seed factors on the microbiome of cotton seeds was studied by Mahafee and Blackman in 1993. The factors affecting the seed include the pH of the surface soil, the plant cultivar, and seed treatments with fungicides which played a role on the spermosphere microbes. Colonization of cotton seeds by *Bacillus subtilis* strain GB03 was conducted. The results indicated a rise in the microbes of the spermosphere up to 5–10 fold in non-neutralized commercial delinted cotton seeds. Though the rhizosphere microbes did not differ much, there was a significant difference in the spermosphere microbes of four different cultivars of cotton. The spermosphere was seen to impact colonization of the root by microbes as the seed treatment with bacteria affected the biological control and plant growth promoting activity of the inoculant microbes.

3.11.3 Treatment of *Celosia* Seeds

Studies were conducted by Saritha and Sreeramulu (2013) on the annual weed *Celosia argentea* L. in the semi-arid soils of Andhra Pradesh. The spermosphere was examined to study the seed germination process involved. The seeds were germinated *in vitro* at pH 0.35 and the spermospheric organisms were observed. The predominant forms obtained were *A. niger, Penicillium notatum, Fusarium solani, and B. subtilis.* Though the individual cultures when grown in laboratory conditions showed growth at a pH of 5.0–6.9, in natural conditions they were shown to lower the pH of the soil thereby impacting the growth of the weed in the fields.

3.11.4 Treatment of Marama Bean Seeds with Bacterial Strains

Studies have been conducted on Marama bean, *Tylosema esculentum*, which is an exclusive legume plant native of southern Africa and shows no nodulation. The bean is consumed widely by Khoisan and Bantu ethnic groups. It has been observed that in spite of the soils limiting in nitrogen content, the Marama bean plant produces seeds with high protein. Studies were aimed to determine the role played by seed-borne bacterial endophytes in promoting plant growth in environmental conditions that are unfavorable for plant growth. The surface sterilized seeds of Marama bean were grown into seedlings in total aseptic conditions. The surface and endophytic bacteria from the roots and shoot regions were cultured *in vitro* on three different media. The bacterial microbes thus isolated were further identified using BOX-PCR fingerprinting and sequence analyses of the 16S rRNA and nifH genes. The phylogenetic analyses of 73 of the isolated bacteria revealed bacteria belonging to 14 genera comprising Proteobacteria (*Rhizobium, Massilia, Kosakonia, Pseudorhodoferax, Caulobacter, Pantoea, Sphingomonas, Burkholderia,* and *Methylobacterium*), Firmicutes (*Bacillus*), Actinobacteria (*Curtobacterium and Microbacterium*), and Bacteroidetes (*Mucilaginibacter and Chitinophaga*). The endophytes isolated from Marama bean showed plant growth

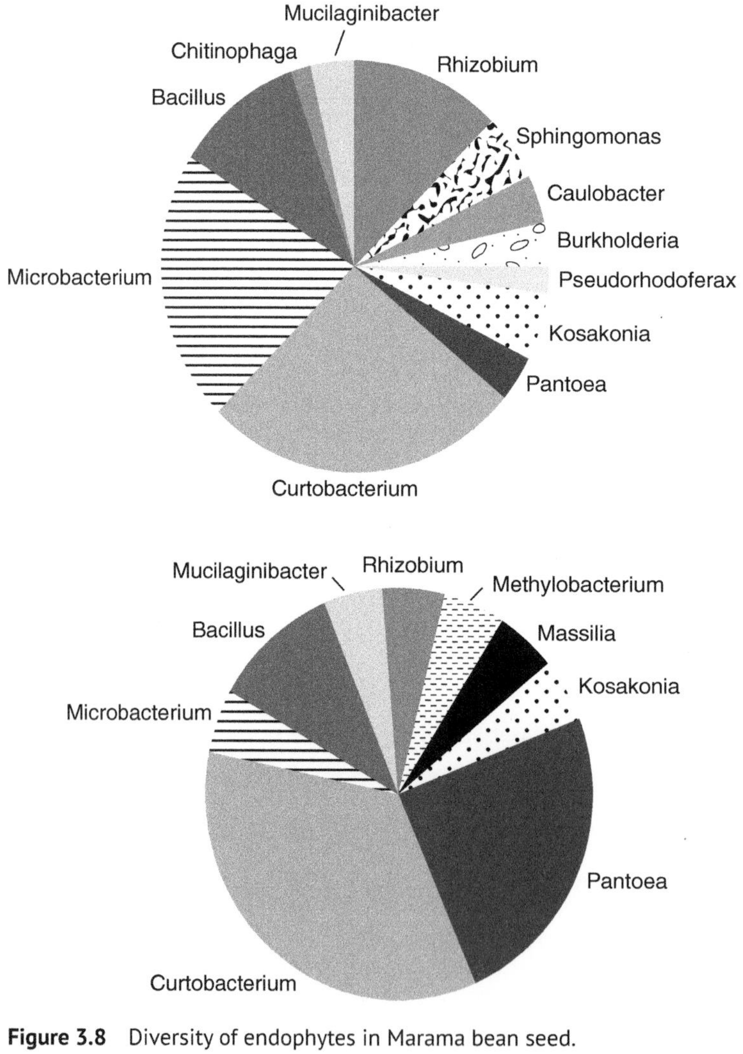

Figure 3.8 Diversity of endophytes in Marama bean seed.

promoting characteristics namely "production of IAA, ACC deaminase, siderophores, endoglucanase, protease, AHLs and capacities to solubilize phosphate and fix nitrogen." Further the 76 isolated strains of bacteria were subjected to 16S rRNA gene sequence analysis which yielded a phylogenetic tree with 29 proteobacteria, of which 14 were alpha-proteobacteria, four belonged to beta-proteobacteria and 11 gamma-proteobacteria; 8 were classified as Firmicutes, 35 to actinobacteria and 4 were grouped under Bacteroidetes (Figure 3.8).

3.12 Conclusions

The seed microbiome is an important milestone in the field of soil and agricultural microbiology which has tremendous application in breeding of crops with enhanced growth characteristics and enables introduction of new traits into the plant varieties. It provides new techniques for introduction of viable microbes into plants in order to enhance crop yield and grow crops with better designed characteristics. The most pertinent objectives of the studies on the seed microbiome are the practical applications done in the agricultural field in order to promote better growth of crops, to create awareness among the farming community to cultivate the beneficial microbes, so that less use of harmful chemical fertilizers paves the way for an ecologically more efficient system of sustainable method of plant cultivation. Research on the plant microbiome is still in its infancy, and needs to be more widely explored so that the growers can benefit. New methods for studying the seed spermosphere and the rhizosphere are being evolved which can promote better understanding of the interactions of the microbes in the soil.

References

Ali, S., Charles, T.C., and Glick, B.R. (2014). Amelioration of high salinity stress damage by plant growth-promoting bacterial endophytes that contain ACC deaminase. *Plant Physiology and Biochemistry* 80: 160–167.

Alvin, A., Miller, K.I., and Neilan, B.A. (2014). Exploring the potential of endophytes from medicinal plants as sources of antimycobacterial compounds. *Microbiological Research* 169 (7–8): 483–495.

Barret, M., Briand, M., Bonneau, S. et al. (2015). Emergence shapes the structure of the seed microbiota. *Applied and Environmental Microbiology* 81 (4): 1257–1266.

Berendsen, R.L., Pieterse, C.M., and Bakker, P.A. (2012). The rhizosphere microbiome and plant health. *Trends in Plant Science* 17 (8): 478–486.

Berg, G., Grube, M., Schloter, M., and Smalla, K. (2014). Unraveling the plant microbiome: looking back and future perspectives. *Frontiers in Microbiology* 5: 148.

Bewley, J.D., Black, M., and Halmer, P. (2006). *The Encyclopedia of Seeds: Science, Technology and Uses*, 828. Wallingford, UK: CABI.

Bewley, J.D., Bradford, K.J., Hilhorst, H.W.M., and Nonogaki, H. (2013). *Seeds: Physiology of Development, Germination and Dormancy*, 392. Springer Science & Business Media.

Bodenhausen, N., Bortfeld-Miller, M., Ackermann, M., and Vorholt, J.A. (2014). A synthetic community approach reveals plant genotypes affecting the phyllosphere microbiota. *PLoS Genetics* 10 (4): e1004283.

Chimwamurombe, P.M., Grönemeyer, J.L., and Reinhold-Hurek, B. (2016). Isolation and characterization of culturable seed-associated bacterial endophytes from gnotobiotically grown Marama bean seedlings. *FEMS Microbiology Ecology* 92 (6): fiw083.

Delaux, P.M., Varala, K., Edger, P.P. et al. (2014). Comparative phylogenomics uncovers the impact of symbiotic associations on host genome evolution. *PLoS Genetics* 10 (7): e1004487.

Edwards, R., Brazier-Hicks, M., Dixon, D.P., and Cummins, I. (2005). Chemical manipulation of antioxidant defences in plants. *Advances in Botanical Research* 42: 1–32.

FAO, IFAD, UNICEF, WFP and WHO. 2018. The State of Food Security and Nutrition in the World 2018. Building climate resilience for food security and nutrition. Rome, FAO.

Hardoim, P.R., Hardoim, C.C., Van Overbeek, L.S., and Van Elsas, J.D. (2012). Dynamics of seed-borne rice endophytes on early plant growth stages. *PLoS One* 7 (2): e30438.

Hardoim, P.R., Van Overbeek, L.S., Berg, G. et al. (2015). The hidden world within plants: ecological and evolutionary considerations for defining functioning of microbial endophytes. *Microbiology and Molecular Biology Reviews* 79 (3): 293–320.

Johnston-Monje, D. and Raizada, M.N. (2011). Conservation and diversity of seed associated endophytes in Zea across boundaries of evolution, ethnography and ecology. *PLoS One* 6 (6): e20396.

Jones, N. (2013). Food fuelled with fungi. *Nature* 504 (7479): 199.

Kanivets, V.I. and Pishchur, I.N. (2001). Bacterial microflora on disinfected sugar beet seeds. *Microbiology* 70 (3): 316–318.

Klaedtke, S., Jacques, M.A., Raggi, L. et al. (2016). Terroir is a key driver of seed-associated microbial assemblages. *Environmental Microbiology* 18 (6): 1792–1804.

Lara-Chavez, A., Lowman, S., Kim, S. et al. (2015). Global gene expression profiling of two switchgrass cultivars following inoculation with Burkholderia phytofirmans strain PsJN. *Journal of Experimental Botany* 66 (14): 4337–4350.

Lau, J.A. and Lennon, J.T. (2012). Rapid responses of soil microorganisms improve plant fitness in novel environments. *Proceedings of the National Academy of Sciences of the United States of America* 109 (35): 14058–14062.

Leuchtmann, A. (1997). Ecological diversity in *Neotyphodium*-infected grasses as influenced by host and fungus characteristics. In: *Neotyphodium/Grass Interactions* (eds. C.W. Bacon and N.S. Hill), 93–108. Boston, MA: Springer.

Mahaffee, W.F. and Backman, P.A. (1993). Effects of seed factors on spermosphere and rhizosphere colonization of cotton by *Bacillus subtilis* GB03. *Phytopathology* 83 (10): 1120–1125.

Mitter, B., Pfaffenbichler, N., Flavell, R. et al. (2017). A new approach to modify plant microbiomes and traits by introducing beneficial bacteria at flowering into progeny seeds. *Frontiers in Microbiology* 8: 11.

Mueller, U.G. and Sachs, J.L. (2015). Engineering microbiomes to improve plant and animal health. *Trends in Microbiology* 23 (10): 606–617.

Mueller, U.G., Gerardo, N.M., Aanen, D.K. et al. (2005). The evolution of agriculture in insects. *Annual Review of Ecology, Evolution, and Systematics* 36: 563–595.

Nelson, E.B. (2004). Microbial dynamics and interactions in the spermosphere. *Annual Review of Phytopathology* 42: 271–309.

Philippot, L., Raaijmakers, J.M., Lemanceau, P., and van der Putten, W.H. (2013). Going Back to the roots: the microbial ecology of the rhizosphere. *Nature Reviews Microbiology* 11 (11): 789–799. https://doi.org/10.1038/nrmicro3109.

Pinedo, I., Ledger, T., Greve, M., and Poupin, M.J. (2015). *Burkholderia phytofirmans* PsJN induces long-term metabolic and transcriptional changes involved in Arabidopsis thaliana salt tolerance. *Frontiers in Plant Science* 6: 466.

Poupin, M.J., Greve, M., Carmona, V., and Pinedo, I. (2016). A complex molecular interplay of auxin and ethylene signaling pathways is involved in *Arabidopsis* growth promotion by *Burkholderia phytofirmans* PsJN. *Frontiers in Plant Science* 7: 492.

Reinhold-Hurek, B. and Hurek, T. (2011). Living inside plants: bacterial endophytes. *Current Opinion in Plant Biology* 14 (4): 435–443.

Rodriguez, R.J., White, J.F. Jr., Arnold, A.E., and Redman, A.R.A. (2009). Fungal endophytes: diversity and functional roles. *New Phytologist* 182 (2): 314–330.

Ronald, P.C. (2014). Lab to farm: applying research on plant genetics and genomics to crop improvement. *PLoS Biology* 12 (6): e1001878.

Saikkonen, K., Wäli, P., Helander, M., and Faeth, S.H. (2004). Evolution of endophyte–plant symbioses. *Trends in Plant Science* 9 (6): 275–280.

Sanchez Marquez, S., Bills, G.F., Herrero, N., and Zabalgogeazcoa, Í. (2012). Non-systemic fungal endophytes of grasses. *Fungal Ecology* 5 (3): 289–297.

Saritha, P. and Sreeramulu, A. (2013). Allelopathic effects of *Celosia argentea* L. on spermosphere microorganisms. *International Journal of Life Sciences Biotechnology and Pharma Research* 2: 77–79.

Schiltz, S., Gaillard, I., Pawlicki-Jullian, N. et al. (2015). A review: what is the spermosphere and how can it be studied? *Journal of Applied Microbiology* 119 (6): 1467–1481.

Schlaeppi, K. and Bulgarelli, D. (2015). The plant microbiome at work. *Molecular Plant-Microbe Interactions* 28 (3): 212–217.

Sheibani-Tezerji, R., Rattei, T., Sessitsch, A. et al. (2015). Transcriptome profiling of the endophyte *Burkholderia phytofirmans* PsJN indicates sensing of the plant environment and drought stress. *MBio* 6 (5): e00621–e00615.

Singh, J.S., Pandey, V.C., and Singh, D.P. (2011). Efficient soil microorganisms: a new dimension for sustainable agriculture and environmental development. *Agriculture, Ecosystems & Environment* 140 (3–4): 339–353.

Slykhuis, J.T. (1947). Studies on *Fusarium culmorum* blight of crested wheat and brome grass seedlings. *Canadian Journal of Research* 25 (5): 155–180.

Swenson, W., Wilson, D.S., and Elias, R. (2000). Artificial ecosystem selection. *Proceedings of the National Academy of Sciences of the United States of America* 97 (16): 9110–9114.

Trognitz, F., Hackl, E., Widhalm, S., and Sessitsch, A. (2016). The role of plant–microbiome interactions in weed establishment and control. *FEMS Microbiology Ecology* 92 (10): pii: fiw138.

Truyens, S., Weyens, N., Cuypers, A., and Vangronsveld, J. (2015). Bacterial seed endophytes: genera, vertical transmission and interaction with plants. *Environmental Microbiology Reports* 7 (1): 40–50.

Turner, T.R., James, E.K., and Poole, P.S. (2013). The plant microbiome. *Genome Biology* 14 (6): 209.

Vaughan, M.J., Mitchell, T., and Gardener, B.B.M. (2015). What's inside that seed we brew? A new approach to mining the coffee microbiome. *Applied and Environmental Microbiology* 81 (19): 6518–6527.

Verona, O. (1958). La spermosphere. *Annales de l'Institut Pasteur de Lille* 95: 795–798.

Vujanovic, V. and Germida, J.J. (2017). Seed endosymbiosis: a vital relationship in providing prenatal care to plants. *Canadian Journal of Plant Science* 97 (6): 972–981.

Weitbrecht, K., Müller, K., and Leubner-Metzger, G. (2011). First off the mark: early seed germination. *Journal of Experimental Botany* 62 (10): 3289–3309.

Whipps, J.M. (1992). Status of biological disease control in horticulture. *Biocontrol Science and Technology* 2 (1): 3–24.

Winston, M.E., Hampton-Marcell, J., Zarraonaindia, I. et al. (2014). Understanding cultivar-specificity and soil determinants of the cannabis microbiome. *PLoS One* 9 (6): e99641.

Zhao, S., Wei, H., Lin, C.Y. et al. (2016). Burkholderia phytofirmans inoculation-induced changes on the shoot cell anatomy and iron accumulation reveal novel components of *Arabidopsis*-endophyte interaction that can benefit downstream biomass deconstruction. *Frontiers in Plant Science* 7: 24.

4

Microbome

The Holobiont, Its Application and Effect on the Plant System

Pragati Sahai[1] *and Vimlendu Bhushan Sinha*[2]

[1] *Amity Institute of Biotechnology, Amity University, Noida, Uttar Pradesh, India*
[2] *Department of Biotechnology, School of Engineering and Technology, Sharda University, Greater Noida, Uttar Pradesh, India*

4.1 Introduction

The world of microbes is spread all over the ecosphere. Their interactions with the environment have been subject to study for more than 100–200 years and still there is scope for further study and research. The symbiotic relationships have shaped the origin, organization, and evolution of all life forms on earth. The term symbiosis has traditionally been applied to associations like mutualism, commensalism, and even parasitism. More recent symbiosis research is expanding this definition to encompass a role of microbial symbiotic relationships in far-reaching themes of biology such as speciation, evolution, and coadaptation. The association and close relationships of organisms that cohabitat are vital for the growth and development of all eukaryotic organisms. These associations and symbiotic networks of microorganisms shape natural landscapes and directly influence the evolutionary trajectory of individual species and entire ecosystems.

The microbial interactions with plants influence plant strength, ecological diversity and nutritional stability to sustain plants in the changing environment. The interaction of the host plant with its inhabitant microbes develops three nutritional zones known as the rhizosphere, phyllosphere, and endosphere (Figure 4.1). The three zones together constitute the Holobiont, the term originally coined by Lynn Marguilis and René Fester (1991) in their book, *Symbiosis as a Source of Evolutionary Innovation* (Margulis and Fester 1991) (Figure 4.1). Before this book, innovatory work on plant–microbe interaction was carried out by Lorenz Hiltner in 1901, later published in the research work of Hartmann et al. (2008).

The association of plants with microbes is very old and ancient. The presence of vesicular-arbuscular mycorrhiza (VAM) fungi is the proof of mutualistic relationship between the host plant and the microbe that gave the nutritional strength and physical stability to the host plant that played a key role in formation of the terrestrial habitat. Their presence also depicted the genetic diversity or changes in their genetic arrangements along with changes in the environment that brought about evolution and diversification in plants

The Plant Microbiome in Sustainable Agriculture, First Edition. Edited by Alok Kumar Srivastava, Prem Lal Kashyap and Madhumita Srivastava.

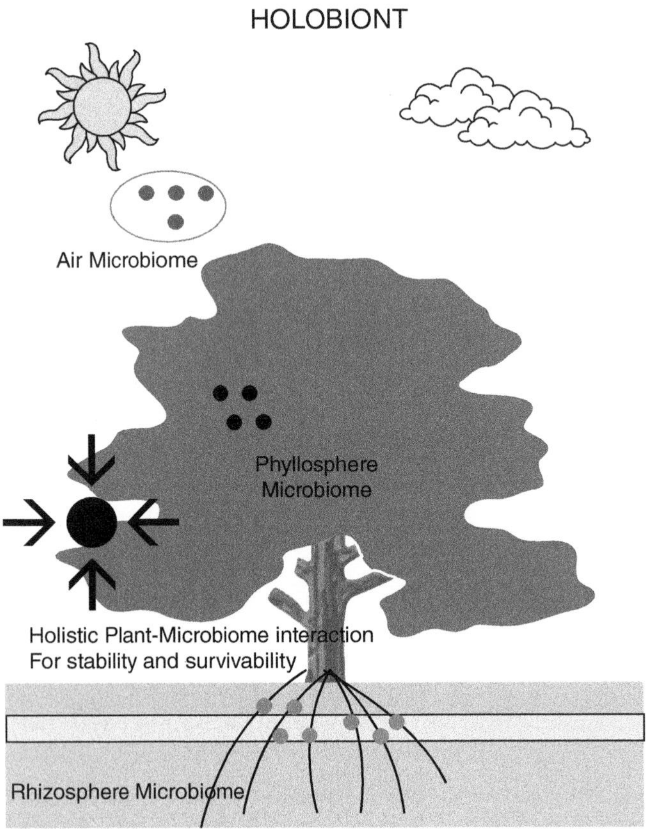

Figure 4.1 The interaction of host plant with its inhabitant microbes at different nutritional zones.

(Heckman et al. 2001). Similarly, the plant model of *Arabidopsis* demonstrated many plant–microbial interactions by different research groups, which further explored the biochemistry of holobionts (Bulgarelli et al. 2013a; Lundberg et al. 2012).

Extensive long research in this field, along with physical parameters, included molecular, proteomics and genomic levels of exploration in order to study the changes in genetic arrangements in the host associated microbes that brought about diversification in the host plant (Prestat et al. 2014). Study at genomic and molecular levels (Fu et al. 2015; Müller et al. 2015) made possible to:

- study suppression of soil-borne pathogens and resistance in plants to such pathogens because of the plant's own rhizosphere (Cook et al. 1995; Mendes et al. 2013);
- study the trans kingdom: different species and genus interactions are a subject of interest because even if the interaction is trans kingdom (plant + microbe), still the rhizosphere specificity may be seen (Berg and Smalla 2009);
- develop molecular markers because of knowledge of the genome of endophytic microorganisms;
- study the signaling behavior.

Thus, the molecular applications based on genomic sequencing and protein analysis have greatly enhanced the knowledge behind the diverse interaction of the plant microbiome (Kiers et al. 2011; Werner et al. 2014); however, this knowledge is as yet insufficient to explain the functionality of the plant microbiome with respect to vast ecological diversification. So, one must accept that despite vast research in this field, the study is still preliminary in the context of Earth as a large biome which opens future prospects of research which may be beneficial for the improvement of crops and enhancing production (Sánchez-Cañizares et al. 2017).

4.2 Effect of the Microbiome on the Plant System

The plant system largely depends on its microbiome for growth and survival. It was reported that in the absence of fungi and bacteria, plant transplants were not successful (Hardoim et al. 2008) and so the concepts of the holobiont came into the picture, stating a single dynamic entity that defines the complete function and interaction between the host plant and its associated microbiome (Figure 4.2). The word holobiont came from word holobiome, which means that when the genome of the host was intercepted, the intervention of its associated microbial genome could be seen as a single dynamic entity (Guerrero et al. 2013). So it was found that whenever there was change in the environment, it brought about a change in holobiont as a whole as well, as it had individual effects on plants and the microbiome also. The stresses faced and the survival achieved in the changing environment

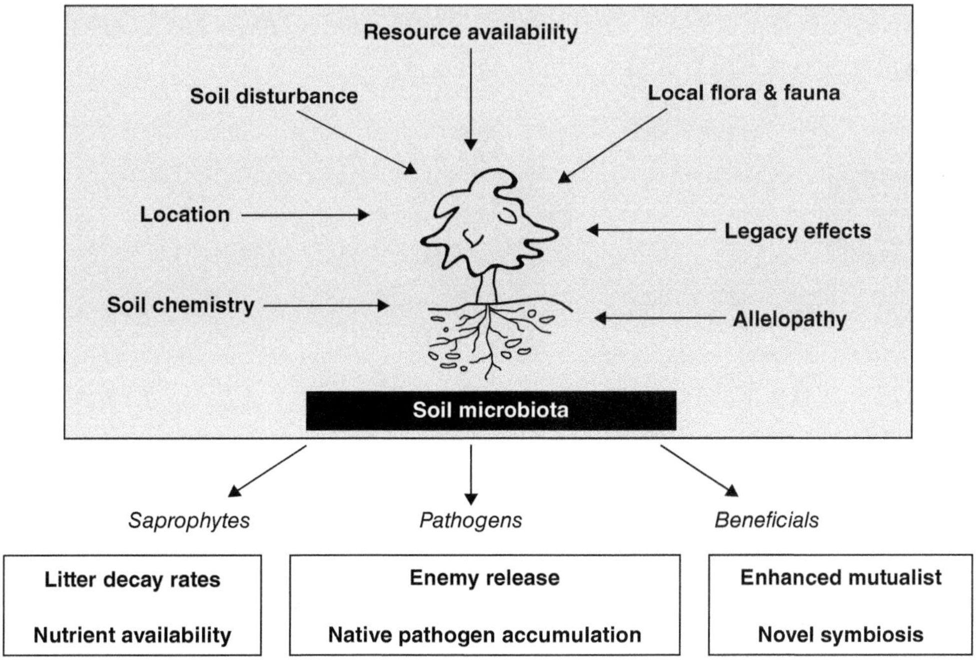

Figure 4.2 Pictorial representation of a holobiont.

largely depend on the strength of the holobiont being the unit of selection and adaptation (Zilber-Rosenberg and Rosenberg 2008). These adaptations bring heterogeneity and a defense system that can be transitional if the relationship is symbiont, or there can be vertical transmission from seed to plant if there are genomic changes in the holobiont that are functionally stable (Berendsen et al. 2012; Cankar et al. 2005).

4.3 Factors that Affect the Holobiont

4.3.1 Rhizosphere and Rhizoplane

The different planes decide the microbial consortia of that plane (Figure 4.3). In the Rhizosphere the root-associated bacteria and soil microorganisms together decide the microbiome of that plane. The resistance to soil-borne pathogens and the defense system of the plant largely depend on this microbiome. There can be a systemic resistance because of temporary association of microbiome with the host plant, or it can be lateral transfer of resistivity from soil to seed and further to the plant where the physical environment such as soil properties, pH, humidity and temperature of the soil act as growth chambers for the microbiome (Bouffaud et al. 2014). It also brings to the light the fact that instead of the plant genome, it is the inhabitant microbial community of the plane that decides the microbiome of that plane, with the host genome playing some role in the selection of the microbiome due to its evolutionary significance (Reinhold-Hurek et al. 2015). Another factor that affects the microbiome is segregation of microbes in different spheres in different planes; the microbes in the rhizosphere are difficult to segregate from the soil microbes

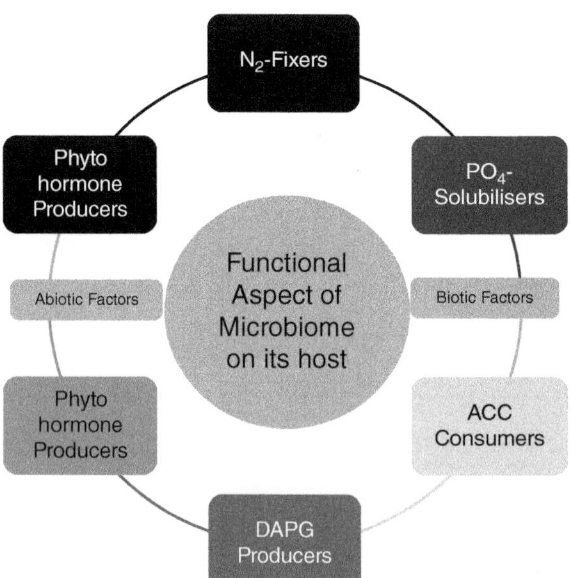

Figure 4.3 The functional aspect of microbiome on its host plant; DAPG: 2,4-diacetylphloroglucinol.

inhabiting the rhizoplane. The 16S rDNA sequencing of these soil inhabitants showed very little difference in their genome, so selection largely depended on their interaction with their host plant, their competitors and soil conditions of pH, temperature and humidity (Bulgarelli et al. 2013b). The reports of CARD-FISH (catalysed reporter deposition fluorescent in situ hybridization) in rice plant also showed the colonization of bacteria in the rhizoplane of rice, thus indicating that the rhizoplane plays an important role in deciding the microbiome of the rhizosphere (Schmidt and Eickhorst 2013).

4.3.2 Endosphere and Host Specificity

After the microbe enters the tissues from the rhizosphere or from the air, they become part of the endosphere. The order of the endosphere microbiome is arbuscular mycorrhizal fungi that penetrate deeply, followed by bacteria and lastly by archaea. The endospore microbiome, unlike the rhizosphere microbiome, is highly specific and chiefly selected by host interaction. There is less diversity seen in the endosphere than in the rhizosphere. Secondly when taxonomic units of the endosphere microbiome and rhizosphere microbiome were studied, it was seen that the microbial diversity decreased as one moved from rhizosphere to endosphere (Schlaeppi et al. 2014). The studies on field-grown poplar trees on structural variability and niche differentiation in the rhizosphere and endosphere bacterial microbiome indicated that the rhizosphere microbiome is less structurally variable, and holds a lower ecological niche than the well-structured endosphere microbiome. Thus, the endosphere microbiome helps in improving adaptability of the host plant and in increasing plant strength (Beckers et al. 2017).

4.3.3 Phyllosphere and Vertical Transmission

The endosphere microbiome can also penetrate leafy tissues and even reproductive parts, probably for vertical transfer or lateral diffusion into the next generation. These microbial denizens are residing either outside or inside the aerial plant tissues (Peñuelas et al. 2014). There have been a few reports of finding the microbiome in reproductive organs like anthers and carpels, giving evidence of vertical transmission from seed to new plant (Lopez-Velasco et al. 2013). The presence of a large amount of microbiome on the surface of host leaf and in the leaf internal tissues shows their growth in different environments, thus showing a very different microbial population in the endosphere and ectosphere of plant tissues above the ground and below the ground (Romero et al. 2014). It was also reported that the microbiomes on the leaf surface or phyllosphere faces can comprise archaea-like methylotrophs as they exist in much harsher conditions than the rhizosphere because of desiccation, UV radiation and oligotrophy. Therefore the microbiome is structurally more defined and has a better ecological niche. It is capable of developing special proteins like microbial rhodopsins that help in absorbing green light and acting as proton pumps, facilitating photosynthesis (Atamna-Ismaeel et al. 2012). These methylotrophs and photoharvesters cause the competitive exclusion of plant pathogens, and create better survival, or have a higher niche as compared to another microbiomes. Thus, the phyllosphere can also affect the holobiont.

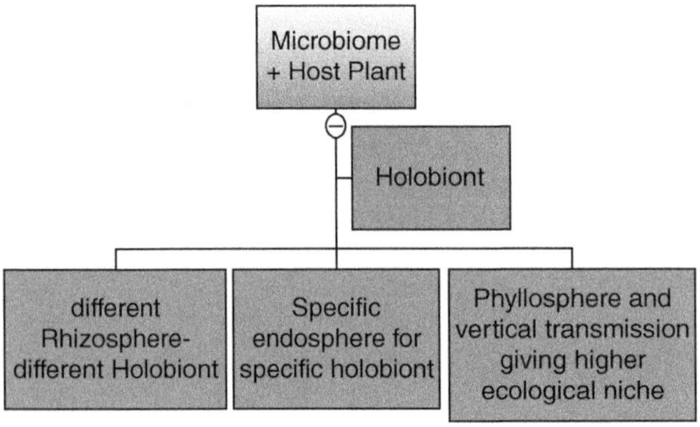

Figure 4.4 Factors that affect the holobiont.

4.4 The Functional Aspect of Microbiome on Its Host Plant

As stated earlier, the microbiome influences the strength of the plant, thus, it participates in the host plant's function (Figure 4.4). It was also reported that these microbiomes provide a defense system to the plant, increasing its immunity. Thirdly it was also reported that because of these microbiomes the plants become more structurally advanced and have a better ecological niche than the plants without their microbiome.

4.5 Core Microbiome and Ecological Niche

Microbiomes provide functional plasticity in plants. This means that the microbiome that is part of the holobiont exists after screening and competitive elimination of microbes that were weak and structurally poor. Such interaction with the environment leads to continuous elimination of weaker populations and the development of better microbial flora on the phyllosphere causing biodiversity. Because of this biodiversity even the same species of plant have different ecological niches resulting in different strains of the plant. The different strains of the same plant species have different holobionts that facilitate different processes in the survival of abiotic stresses, heterogeneity, mutualism sharing, resource accessibility and creating a novel habitat (McIntire and Fajardo 2014; Wagner et al. 2014). Here, it is important to mention the term *core microbiome* that is generated out of a plant community, for example if leguminoceae have a population of bacteria as their microbiome, then all the plants belonging to this family will have at least that population set, apart from some different ones. Again, the different analysis at genomic and transcriptome levels will help to understand the functionality of the core microbiome. That is, the microbes of different holobionts of the same family with similar genomic and transcriptome analysis will constitute a core microbiome. It is this core microbiome that will place that particular family on the ecological map, so it becomes very important for a plant community (Finkel et al. 2017; Lemanceau et al. 2017).

4.6 Developing the Immunity in Plants

Plants have two types of innate immunity: horizontal resistance and vertical resistance. Horizontal resistance is one in which the immunity is triggered due to pathogenic molecular patterns and so it is also called pathogen-associated molecular pattern triggered immunity or PTI. Similarly, vertical resistance or R-gene based immunity is triggered by effector cells, also called effector cell triggered immunity or ETI (Jones and Dangl 2006). The mechanism of defense in the former case depends on two types of molecular pattern; one that is generated by microbes and easily identified by the plants' microbiome, and second type of pattern is generated by cellular damage or any kind of biotic stress to the cell that is identified by host plant. So, in this mechanism the microbiome sends a signal to its host and the entire holobiont prepares its defense system against its microbial enemy. The defense system and hypersensitive reaction is also triggered by effector cells that are produced by the microbiome of the host plant. This helps in signaling cell apoptosis or local necrosis in case of danger (Boller and Felix 2009). Thus, both types of immunity because of microbiome help the plant to gain strength and survival. Many fungi have been reported to provide fitness and strength to tall trees; similarly numerous bacteria have provided resistance against many soil-borne bacteria (Hacquard et al. 2017).

4.7 Current Application of Microbiome and Their Future Aspects

The recent insight into the application of the microbiome revolves around the three habitats in the holobiont, *viz.*, rhizosphere, phyllosphere, and endosphere that harbor different microbes (Hardoim et al. 2015; Philippot et al. 2013; Vorholt 2012; Wagner et al. 2015). The rhizosphere apart from providing nutritional benefits to host plant, also provides effective biocontrol to the plant against phytopathogens. This biocontrol activity of the microbiome is being extensively used to replace the existing pesticidal activities that are harmful to the environment. The specificity of the microbiome to its host helps in defending the host from soil-borne pathogens (Cole et al. 2015; Zachow et al. 2016). These associations form a holobiont that can be considered as evolved microbial species, and a host assembly with better survival in the environment. The microbiome helps in preventing diseases because of its rich diversity acting as biomarkers in plant protection strategies; thus by identifying the biocontrol agents in the rhizosphere, targeted and predictive protection strategies can be designed. These microbiome-based solutions are key to targeted and predictive biocontrol approaches that are required for effective plant breeding and ecosystem health (Berg et al. 2017). However, in this symbiotic behavior there are some cheaters too, which are those microbes which take the gain from the host plant at a rate that is so rapid that it makes the mutualism unstable. There are many hypothesis that support this fact. The first hypothesis is the luxury resource exchange that allows both the host and symbiont to enjoy the surplus nutrients in the environment; however if this utilization shift tends more towards the symbiont, then it will be stronger than the host, destroying the mutualism (Kiers and Van Der Heijden 2006). But this hypothesis does not hold true as the nutritional availability in the environment is very limited, so having a surplus is out of question. The second hypothesis states the partner choice, according to which the partner with better strength for survival will be selected by the symbiont leading to disruption in mutualism with the weaker partner. Such processes occur on the basis of molecular signaling that sends a message to the symbiont to continue with the interaction or start another (Sachs and Simms 2006; Sachs et al. 2004). Again, by studying this molecular signaling, a better plant symbiont relationship can be researched and experimented upon. This molecular signaling is called plant–microbe communication. This communication is largely seen in the rhizosphere of the holobiont, and is extensively studied (Van Overbeek and Van Elsas 2008; Venturi and Keel 2016). The third hypothesis is of forced cooperative behavior that sounds very interesting in the way that the strong symbiont does the resource allocation and provides lesser resource to the weaker symbiont, thereby forcing the weaker one to be always dependent on the host and the stronger symbiont. Prominently such behavior is seen in legumes where N_2 fixing organisms penalize the rhizobia that do not participate in fixation (Kiers et al. 2003); it is similarly seen in AM fungi where the stronger ones reallocate carbon and phosphorous uptake of the cheater AM fungi and transfer the phosphorous and carbon to the host plants forcing them to be a part of the symbiont (Kiers et al. 2011). Apart from the fact that the cheaters are in symbiosis, there are reports of them being free riders, such as rhizobia in legumes which after some time does not benefit its host but starts reallocating its lipid contents by producing rhizobitoxine, an ethylene inhibitor (Ratcliff and Denison 2009). This symbiont has recently been reported to be selective for its partner (Porter and Simms 2014). The recent advances in cropping have largely eliminated the growth of these cheaters and free riders as the soil these days is nutrient enriched, so reallocation of resources or selecting stronger hosts for symbiosis is no longer required, hence the cheater population is high in conventional farming methods, but not in modern methods of agriculture (Duhamel and Vandenkoornhuyse 2013; Johnson 1993). Thus these plant microbiomes which act as free riders suggest their adaptation to changing environmental needs, thus contributing to the plant health mechanism.

4.8 Plant Fitness and Health Mechanism

The plant fitness and health mechanism revolves around its endosphere and phyllosphere after the rhizosphere, as these symbionts pass from the soil into the plant tissue or inhabit the leaf surface from the external environment. This arrangement is basically the collaboration of detoxifying enzymes, nutrients to and from the host and its microbe, and gene uptake and expression (Kemen 2014). According to the Black Queen hypothesis, this symbiosis is automatic and for selfish needs. The hypothesis was supported by the fact that the microbial interactions involve two types of population: helper (host and strong microbe) and beneficiary (other microbe or host) where helpers are responsible for major productions and beneficiaries are dependent on these helpers. This results in the loss of genes and functions in the latter case, because of non-usability and this shows the selfish trajectory that makes powerful symbiosis and accounts for plant health (Morris et al. 2012). The plants are highly adaptable and can survive with changing environmental conditions by bringing about the change in their microbiome. They can select their ectospheric and endospheric microbiome by mutual cooperation for better survival in the environment. This mutualism helps the holobiont as a whole to improve its fitness and promote evolution in ecology.

Vertical transmission is also the cause of plant fitness. Not all microbes of the microbiome can be part of the vertical transmission process that involves the integration of the microbial genome with the host genome so that the trait can be transferred to next generation, which otherwise is temporary and only in the presence of that symbiont. Once the symbiont is removed, that trait is also lost, however in the case of vertical transmission the trait enters the progeny and enables the plant to be better adapted to the environment. Various fungi like *Epichole festucae* in *Festuca rubra* because of vertical transmission affect the plant's nutrient balance (Vázquez-de-Aldana et al. 2013). This also allows the partners to have a fidelity check on each other (Frederickson 2013) and also to have an automatic fitness feedback (Sachs et al. 2011). Thus, on the broader scale the partner feedback is largely based on cooperation, so it completely collaborates with their personal feedback too, stating the relationship to be positive–positive (Friesen 2012). The recent approaches to improve plant fitness thus largely revolve on this feedback, which indicates that the plant works as a holobiont and this community dynamics help in modifying the intensity and nature of plant competition to ensure the survival rate of the plant (Pendergast et al. 2013). The future of effective biocontrol and production of the next generation of bioproducts are part of a sustainable agriculture that visualizes the plant as a meta-organism because of its ectosphere and endosphere mutualistic interactions, which are transferred to its seed and via the spermosphere to the next generation. Hence, there is the possibility of designing better agricultural strategies with the help of experimental botany tools and knowing the microbiome of the plant that is beneficial to it, the rate of mutualism, the nature of collaboration, the molecular signaling and the developed defense system (Berg et al. 2013; Rybakova et al. 2016a, 2016b).

4.9 Plant–Microbe Communication

The plant–microbe communication is another aspect that is of great interest for designing effective agricultural practices. As from the experiments on plants in axenic environments, the most common molecular signal that was identified was IAA (indole acetic acid). The phylogenetic evidence suggests that IAA biosynthesis has an evolutionary significance on natural selection. Though the biosynthesis of IAA in bacteria to plants is independent of each other, natural selection makes it a widespread physiological code in the microbiome and its participation in their interactions (Fu et al. 2015). However, the recent research in plant–microbe communication revolves around secondary metabolites like volatile organic compounds (VOCs) and their involvement in communication between the symbiont and the host. It was reported that microbes use widespread VOCs to communicate among themselves and with the host (Bitas et al. 2013; Peñuelas et al. 2014). Besides VOCs, the structural components of microbes like LPS (lipopolysaccharides), siderophores, flagella, quorum-sensing molecules in Gram-negative molecules are found to activate the signaling pathway for the plant's induced systemic resistance of the defense system. Thus, the continuous interaction and communication of symbionts with each other and with the hosts of the entire microbiome will help in developing effective defense systems for plants that will lead to better crop productivity in future (Hartmann and Schikora 2012; Lugtenberg and Kamilova 2009; Lee et al. 2012; Ryu et al. 2003; Schenk and Schikora 2015).

The Earth Microbiome Project helped in microbiome profiling of soil that has enhanced the identification of microbiomes in high and low susceptibility regions (Gilbert et al. 2010). Such investigations assist in the development of microbial probiotics and developing targeted approaches for habitat reclamation that are heavily invaded because of land pollution and bad agricultural practices. Efforts are now focused toward sustainable development (Berlec 2012).

4.10 Plant and Soil Microbiome Engineering

After the advances in genome sequencing and metagenomic tools, microbiome analysis without culturing has become feasible, thus further accelerating microbiome research influencing the ecosystem (Mueller and Sachs 2015). The recent advance in this field has been microbiome engineering which helps in shaping the microbiome to alter the ecosystem of interest.

Plant microbiome engineering is basically engineering of the root associated microbes that include endosphere and rhizosphere microbiome, since it has been proven that they play a crucial role in plants' wellbeing and fitness (Agler et al. 2016). Thus by altering their composition and signaling pathways, improved phenotypes can be generated, and when these phenotypes become the part of the plant microbiome, they benefit the plant in its growth, fitness and health. The microbiome is engineered by simple molecular techniques of microbiome transfer (Zolla et al. 2013), formation of synthetic microbiomes (Glick 2012; Santhanam et al. 2015) and host mediated artificial selection of developing evolved microbiomes (Edwards et al. 2015; Swenson et al. 2000). However, even the best molecular technique cannot guarantee the successful microbiome modification, due to lack of functional native microbiome that can be cultured easily. So, the concept of soil microbiome engineering came into the picture. The soil microbiome refers to the entire population outside the rhizosphere that is mainly influenced by agricultural management practices than by the host plant. So the engineering of the soil microbiome is basically using healthier green practices of agriculture like organic farming that bans the use of chemical fertilizers and pesticides (Hartmann et al. 2015) and other agricultural practices like the cropping system (Xiong et al. 2015), tillage (Souza et al. 2016), and rotational cropping (Frampton et al. 2012).

This balances the soil composition and the residents of the soil, bringing back the soil health and thereby plant fitness. Microbiome engineering was further facilitated by use of phages for microbiome transfer; however the utilization of phageome is very limited because of their limited host specificity. However, phages are still part of agricultural practices as antibacterial agents and for antibacterial purposes (Andreote and de Silva 2017).

4.11 Conclusion

The conclusion of the study is to consider the plant as a metaorganism which consists of different types of microbial community and plant cells, and which is influenced by biotic and abiotic stresses, following the rules of natural selection. Thus, a plant is a holobiont where fitness and health are interrelated to its microbiome, sharing a mutualistic relationship. The microbiome and the holobiont are connected, and together they help in plant growth functioning. The above studies have clearly stated the importance and contribution of the microbiome to the host physiology and ecological characteristics. The recent metagenomic tools and transcriptome analysis have shown recent advances in microbiome research that have helped in controlling the pathogen burden. It was reported that the developed phenotype of the plant (agriculture probiotic) had resistant genes for the plant pathogen (innate immune system) that led to chemical-free biocontrol of the plant as a part of sustainable agriculture (Dangl et al. 2013). Another concept which came into the picture was growing plants under axenic conditions or gnotobiotic conditions also called neodomestication of the plant. These helped in understanding microbe and plant communication and their collaboration, which was instrumental in developing innovative and ecologically intensive agriculture (Duhamel and Vandenkoornhuyse 2013). Further, the implementation of agricultural practices has recently seen advances which have enabled rational engineering of the soil (soil microbiome engineering) for crop improvement. Similarly, capitalizing on the newfound knowledge on the microbiome led to the start of plant microbiome engineering which has helped in the development of plant protection strategies and reclamation of habitat. Microbiome engineering has largely been done on *Arabidopsis thaliana* as the host plant, however, it raises the possibility of using other crops also as host plants. Therefore, microbiome engineering is in its early stages, and a great deal

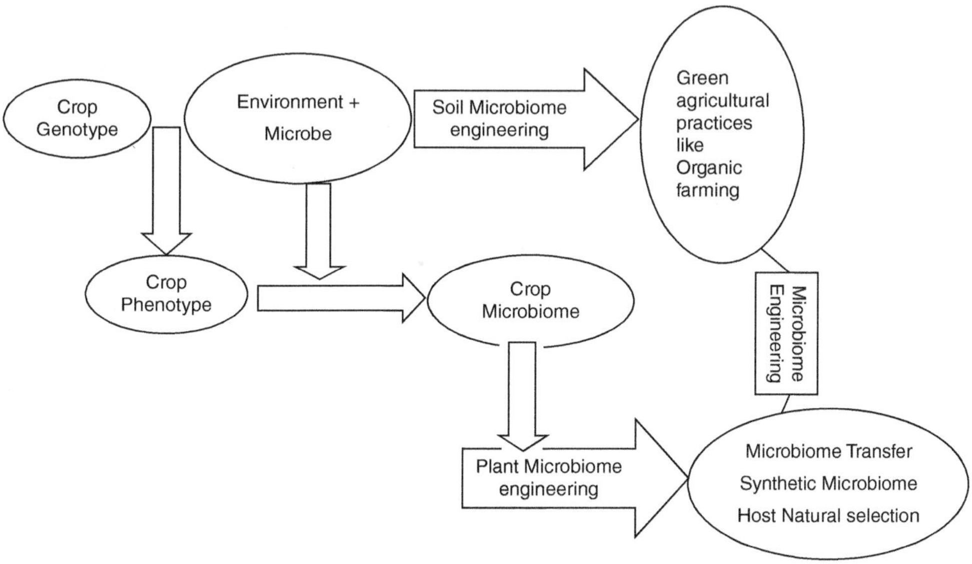

Figure 4.5 Approaches for plant and soil microbiome engineering.

of research is required on different plants with different microbiomes, so that stronger crops and drought, wilt, and rust resistant crops can be produced using green approaches. If synthetic microbes are engineered, they will also contribute to sustainable agriculture as they possess novel functionalities (Figure 4.5).

So, to conclude, with continuous efforts in gaining microbiome knowledge and its manipulation, microbiome engineering will soon become an important tool in agriculture for sustainable development.

References

Agler, M.T., Ruhe, J., Kroll, S. et al. (2016). Microbial hub taxa link host and abiotic factors to plant microbiome variation. *PLoS Biology* 14 (1): e1002352.

Andreote, F.D. and de Silva, M.D.C.P. (2017). Microbial communities associated with plants: learning from nature to apply it in agriculture. *Current Opinion in Microbiology* 37: 29–34.

Atamna-Ismaeel, N., Finkel, O.M., Glaser, F. et al. (2012). Microbial rhodopsins on leaf surfaces of terrestrial plants. *Environmental Microbiology* 14 (1): 140–146.

Beckers, B., De Beeck, M.O., Weyens, N. et al. (2017). Structural variability and niche differentiation in the rhizosphere and endosphere bacterial microbiome of field-grown poplar trees. *Microbiome* 5 (1): 25.

Berendsen, R.L., Pieterse, C.M., and Bakker, P.A. (2012). The rhizosphere microbiome and plant health. *Trends in Plant Science* 17 (8): 478–486.

Berg, G. and Smalla, K. (2009). Plant species and soil type cooperatively shape the structure and function of microbial communities in the rhizosphere. *FEMS Microbiology Ecology* 68 (1): 1–13.

Berg, G., Zachow, C., Müller, H. et al. (2013). Next-generation bio-products sowing the seeds of success for sustainable agriculture. *Agronomy* 3: 648–656.

Berg, G., Köberl, M., Rybakova, D. et al. (2017). Plant microbial diversity is suggested as the key to future biocontrol and health trends. *FEMS Microbiology Ecology* 93 (5): fix050.

Berlec, A. (2012). Novel techniques and findings in the study of plant microbiota: search for plant probiotics. *Plant Science* 193: 96–102.

Bitas, V., Kim, H.S., Bennett, J.W., and Kang, S. (2013). Sniffing on microbes: diverse roles of microbial volatile organic compounds in plant health. *Molecular Plant-Microbe Interactions* 26 (8): 835–843.

Boller, T. and Felix, G. (2009). A renaissance of elicitors: perception of microbe-associated molecular patterns and danger signals by pattern-recognition receptors. *Annual Review of Plant Biology* 60: 379–406.

Bouffaud, M.L., Poirier, M.A., Muller, D., and Moënne-Loccoz, Y. (2014). Root microbiome relates to plant host evolution in maize and other Poaceae. *Environmental Microbiology* 16 (9): 2804–2814.

Bulgarelli, D., Rott, M., Schlaeppi, K. et al. (2013a). Revealing structure and assembly cues for *Arabidopsis* root-inhabiting bacterial microbiota. *Nature* 501 (7468): S25.

Bulgarelli, D., Schlaeppi, K., Spaepen, S. et al. (2013b). Structure and functions of the bacterial microbiota of plants. *Annual Review of Plant Biology* 64: 807–838.

Cankar, K., Kraigher, H., Ravnikar, M., and Rupnik, M. (2005). Bacterial endophytes from seeds of Norway spruce (*Picea abies* L. Karst). *FEMS Microbiology Letters* 244 (2): 341–345.

Cole, B.J., Fletcher, M., Waters, J., Wetmore, K., Blow, M.J., Deutschbauer, A.M., Dangl, J.L. and Visel, A. 2015. Genetic control of plant root colonization by the biocontrol agent, *Pseudomonas fluorescens* (No. LBNL-178476). Lawrence Berkeley National Laboratory (LBNL), Berkeley, CA, United States.

Cook, R.J., Thomashow, L.S., Weller, D.M. et al. (1995). Molecular mechanisms of defense by rhizobacteria against root disease. *Proceedings of the National Academy of Sciences USA* 92 (10): 4197–4201.

Dangl, J.L., Horvath, D.M., and Staskawicz, B.J. (2013). Pivoting the plant immune system from dissection to deployment. *Science* 341 (6147): 746–751.

Duhamel, M. and Vandenkoornhuyse, P. (2013). Sustainable agriculture: possible trajectories from mutualistic symbiosis and plant neodomestication. *Trends in Plant Science* 18 (11): 597–600.

Edwards, J., Johnson, C., Santos-Medellín, C. et al. (2015). Structure, variation, and assembly of the root-associated microbiomes of rice. *Proceedings of the National Academy of Sciences USA* 112 (8): E911–E920.

Finkel, O.M., Castrillo, G., Paredes, S.H. et al. (2017). Understanding and exploiting plant beneficial microbes. *Current Opinion in Plant Biology* 38: 155–163.

Frampton, R.A., Pitman, A.R., and Fineran, P.C. (2012). Advances in bacteriophage-mediated control of plant pathogens. *International journal of microbiology* 2012: 326452.

Frederickson, M.E. (2013). Rethinking mutualism stability: cheaters and the evolution of sanctions. *The Quarterly Review of Biology* 88 (4): 269–295.

Friesen, M.L. (2012). Widespread fitness alignment in the legume–rhizobium symbiosis. *New Phytologist* 194 (4): 1096–1111.

Fu, S.F., Wei, J.Y., Chen, H.W. et al. (2015). Indole-3-acetic acid: a widespread physiological code in interactions of fungi with other organisms. *Plant Signaling and Behavior* 10 (8): e1048052.

Gilbert, J.A., Meyer, F., Jansson, J. et al. (2010). The Earth Microbiome Project: meeting report of the "1st EMP meeting on sample selection and acquisition" at Argonne National Laboratory October 6th 2010. *Standards in Genomic Sciences* 3 (3): 249.

Glick, B.R. (2012). Plant growth-promoting bacteria: mechanisms and applications. *Scientifica* 2012: 963401. https://doi.org/10.6064/2012/963401.

Guerrero, R., Margulis, L., and Berlanga, M. (2013). Symbiogenesis: the holobiont as a unit of evolution. *International Microbiology* 16 (3): 133–143.

Hacquard, S., Spaepen, S., Garrido-Oter, R., and Schulze-Lefert, P. (2017). Interplay between innate immunity and the plant microbiota. *Annual Review of Phytopathology* 55 (1): 565–589.

Hardoim, P.R., van Overbeek, L.S., and van Elsas, J.D. (2008). Properties of bacterial endophytes and their proposed role in plant growth. *Trends in Microbiology* 16 (10): 463–471.

Hardoim, P.R., Van Overbeek, L.S., Berg, G. et al. (2015). The hidden world within plants: ecological and evolutionary considerations for defining functioning of microbial endophytes. *Microbiology and Molecular Biology Reviews* 79 (3): 293–320.

Hartmann, A. and Schikora, A. (2012). Quorum sensing of bacteria and trans-kingdom interactions of N-acyl homoserine lactones with eukaryotes. *Journal of Chemical Ecology* 38 (6): 704–713.

Hartmann, A., Rothballer, M., and Schmid, M. (2008). Lorenz Hiltner, a pioneer in rhizosphere microbial ecology and soil bacteriology research. *Plant and Soil* 312 (1–2): 7–14.

Hartmann, M., Frey, B., Mayer, J. et al. (2015). Distinct soil microbial diversity under long-term organic and conventional farming. *The ISME Journal* 9 (5): 1177.

Heckman, D.S., Geiser, D.M., Eidell, B.R. et al. (2001). Molecular evidence for the early colonization of land by fungi and plants. *Science* 293 (5532): 1129–1133.

Johnson, B.J. (1993). Response of tall fescue to plant growth regulators and mowing frequencies. *Journal of Environmental Horticulture* 11 (4): 163–167.

Jones, J.D. and Dangl, J.L. (2006). The plant immune system. *Nature* 444 (7117): 323.

Kemen, E. (2014). Microbe–microbe interactions determine oomycete and fungal host colonization. *Current Opinion in Plant Biology* 20: 75–81.

Kiers, E.T. and Van Der Heijden, M.G. (2006). Mutualistic stability in the arbuscular mycorrhizal symbiosis: exploring hypotheses of evolutionary cooperation. *Ecology* 87 (7): 1627–1636.

Kiers, E.T., Rousseau, R.A., West, S.A., and Denison, R.F. (2003). Host sanctions and the legume-rhizobium mutualism. *Nature* 425 (6953): 78.

Kiers, E.T., Duhamel, M., Beesetty, Y. et al. (2011). Reciprocal rewards stabilize cooperation in the mycorrhizal symbiosis. *Science* 333 (6044): 880–882.

Lee, B., Farag, M.A., Park, H.B. et al. (2012). Induced resistance by a long-chain bacterial volatile: elicitation of plant systemic defense by a C13 volatile produced by *Paenibacillus polymyxa*. *PloS One* 7 (11): e48744.

Lemanceau, P., Blouin, M., Muller, D., and Moënne-Loccoz, Y. (2017). Let the core microbiota be functional. *Trends in Plant Science* 22 (7): 583–595.

Lopez-Velasco, G., Carder, P.A., Welbaum, G.E., and Ponder, M.A. (2013). Diversity of the spinach (*Spinacia oleracea*) spermosphere and phyllosphere bacterial communities. *FEMS Microbiology Letters* 346 (2): 146–154.

Lugtenberg, B. and Kamilova, F. (2009). Plant-growth-promoting rhizobacteria. *Annual Review of Microbiology* 63: 541–556.

Lundberg, D.S., Lebeis, S.L., Paredes, S.H. et al. (2012). Defining the core *Arabidopsis thaliana* root microbiome. *Nature* 488 (7409): 86.

Margulis, L. and Fester, R. (eds.) (1991). *Symbiosis as a Source of Evolutionary Innovation: Speciation and Morphogenesis*, 482. MIT Press.

McIntire, E.J. and Fajardo, A. (2014). Facilitation as a ubiquitous driver of biodiversity. *New Phytologist* 201 (2): 403–416.

Mendes, R., Garbeva, P., and Raaijmakers, J.M. (2013). The rhizosphere microbiome: significance of plant beneficial, plant pathogenic, and human pathogenic microorganisms. *FEMS Microbiology Reviews* 37 (5): 634–663.

Morris, J.J., Lenski, R.E., and Zinser, E.R. (2012). The Black Queen Hypothesis: evolution of dependencies through adaptive gene loss. *mBio* 3 (2): e00036–e00012.

Mueller, U.G. and Sachs, J.L. (2015). Engineering microbiomes to improve plant and animal health. *Trends in Microbiology* 23 (10): 606–617.

Müller, H., Berg, C., Landa, B.B. et al. (2015). Plant genotype-specific archaeal and bacterial endophytes but similar *Bacillus* antagonists colonize Mediterranean olive trees. *Frontiers in Microbiology* 6: 138.

Pendergast, T.H., Burke, D.J., and Carson, W.P. (2013). Belowground biotic complexity drives aboveground dynamics: a test of the soil community feedback model. *New Phytologist* 197 (4): 1300–1310.

Peñuelas, J., Asensio, D., Tholl, D. et al. (2014). Biogenic volatile emissions from the soil. *Plant, Cell & Environment* 37 (8): 1866–1891.

Philippot, L., Raaijmakers, J.M., Lemanceau, P., and Van Der Putten, W.H. (2013). Going back to the roots: the microbial ecology of the rhizosphere. *Nature reviews. Microbiology* 11 (11): 789–799.

Porter, S.S. and Simms, E.L. (2014). Selection for cheating across disparate environments in the legume-rhizobium mutualism. *Ecology Letters* 17 (9): 1121–1129.

Prestat, E., David, M.M., Hultman, J. et al. (2014). FOAM (functional ontology assignments for metagenomes): a hidden Markov model (HMM) database with environmental focus. *Nucleic Acids Research* 42 (19) e145–e145.

Ratcliff, W.C. and Denison, R.F. (2009). Rhizobitoxine producers gain more poly-3-hydroxybutyrate in symbiosis than do competing rhizobia, but reduce plant growth. *The ISME Journal* 3 (7): 870.

Reinhold-Hurek, B., Bünger, W., Burbano, C.S. et al. (2015). Roots shaping their microbiome: global hotspots for microbial activity. *Annual Review of Phytopathology* 53: 403–424.

Romero, F.M., Marina, M., and Pieckenstain, F.L. (2014). The communities of tomato (*Solanum lycopersicum* L.) leaf endophytic bacteria, analyzed by 16S-ribosomal RNA gene pyrosequencing. *FEMS Microbiology Letters* 351 (2): 187–194.

Rybakova, D., Cernava, T., Köberl, M. et al. (2016a). Endophytes-assisted biocontrol: novel insights in ecology and the mode of action of *Paenibacillus*. *Plant and Soil* 405 (1–2): 125–140.

Rybakova, D., Schmuck, M., Wetzlinger, U. et al. (2016b). Kill or cure? The interaction between endophytic *Paenibacillus* and *Serratia* strains and the host plant is shaped by plant growth conditions. *Plant and Soil* 405 (1–2): 65–79.

Ryu, C.M., Farag, M.A., Hu, C.H. et al. (2003). Bacterial volatiles promote growth in *Arabidopsis*. *Proceedings of the National Academy of Sciences* 100 (8): 4927–4932.

Sachs, J.L. and Simms, E.L. (2006). Pathways to mutualism breakdown. *Trends in Ecology & Evolution* 21 (10): 585–592.

Sachs, J.L., Mueller, U.G., Wilcox, T.P., and Bull, J.J. (2004). The evolution of cooperation. *The Quarterly Review of Biology* 79 (2): 135–160.

Sachs, J.L., Skophammer, R.G., and Regus, J.U. (2011). Evolutionary transitions in bacterial symbiosis. *Proceedings of the National Academy of Sciences* 108 (Supplement 2): 10800–10807.

Sánchez-Cañizares, C., Jorrín, B., Poole, P.S., and Tkacz, A. (2017). Understanding the holobiont: the interdependence of plants and their microbiome. *Current Opinion in Microbiology* 38: 188–196.

Santhanam, R., Weinhold, A., Goldberg, J. et al. (2015). Native root-associated bacteria rescue a plant from a sudden-wilt disease that emerged during continuous cropping. *Proceedings of the National Academy of Sciences* 112 (36): E5013–E5020.

Schenk, S.T. and Schikora, A. (2015). AHL-priming functions via oxylipin and salicylic acid. *Frontiers in Plant Science* 5: 784.

Schlaeppi, K., Dombrowski, N., Oter, R.G. et al. (2014). Quantitative divergence of the bacterial root microbiota in Arabidopsis thaliana relatives. *Proceedings of the National Academy of Sciences* 111 (2): 585–592.

Schmidt, H. and Eickhorst, T. (2013). Spatio-temporal variability of microbial abundance and community structure in the puddled layer of a paddy soil cultivated with wetland rice (*Oryza sativa* L.). *Applied Soil Ecology* 72: 93–102.

Souza, R.C., Mendes, I.C., Reis-Junior, F.B. et al. (2016). Shifts in taxonomic and functional microbial diversity with agriculture: how fragile is the Brazilian Cerrado? *BMC Microbiology* 16 (1): 42.

Swenson, W., Wilson, D.S., and Elias, R. (2000). Artificial ecosystem selection. *Proceedings of the National Academy of Sciences* 97 (16): 9110–9114.

Van Overbeek, L. and Van Elsas, J.D. (2008). Effects of plant genotype and growth stage on the structure of bacterial communities associated with potato (*Solanum tuberosum* L.). *FEMS Microbiology Ecology* 64 (2): 283–296.

Vázquez-de-Aldana, B.R., García-Criado, B., Vicente-Tavera, S., and Zabalgogeazcoa, I. (2013). Fungal endophyte (*Epichloë festucae*) alters the nutrient content of *Festuca rubra* regardless of water availability. *PloS One* 8 (12): e84539.

Venturi, V. and Keel, C. (2016). Signaling in the rhizosphere. *Trends in Plant Science* 21 (3): 187–198.

Vorholt, J.A. (2012). Microbial life in the phyllosphere. Nature reviews. *Microbiology* 10 (12): 828–840.

Wagner, M.R., Lundberg, D.S., Coleman-Derr, D. et al. (2014). Natural soil microbes alter flowering phenology and the intensity of selection on flowering time in a wild *Arabidopsis relative*. *Ecology Letters* 17 (6): 717–726.

Wagner, M.R., Lundberg, D.S., Coleman-Derr, D. et al. (2015). Corrigendum to Wagner et al.: natural soil microbes alter flowering phenology and the intensity of selection on flowering time in a wild *Arabidopsis* relative. *Ecology Letters* 18 (2): 218–220.

Werner, G.D., Strassmann, J.E., Ivens, A.B. et al. (2014). Evolution of microbial markets. *Proceedings of the National Academy of Sciences USA* 111 (4): 1237–1244.

Xiong, W., Li, Z., Liu, H. et al. (2015). The effect of long-term continuous cropping of black pepper on soil bacterial communities as determined by 454 pyrosequencing. *PloS One* 10 (8): e0136946.

Zachow, C., Berg, C., Müller, H. et al. (2016). Endemic plants harbour specific *Trichoderma* communities with an exceptional potential for biocontrol of phytopathogens. *Journal of Biotechnology* 235: 162–170.

Zilber-Rosenberg, I. and Rosenberg, E. (2008). Role of microorganisms in the evolution of animals and plants: the hologenome theory of evolution. *FEMS Microbiology Reviews* 32 (5): 723–735.

Zolla, G., Badri, D.V., Bakker, M.G. et al. (2013). Soil microbiomes vary in their ability to confer drought tolerance to Arabidopsis. *Applied Soil Ecology* 68: 1–9.

5

Ecology of the Diazotrophic Microbiome

Preeti Singh[1], *Rahul Kunwar Singh*[1], *Dhananjay Kumar*[1], *and Shree Prakash Tiwari*[2]

[1] *Department of Botany and Microbiology, HNB Garhwal University, Srinagar Garhwal, Uttarakhand, India*
[2] *Department of Microbiology, VBS Purvanchal University, Jaunpur, Uttar Pradesh, India*

5.1 Introduction

Nitrogen (N) is a crucial element for all forms of life on the earth; required for synthesis of nucleotides and amino acids. Though N is available in the atmosphere as a major component (78%), it is in the form of diazo nitrogen which cannot be utilized by the producers in any ecosystem due to its chemical inertness. So, limited availability of nitrogen is usually linked to nitrogen phosphorous ratio, productivity, as well as other processes, in several ecosystems (Hou et al. 2012). The conversion of atmospheric diazo nitrogen into biologically utilizable forms like nitrates (NO_3^-) and ammonia (NH_3) is known as nitrogen fixation and the microbial communities involved in the phenomenon are called nitrogen fixers or diazotrophic microbiome. There are several diazotrophs maintaining a symbiotic relationship with legumes and non-legumes; the majority of the diazotrophic microbiome is free living and found in the rhizospheric region of plants. Hence, the diversity and composition of the diazotrophic microbiome varies with the physico-chemical parameters and the vegetation of the ecosystem (Chen et al. 2019). This chapter focuses on the basics of biological nitrogen fixation (BNF) along with diversity and ecology of the diazotrophic microbiome in different ecosystems.

5.2 Different Modes of Nitrogen Fixation

Nitrogen fixation is done artificially in industries for commercial production of chemical nitrogen fertilizers. In nature, nitrogen fixation may take place either naturally during thunderstorms and lightning, or biologically by a unique group of microorganisms, nitrogen fixers or diazotrophs, which utilize the atmospheric bimolecular nitrogen and convert it into ammonia using nitrogenase enzyme complex (Wagner 2011). The BNF accounts for about 50 percent of fixed nitrogen available on our planet and produces about 210 teragram

The Plant Microbiome in Sustainable Agriculture, First Edition. Edited by Alok Kumar Srivastava, Prem Lal Kashyap and Madhumita Srivastava.

(Tg) fixed nitrogen per year globally (Fowler et al. 2013). Thus, BNF acts as an eco-friendly and cost-effective substitute for extensive use of chemical fertilizers (Adesemoye et al. 2009; Bargaz et al. 2018). Types of nitrogen fixation are shown in Figure 5.1.

Nitrogen fixing microbiome may be defined as microbial flora able to fix the dimolecular nitrogen and are generally categorized as symbiotic and non-symbiotic.

5.2.1 Symbiotic Diazotrophic Microbiome

Symbiotic diazotrophic microbiome includes the rhizobia (*Rhizobium, Sinorhizobium, Bradyrhizobium, Mesorhizobium,* and *Azorhizobium*) living as symbionts with leguminous plants and a non-legume, *Parasponia*, the actinobacteria (*Frankia*) associated with actinorhizal plants, as well as cyanobacterium–*Gunnera* associations (Mus et al. 2016). However, a few studies have revealed that a diverse range of microorganisms have the ability of BNF, including archaea and unknown bacteria (Reed et al. 2011). Even the new microbial partners are being added to the arena of the rhizobia/legume association, the most studied systems that can nodulate legumes (Gyaneshwar et al. 2011; Vitousek et al. 2013).

The rate of fixed nitrogen addition by this type of microbiome is reported up to $>150\,kg\,N\,ha^{-1}\,year^{-1}$ in some unmanaged ecosystems which is very high in comparison to the rate of nitrogen added by the use of fertilizers, i.e. $100–500\,kg\,N\,ha^{-1}\,year^{-1}$ (Reed et al. 2011). However, the rate of BNF by symbiotic microbiomes in several terrestrial ecosystems reported in the literature (Table 5.1), were actually the point measurements which have been generalized to the biome scale (Cleveland et al. 1999).

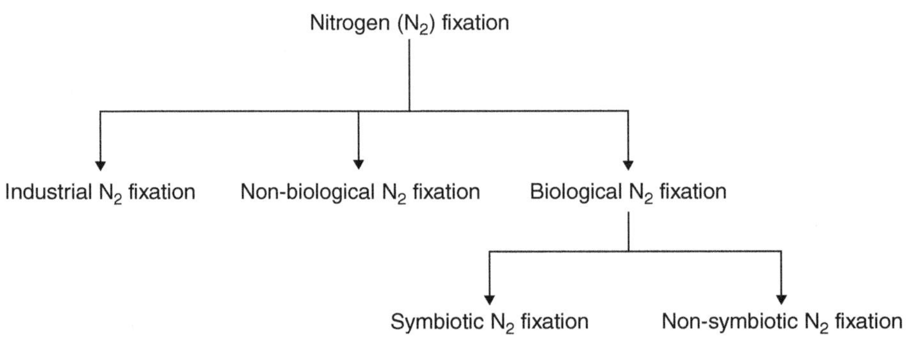

Figure 5.1 Types of nitrogen fixation.

Table 5.1 Rate of nitrogen fixation in different ecosystems.

S. No.	Ecosystem	Rate of nitrogen fixation	References
1	Terrestrial ecosystem	$128\,Tg\,N\,year^{-1}$	Galloway et al. (2008)
1A	Tropical Forest	$26 \pm 10\,kg\,N\,ha^{-1}\,year^{-1}$	Cleveland et al. (1999)
1B	Cropland	$60 \pm 10\,Tg\,N\,year^{-1}$	Fowler et al. (2013), Herridge et al. (2008)
2	Marine ecosystem	$140 \pm 50\,Tg\,N\,year^{-1}$	Voss et al. (2013)
3	Freshwater ecosystem	$0.2–9.2\,kg\,N\,ha^{-1}\,year^{-1}$	Haworth et al. (1988)

5.2.2 Non-symbiotic Diazotrophic Microbiome

Non-symbiotic diazotrophic microbiome (free living, associative, and endophytes) include cyanobacteria (*Anabaena and Nostoc*), *Azospirillum, Azotobacter, Gluconacetobacter diazotrophicus* and *Azocarus,* etc. (Bhattacharyya and Jha 2012), and may be defined as the nitrogen fixing microflora that do not form a clear-cut symbiotic relationship with plants; they include free living and associative microbiomes. This type of microbiome inhabit almost all the ecosystems including aerobic and anaerobic niches of terrestrial, freshwater and marine ecosystems (Barron et al. 2009; Benner and Vitousek 2007). Several reports suggest very low rate of BNF (1–20 kg N ha^{-1} $year^{-1}$) by non-symbiotic microbiomes in terrestrial ecosystems (Cleveland et al. 1999; Son 2001). However, recent studies suggest that non-symbiotic microbiomes contribute significant amounts of fixed nitrogen in most terrestrial ecosystems, particularly those lacking symbiotic microbiome (Hedin et al. 2009; Reed et al. 2011; Wagner 2011).

The rate and timing of nitrogen fixation varies with changes in diversity and abundance of diazotrophic microbiome, which may change in response to variations in abiotic characteristics even on a single site (Lindsay et al. 2010; Teng et al. 2009). Particularly the quantity and quality of carbon is related to the structure of the microbiome. Studies suggest that availability of carbon stock, its chemistry and input rate are also factors which control the diversity, abundance and structure and activity of the diazotrophic microbiome (Forbes et al. 2009; Reed et al. 2011; Wakelin et al. 2010).

In addition, the changes in availability of nutrients may also alter the composition of the diazotrophic microbiome (Table 5.2) thereby changing the rate of nitrogen fixation (Lindsay et al. 2010; Reed et al. 2010).

Table 5.2 Major constituents of diazotrophic microbiomes in different ecosystems.

S. No.	Ecosystem	Dominant diazotrophs	References
1	Tropical forest	Firmicutes, α-proteobacteria	Reed et al. (2010)
		α-proteobacteria, β-proteobacteria, methanogenic archaea	Mirza et al. (2014)
2	Temperate forest	Geobacteriaceae, α-proteobacteria	Berthrong et al. (2014)
3	Legume cropland	α-proteobacteria	Laranjo et al. (2014)
4	Non-legume cropland	*Azosprillum, Azotobacter, Pseudomonas*, Cyanobacteria, Actinobacteria	Bhattacharyya and Jha (2012)
5	Marine ecosystem	*Trichodesmium, Richella*	LaRoche and Breitbarth (2005)
6	Freshwater ecosystem	Cyanobacteria	Haworth et al. (1988)

5.3 Physiology and Biochemistry of Nitrogen Fixation

BNF is an energy expensive anabolic process operated in a variety of prokaryotes known as diazotrophs, belonging to archaeal and bacterial domains (Raymond et al. 2004). Diazotrophs may derive energy required for the process either by autotrophy or chemotrophy. They may be aerobic or anaerobic. Despite a high degree of diversity among diazotrophs, the physiology of BNF is quite similar in all the organisms (Darnajoux et al. 2017). The phenomenon is catalyzed by an enzyme complex called the nitrogenase enzyme complex. This enzyme complex consists of two different enzymes, dinitrogenase reductase and dinitrogenase. The latter enzyme usually has an FeMo cofactor, where the actual nitrogen fixation takes place, and is called the molybdenum dependent enzyme. Reduction of dimolecular nitrogen into ammonia involves transfer of six electrons in three successive reduction steps taking place on nitrogenase enzyme. Due to chemical inertness of the diazo bond in dimolecular nitrogen, the process requires a high amount (16–24 ATP) of energy. The electrons for reduction of dimolecular nitrogen are supplied to nitrogenase enzyme by dinitrogenase reductase, which actually receives these electrons from oxidation of pyruvate through flavodoxin or ferredoxin. With each cycle of electron transfer, two molecules of ATP bind with dinitrogenase reductase to change its conformation, lower its reduction potential and thereby allow it to interact with and reduce nitrogenase enzyme. After transfer of electrons to nitrogenase, ATP hydrolyzes; dinitrogenase reductase dissociates from nitrogenase and initiates another cycle of electron transfer. Upon complete reduction of nitrogenase, the actual reduction of dimolecular nitrogen takes place at the FeMo cofactor of the nitrogenase enzyme. Despite only six electrons being required for the reduction of dimolecular nitrogen into ammonia, a total of eight electrons are actually used in the process, two electrons being utilized in the process of simultaneous production of hydrogen. The biochemistry of nitrogen fixation is shown in Figure 5.2.

Earlier, it was assumed that only FeMo nitrogenase contributes to the nitrogen fixation. Recent studies have revealed the presence of alternative nitrogenases in various ecosystems and its contribution to BNF in soils (Bellenger et al. 2014; Betancourt et al. 2008; Hodkinson et al. 2014). These alternative nitrogenases are actually the isozymes of FeMo nitrogenase: one is vanadium dependent nitrogenase having FeVa cofactor, and the other is iron-only nitrogenase having FeFe cofactor (Mus et al. 2018; McRose et al. 2017). However, FeMo nitrogenase is the preferred one among all, and synthesized first in nitrogen limiting conditions. The synthesis of alternative nitrogenases occurs only due to insufficient amount of molybdenum in the environment and acts as a backup system for nitrogen fixation. When there is sufficient molybdenum in the environment, it suppresses the synthesis of alternative nitrogenases.

An entirely distinct FeMo nitrogenase complex is reported from a thermophilic, filamentous actinobacterium, *Streptomyces thermoautotrophicus*. This nitrogenase is insensitive to oxygen. This nitrogenase complex has two proteins, Str1 and Str2. Str1 is analogous, and structurally similar to dinitrogenase enzyme of other diazotrophs. The reduction of dimolecular nitrogen takes place on this protein only. However, the Str2 protein does not show any structural similarity to dinitrogenase reductase enzyme, but is analogous to that. Str2 protein actually acts like superoxide dismutase enzyme and supplies electrons to Str1 from superoxide radical, which is generated from the reduction of oxygen during the oxidation

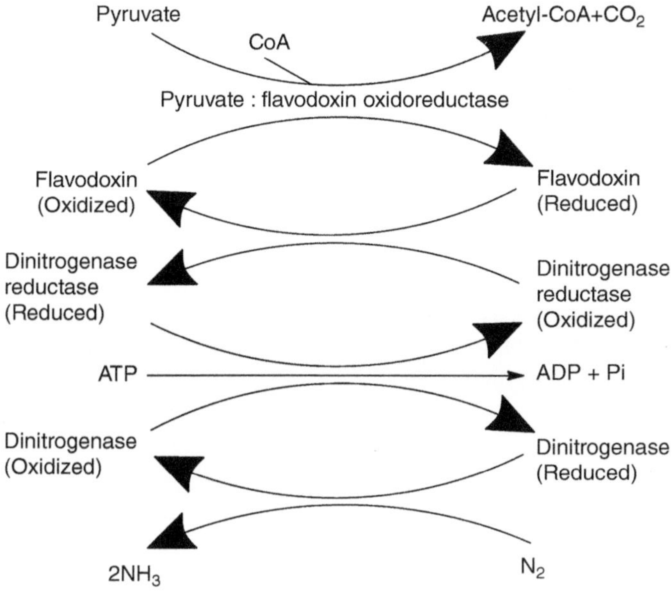

Figure 5.2 Biochemistry of nitrogen fixation. *Source:* Adapted from *Brock Biology of Microorganisms*, 13th edition.

of carbon monoxide. However, MacKellar et al. (2016) could not find the gene sequences, reported as putative nitrogenase of *S. thermoautotrophicus*, in the draft genome sequence of a few strains of the same organism.

5.3.1 Protection of Nitrogenase from Oxygen

Oxygen has the capability to irreversibly inactivate nitrogenase enzyme and thereby affects the nitrogen fixation process. Some diazotrophs live in an anaerobic environment to avoid the oxygen and get their energy from anaerobic respiration. However, the aerobic diazotrophs have developed different mechanisms to protect their nitrogenase enzyme from the oxygen, as well as permitting oxygen supply for respiration. One mechanism is to limit nitrogenase in a space like heterocyst, developing a thick cell wall around it, thereby limiting oxygen diffusion. The second mechanism is to isolate nitrogenase activity in time from oxygen generating metabolic process like photosynthesis. The third mechanism is to reduce oxygen concentration around nitrogenase by metabolic regulation such as increasing the rate of respiration (Oelze 2000). The fourth mechanism is to modify the nitrogenase enzyme by conformational change to make it resistant to oxygen (Schlesier et al. 2015; Soto-Urzúa and Baca 2001).

5.3.2 Metals

FeMo nitrogenase is the major contributor in the BNF process. Studies have revealed that few diazotrophs, i.e. *Azotobacter vinelandii, Escherichia coli, Azomonas macrocytogenes and Rhodopseudomonas palustris*, are able to synthesize alternative nitrogenases in addition to

FeMo nitrogenase (Betancourt et al. 2008; McRose et al. 2017). However, synthesis of a particular type of nitrogenase on a particular time point depends on genetic as well as environmental factors like availability of Mo, V, light and oxygen concentration (Bellenger et al. 2014). However, it has been widely accepted that FeMo-nitrogenase is the preferred and most efficient nitrogenase enzyme when molybdenum (Mo), vanadium (V) and iron (Fe) all are present in sufficient amount (Bellenger et al. 2014).

5.3.3 ATP and Phosphorus

BNF is an energy expensive metabolic process (Halbleib and Ludden 2000). For example *A. vinelandii* use 100 g glucose and 16 ATP for fixation of 1 mol of nitrogen. This large amount of energy may be supplied either by phototrophy or chemotrophy (Desai et al. 2013; Wagner 2011). However, the efficiency of nitrogenase enzyme is affected by the type of energy source being used (Hill 1992). This extensive need for ATP necessitates a large amount of phosphorous also (Vitousek et al. 2002). Besides, bioavailable phosphorus is needed for growth of the host plant, as well as development and formation of root nodules, necessary for symbiotic nitrogen fixation.

5.3.4 Nitrogen

The diazotrophs fix nitrogen only in conditions when fixed forms of nitrogen such as ammonia, nitrate and nitrite are limited in the environment. Where an ample amount of fixed N is present, the diazotrophs usually switch-off the process of nitrogen fixation either by inhibition of the activity of nitrogenase enzyme, or by inhibiting its synthesis. However, in several cases, the efficiency of this switch depends on concentration of dissolved organic carbon, pH, and the carbon–nitrogen ratio of the cell (Cejudo and Paneque 1988). Thus, addition of chemical nitrogen fertilizers to any ecosystem usually represses the rate of nitrogen fixation as the free living diazotrophic microbiomes are facultative nitrogen fixers and do not fix nitrogen, when the fixed nitrogen is already available in the environment (Galloway et al. 2008). Even the diversity of nitrogen fixers in the soil shifts quickly in response to the amount of fixed nitrogen available (Vitousek and Field 1999).

5.3.5 Carbon Dioxide

The diazotrophic microbiomes may be affected by climate change drivers. They are able to alter their diversity, abundance, and nitrogen fixation rates in response to change in climate. The increased level of CO_2 stimulates the rate of nitrogen fixation by increasing the carbon nitrogen (C/N) ratio of plant input to soil (LeBauer and Treseder 2008).

5.4 Genes Involved in Nitrogen Fixation and Its Regulation

Several studies have been performed on genetics of the nitrogenase enzyme complex, as the structure of nitrogenase and dinitrogenase reductase enzymes is highly conserved throughout the majority of diazotrophic organisms (Gaby and Buckley 2012; Ludden 2001). The FeMo nitrogenase complex is encoded by *nifHDK* genes, whereas the alternative nitrogenases,

Fe-Va nitrogenase and Fe only nitrogenase are encoded by *vnfHDK* and *anfHDk* genes, respectively. The sequences of *vnfHDK* and *anfHDK* genes also have significant similarity with that of *nifHDK* genes (Hu et al. 2012; McRose et al. 2017; Yang et al. 2014).

The *nifH* gene has been widely used as a genetic probe to determine the presence and activity of the diazotrophic microbiome as well as its composition (Raymond et al. 2004). The studies involving *nifH* genes have significantly increased the understanding about the biology and ecology of the diazotrophic microbiome (Gaby and Buckley 2011; Penton et al. 2016). The *nifH* gene has been chosen for the purpose as it has highly conserved sequences and gene distribution similar to 16S rRNA sequences (Yeager et al. 2004).

5.4.1 Regulation of Nitrogenase Synthesis

The synthesis of nitrogenase enzyme is suppressed by oxygen and fixed forms of nitrogen like ammonia and nitrate. The inhibition of nitrogenase synthesis in the presence of oxygen is achieved by inhibiting the synthesis of nifA protein which is actually a positive regulator for expression of structural genes, *nifHDK*. By contrast, NifL protein is a negative regulator of structural gene expression and is activated by the availability of sufficient oxygen.

Availability of ammonia or other fixed forms of nitrogen inactivates another protein NtrC, which in turn represses the transcription of nifA, the positive regulator of nifHDK. When fixed nitrogen is limited in the cell, NtrC becomes activated and thereby activates the expression of nifHDK through nifA protein. The ammonia produced by nitrogenase itself does not inhibit the expression of nifHDK as it is utilized in biosynthesis of amino acids immediately.

5.4.2 Regulation of Nitrogenase Activity

The nitrogen fixation process can also be regulated by shutting down the activity of existing nitrogenase enzyme, when fixed nitrogen (i.e. ammonia) is present in excess. The mechanism is known as the ammonia switch-off effect and is operated in the majority of diazotrophs. In this mechanism, excess ammonia causes addition of an ADP molecule to dinitrogenase reductase enzyme, resulting in loss of enzymatic activity. The enzyme activity resumes when ammonia will be limited in the cell. Thus, this regulatory mechanism is a reversible and a rapid method for control of nitrogen fixation activity.

In archaebacteria, the regulation of nitrogen fixation activity in the presence of excess ammonia takes place in another way. There is an ammonia sensing protein in these organisms which bind to the nitrogenase and inactivate it. Interestingly, these organisms sense the excess of ammonia indirectly by sensing the cytoplasmic level of alpha-ketoglutarate, a precursor of glutamate. A scarcity of α-ketoglutarate in the cytoplasm is considered as a sign of excess of glutamate which in turn indicates excess of ammonia in the cell as ammonia also favors glutamate synthesis.

5.5 Nitrogen Fixation in Termites

Termites have diazotrophic microbiome in their gut as normal flora and are distributed worldwide (Thanganathan and Hasan 2018). The fixed nitrogen by termite gut microbiome is first integrated into termite tissues and can be disseminated all over the colony by social feeding

(Sapountzis et al. 2016). The process provides the nitrogen supplementation to the nitrogen deficient diet of termites and also contributes to the nitrogen cycle of terrestrial ecosystems (Vecherskii et al. 2008). The rate of nitrogen fixation by termites varies from species to species and also in members of same species due to variation in termite size, caste, food quality and season (Curtis and Waller 1995; Sapountzis et al. 2016). The diazotrophic microbiomes of the termite gut are mainly composed of Bacteroidetes, Spirochaetes, and Clostridia (Brune 2014). According to an estimate by Pandey et al. (1992), *Reticulitemes* spp. of termites fix 0.12–0.44 kg N ha^{-1} $year^{-1}$ in a forest ecosystem, whereas Yamada et al. (2006) estimated the contribution of 0.21 kg ha^{-1} $year^{-1}$ fixed in the dry deciduous forest by *Microcerotermes crassus* and 0.28 kg ha^{-1} $year^{-1}$ by *Globitermes sulphureus*, in the dry evergreen forest of Thailand.

Termites inhabit fallen logs which occur in patches all over the forest ecosystems. The nitrogen contribution of termites may be seen as a montage of nitrogen "hot-spots" concentrated in termite-infested logs. Curtis and Waller (1998) estimated that termites might contribute up to 0.015 g N log^{-1} day^{-1} to logs, which may be accrued over a season and be released during the other season when the rate of termite nitrogen fixation is slow. The enrichment of logs with nitrogen may help in its decomposition by fungi and thus speed up the nutrient release from infested wood. Ohkuma et al. (1999) have analyzed the *nifH* sequences from six different lower termites showing nitrogen fixation, and assigned the majority of them to either the anaerobic *nif* group or the alternate *nif* methanogen group.

5.6 Nitrogen Fixation in the Aquatic Ecosystem

The diazotrophic microbiome in the aquatic system consists of a variety of prokaryotes such as planktonic as well as benthic organisms, including autotrophs and heterotrophs. The rate of nitrogen fixation by planktonic organisms differs significantly in freshwater lakes based on the level of nutrients present in them. Haworth et al. (1988) reported higher nitrogen fixation rates (02–92 kg N ha^{-1} $year^{-1}$) in eutrophic lakes containing cyanobacteria as a major constituent of its planktonic biomass in comparison to that in the oligotrophic and mesotrophic lakes (<01 kg N ha^{-1} $year^{-1}$). However, the rates of benthic nitrogen fixation in the sediments of most lakes are much lower (<2.5 kg N ha^{-1} $year^{-1}$) than that of planktonic fixation in oligotrophic lakes.

As far as the marine ecosystem is concerned, the global rate of N_2 fixation in the ocean is quite similar to that of the terrestrial N_2 fixation rate and is in the range of 100–200 Tg N $year^{-1}$ with large uncertainties (Bostrom 2006; Karl et al. 2002). This is due to the fact that nitrogen is often limited in most of the marine ecosystem and the nitrogen fixation process is the only key to maintain the productivity of these ecosystems (Karl et al. 2002). *Trichodesmium*, a filamentous, non-heterocystous cyanobacterium and *Richelia*, a heterocystous cyanobacterium living in endosymbiotic association with diatoms such as *Epithemia* and *Rhapalodia*, are the major constituents of the diazotrophic microbiome in marine ecosystems (LaRoche and Breitbarth 2005; Mahaffey et al. 2005). However, recent studies have recognized the noteworthy contribution of unicellular cyanobacteria such as *Synechococcus*, *Gloeocapsa* and few unculturable species in marine nitrogen fixation (Bergman 2001; Zehr 2011). Moreover, heterotrophic bacterioplankton having the *nifH* gene have also been reported in the marine ecosystem (Bird et al. 2005; Church et al. 2005).

As the activity of nitrogenase enzyme is usually inhibited by oxygen, most of the diazotrophic microbiome in the marine ecosystem are either anaerobes or facultative anaerobes. However, some marine cyanobacteria have limited their nitrogenase only in heterocysts, protecting nitrogenase from oxygen being evolved during photosynthesis (Paerl and Zehr 2000). Cyanobacterial heterocysts do not have PSII and are not able to produce oxygen during photosynthesis. In addition, these cells have a thick cell wall and thus are able to separate photosynthesis and the N_2-fixation process. Similarly, *Trichodesmium* protects the nitrogenase by allocating it to only few non-photosynthetic cells and thus maintains a high rate of N_2 fixation as well as photosynthesis during the day time (Berman-Frank et al. 2007). This phenomenon is advantageous for the cell as photosynthesis provides the energy for the energy expensive nitrogen fixation process. The non-heterocystous cyanobacteria protect the nitrogenase by managing nitrogen fixation during the night (Church et al. 2005). Moreover, an alternative mechanism, i.e. maintenance of low oxygen concentration in the cell through high respiration rate, is also well-studied in the bacterium *A. vinelandii*.

The activity of nitrogenase enzyme and thus the rate of BNF in the aquatic ecosystem is controlled by the diazotrophic microbiome as well as several physicochemical factors (Mahaffey et al. 2005). The nitrogenase enzyme needs iron as a cofactor for its activity (Kustka et al. 2002). The availability of both iron as well as phosphorus may act as limiting factors for the nitrogen fixation process in some parts of the marine ecosystem (Karl et al. 2002; Mills et al. 2004; Sañudo-Wilhelmy et al. 2001). Turk et al. (2011) have reported a positive correlation between the nitrogen fixation rate and concentration of dissolved iron in the ocean near the Cape Verde Islands, in the eastern North Atlantic. Recently, highest abundance of heterocystous cyanobacteria in a stream having high phosphorus and iron content but low nitrogen has been reported by Larson et al. (2018).

5.7 Nitrogen Fixation in the Terrestrial Ecosystem

BNF is the major source of nitrogen input to terrestrial ecosystems throughout the world (Cleveland et al. 1999; Galloway et al. 2008; Vitousek et al. 2002). The relation between diazotrophic microbiome and rate of BNF has been worked out by several workers in different type of terrestrial ecosystems, such as tropical and temperate forest soils (Chen et al. 2019; Mirza et al. 2014), wetlands (Rejmánková et al. 2018), desert systems (Wang et al. 2016; Yeager et al. 2004), grasslands (Tu et al. 2016), and agricultural lands (Pereira E Silva et al. 2013). Most of these studies were based on genetic techniques which have provided significant advances in BNF research, and assessments of diazotrophic microbiome composition.

5.8 Nitrogen Fixation in the Tropical Forest

In tropical ecosystems, soil are generally poor in nutrients; there are zones where nutrient availability is higher to a larger extent than average, even on highly battered soils (Townsend and Howarth 2010). These zones have a high number of diazotrophic organisms and high nitrogen fixation rates, and are known as nitrogen fixing hotspots.

The hotspots reflect that the composition of abiotic factors, which regulate the dynamics of nitrogen fixers, changes on a very small scale (Reed et al. 2010). The major abiotic factors affecting nitrogen fixation include moisture, availability of phosphorous and molybdenum, and C/N ratio (Barron et al. 2009; Cusack et al. 2009; Vitousek et al. 2002; Wurzburger et al. 2012).

Due to presence of N-fixing hotspots, the tropical forest ecosystem upholds a significant amount of BNF which varies from 15 to 36 $kg\,N\,ha^{-1}\,year^{-1}$ and is higher than its temporal counterpart (Cleveland et al. 1999). This high rate of nitrogen fixation is maintained in tropical forests by the nitrogen fixing organisms which obtain an adequate amount of nitrogen to maintain its extracellular phosphatase activity. This is essential to overcome phosphorous restraint in these ecosystems, and because of optimum temperature for BNF (Houlton et al. 2008). Earlier, tropical forests were thought to be dominated by symbiotic nitrogen fixing microbiome (Cleveland et al. 1999). However, several workers have shown that free living diazotrophic bacteria inhabiting the litter, soil as well as forest canopies also contribute significantly in nitrogen input to these forests (Benner and Vitousek 2007; Hedin et al. 2009; Reed et al. 2013). Moreover, archaebacteria may also play an important role in nitrogen fixation in tropical soils (Mirza et al. 2014).

The diazotrophic microbiome of tropical rainforest in Costa Rica was found to be dominated by Firmicutes (*Heliobacterium*) and alpha-proteobacteria (*Gluconacetobacter*, *Methylobacterium*, *Azospirillum*, and *Zymomonas*) in a metagenomic study by Reed et al. (2010). In a similar study in rainforest of the western Amazon basin, Mirza et al. (2014) revealed the presence of alpha and beta-proteobacteria (*Azospirillum*, *Azorhizobium*, *Bradyrhizobium*, *Methylobacterium*, and *Burkholderia*), Firmicutes (e.g. *Paenibacillus* and *Heliobacterium*), Cyanobacteria (*Nostoc* and *Anabaena*) as well as methanogenic Archaea (*Methanoregula*, *Methanosphaerula* and *Methanocella*). Interestingly, associated *nifH* sequences were also found in these soils.

The BNF in tropical forests by non-symbiotic microbiome is affected by nutrient level as well as the soil characteristics. Cusack et al. (2009) reported that addition of nitrogen fertilizers in tropical soil could significantly inhibit the nitrogen fixation process. However, increase in soil moisture level could enhance the rate of nitrogen fixation (Barron et al. 2009). Similarly, a high level of primary minerals in the soil provides sympathetic conditions for growth of diazotrophic microbiome in comparison to soil containing secondary minerals (Stone et al. 2015).

Moreover, the rate of nitrogen fixation in tropical forest may also be modulated by phosphorus and micronutrients.

A high amount of phosphorous is needed by diazotrophs to fulfill the high energy demand of energy expensive nitrogen fixation process. To get this high amount of phosphorus, diazotrophs secrete phosphatase enzyme which liberates the phosphorous bound to soil minerals and organic matter in most of the tropical forests. Studies by Benner and Vitousek (2007) and Reed et al. (2007) have revealed that the addition of phosphorus in soil stimulates the nitrogen fixation process and also increases the diversity and abundance of non-symbiotic diazotrophic microbiome. Similarly, Mo is also a limiting element for nitrogen fixation and is a common constituent of nitrogenase enzyme (Wurzburger et al. 2012). A recent study by Reed et al. (2013), to explore P and Mo controls over N fixation rates in a tropical rainforest revealed the positive correlation between phosphorous concentration and nitrogen fixation rate of the non-symbiotic microbiome. The workers could not draw any conclusion regarding Mo concentration.

5.9 Nitrogen Fixation in the Temperate Forest

The diazotrophic microbiome of the temperate pine forest ecosystem is dominated by free living diazotrophs (Cleveland et al. 1999). Berthrong et al. (2014) on the basis of *nifH* sequences obtained from three different temperate forests reported that the diazotrophic microbiome of temperate forests is dominated mainly by Geobacteriaceae followed by alpha-proteobacteria. The composition of the microbiome was observed to be changed in a period of 10 years with an increased concentration of atmospheric carbon dioxide. Moreover, the structure of the diazotrophic microbiome was strongly affected by nitrogen fertilization.

5.10 Nitrogen Fixation in Crop Land

Nowadays, the major sources of nitrogen input in crop lands are chemical nitrogen fertilizer, which accounts for about half of the global inorganic nitrogen, throughout the world (Liu et al. 2010). The BNF, which was the major source of nitrogen input before the industrial era, now accounts for only 16 percent of total nitrogen input to the cropland (Yuan and Peng 2017). The use of chemical nitrogen fertilizers has increased up to sevenfold in the last five decades, which undoubtedly doubled global food production (Tilman et al. 2001). However, extensive use of chemical nitrogenous fertilizers has led to an astonishing perturbation of the nitrogen cycle as evidenced by the increasing accretion of nitrates in soils and waters and of nitrogen oxides in the atmosphere (Green et al. 2004; Grote et al. 2005). These anthropogenic changes in global nitrogen flows suppress the BNF process. Uddin et al. (2008) revealed that the high levels of N fertilizer in the soil reduce the number and size of nitrogen fixing root nodules which in turn inhibit the growth of seedlings. Moreover, they affect sustainable food production as well as environmental health (Olivares et al. 2013).

The quantity of nitrogen needed for agriculture is predicted to increase in the next decades, which could lead to greater environmental pollution and a deterioration of soil health. Hence, the only way to sustain crop productivity is to reduce our dependence on chemical fertilizers and to rely on methods of traditional agriculture such as BNF (Huhe et al. 2014).

Moreover, climate change is also responsible for changes in nitrogen fixation rate, abundance and diversity of diazotrophic microbiome. The increased amount of carbon dioxide usually enhances nitrogen fixation through a high C/N ratio in the soil. Olivares et al. (2013) have hypothesized that a high level of carbon dioxide in the atmosphere for a long duration would increase the abundance of N-fixing bacteria and change the composition of the diazotrophic community throughout the world.

The diversity of the nitrogen fixing microbiome in cropland is distributed across the Bacteria and Achaea including the symbiotic and non-symbiotic diazotrophs (Dixon and Kahn 2004). The major portion (about 80%) of biologically fixed nitrogen in croplands comes from the diazotrophs living in symbiosis with legumes (Peoples et al. 1995). However, in specific conditions such as in neglected croplands, where no legumes have been cultivated or few symbiotic plants growing, the free-living diazotrophs like *Azospirillum*, *Azotobacter*, *Pseudomonas* and Cyanobacteria, may contribute significantly (Burgmann et al. 2004; Kahindi et al. 1997). However, several actions are required to obtain better deployment of this biological process before completely relying on it for sustainable agriculture. These actions are: (i) the optimal use of known nitrogen-fixing systems; (ii) development of new nitrogen fixing plant microbe associations; and (iii) transfer of the nitrogen-fixing ability to non-fixing organisms.

5.11 Genetic Modification of Crops for Nitrogen Fixation

The generation of genetically modified plants that could fix their own nitrogen has been continuously discussed and attempted by scientists since the beginning of genetic engineering technologies (Ivleva et al. 2016). To achieve this, three different mechanisms are being considered by workers. One mechanism is to engineer the legume plants and/or the symbiotic diazotrophic organisms for improvement of the nodule formation activity of plants (Ling et al. 2016; Rogers and Oldroyd 2014). The second approach is focused on improvement of nitrogen fixation activity or introduction of the nitrogen fixation pathway in bacterial endophytes (Geddes et al. 2015; Oldroyd and Dixon 2014). The third mechanism is to incorporate the nitrogen fixation pathway in the plant itself. All these mechanism attempted by several workers are facing challenges in execution because of the complications of the nitrogen fixation pathway, sensitivity of nitrogenase enzyme to the oxygen, as well as complex interaction between plant and microorganisms (Beatty and Good 2011; Curatti and Rubio 2014).

Though several non-nitrogen fixing bacterial strains have been engineered to produce functional nitrogenase enzymes in recent years (Temme et al. 2012; Wang et al. 2013), engineering of eukaryotic cells to produce functional nitrogenase is still not successful (Curatti and Rubio 2014). The number of *nif* genes and their ratio of expression involved in nitrogenase production is a major challenge to biotechnologists (Poza-Carrión et al. 2014; Wang et al. 2013). In addition, the management of oxygen during the expression of nitrogenase in plant cells is another challenge as the enzyme is extremely sensitive to oxygen (Ivelva et al. 2016).

5.12 Conclusions and Future Prospective

The intense application of chemical nitrogen fertilizers in crops to meet the global food demand is creating an economic as well as an environmental burden throughout the globe. Hence, to fulfill the nitrogen demand of soil, and thereby global food demand, it is necessary to avoid the use of chemical fertilizers and shift to BNF, which is an alternative sustainable way to achieve the same. This biological phenomenon offers an eco-friendly and cost effective way for improvement of the nitrogen health of soil.

However, social as well as technical constraints are still there which limit the effective use of BNF in agriculture. In most cases, the diazotrophic microbiome, when applied to cropland, faces limitations of bio-available phosphorus which ultimately affects the rate of nitrogen fixation. Molybdenum and iron are the other nutrients required in larger amounts for sustainable BNF in cropland. In addition, data on the response of a diazotrophic organism to a given set of natural conditions is still lacking.

Though the ecological models predicting the magnitude of response to rhizobial inoculation are now becoming available, inadequate information on their interaction with the environment is a major limitation to use them in the field. Moreover, the use of rhizobia is limited only to the legumes. The ecological models able to predict the response of plant and environment to inoculation of free living diazotrophs will be more useful, because these organisms have a broader range of host diversity.

Application of recombinant DNA technology to extend the symbiotic capability of rhizobia to non-symbiotic organisms and extension of nitrogen fixation activity to non-nitrogen fixing microorganisms as well as plants could be an alternative to improve the magnitude of BNF. However, these approaches are facing a lot of obstacles, like complexity of genes involved in nitrogen fixation, management of oxygen toxicity to nitrogenase, and are still in infancy.

Furthermore, the detection of introduced organisms in the field, and production level problems are the other issues, which need to be addressed before the incorporation of BNF into actual farming situations.

References

Adesemoye, A.O., Torbert, H.A., and Kloepper, J.W. (2009). Plant growth-promoting rhizobacteria allow reduced application rates of chemical fertilizers. *Microbial Ecology* 58 (4): 921–929.

Bargaz, A., Lyamlouli, K., Chtouki, M. et al. (2018). Soil microbial resources for improving fertilizers efficiency in an integrated plant nutrient management system. *Frontiers in Microbiology* 9: 1606. https://doi.org/10.3389/fmicb.01606.

Barron, A.R., Wurzburger, N., Bellenger, J.P. et al. (2009). Molybdenum limitation of asymbiotic nitrogen fixation in tropical forest soils. *Nature Geoscience* 2 (1): 42.

Beatty, P.H. and Good, A.G. (2011). Future prospects for cereals that fix nitrogen. *Science* 333 (6041): 416–417.

Bellenger, J.P., Xu, Y., Zhang, X. et al. (2014). Possible contribution of alternative nitrogenases to nitrogen fixation by asymbiotic N_2-fixing bacteria in soils. *Soil Biology and Biochemistry* 69: 413–420.

Benner, J.W. and Vitousek, P.M. (2007). Development of a diverse epiphyte community in response to phosphorus fertilization. *Ecology Letters* 10: 628–636.

Bergman, B. (2001). Nitrogen-fixing cyanobacteria in tropical oceans, with emphasis on the Western Indian Ocean. *South African Journal of Botany* 63 (3): 426–432.

Berman-Frank, I., Quigg, A., Finkel, Z.V. et al. (2007). Nitrogen-fixation strategies and Fe requirements in cyanobacteria. *Limnology and Oceanography* 52: 2260–2269.

Berthrong, S.T., Yeager, C.M., Gallegos-Graves, L. et al. (2014). Nitrogen fertilization has a stronger effect on soil nitrogen-fixing bacterial communities than elevated atmospheric CO_2. *Applied and Environmental Microbiology* 80 (10): 3103–3112.

Betancourt, D.A., Loveless, T.M., Brown, J.W., and Bishop, P.E. (2008). Characterization of diazotrophs containing Mo-independent nitrogenases, isolated from diverse natural environments. *Applied and Environmental Microbiology* 74: 3471–3480.

Bhattacharyya, P.N. and Jha, D.K. (2012). Plant growth-promoting rhizobacteria (PGPR): emergence in agriculture. *World Journal of Microbiology and Biotechnology* 28: 1327–1350.

Bird, C., Martinez, J., O'Donnell, A.G., and Wyman, M. (2005). Spatial distribution and transcriptional activity of an uncultured clade of planktonic diazotrophic gamma-proteobacteria in the Arabian Sea. *Applied and Environmental Microbiology* 71: 2079–2085.

Bostrom, K.H. (2006). Nitrogen fixation among marine bacterioplankton. Doctoral Thesis, University of Kalmar, Sweden. pp. 26.

Brune, A. (2014). Symbiotic digestion of lignocellulose in termite guts. *Nature Reviews Microbiology* 12: 168–180.

Burgmann, H., Widmer, F., Von Sigler, W., and Zeyer, J. (2004). New molecular screening tools for analysis of free-living diazotrophs in soil. *Applied and Environmental Microbiology* 70: 240–247.

Cejudo, F.J. and Paneque, A. (1988). Effect of nitrogen starvation on ammonium-inhibition of nitrogenase activity in *Azotobacter chroococcum*. *Archives of Microbiology* 149: 481–484.

Chen, J., Shen, W., Xu, H. et al. (2019). The composition of nitrogen-fixing microorganisms correlates with soil nitrogen content during reforestation: a comparison between legume and non-legume plantations. *Frontiers in Microbiology* 10: 508. https://doi.org/10.3389/fmicb.2019.00508.

Church, M.J., Short, C.M., Jenkins, B.D. et al. (2005). Temporal patterns of nitrogenase gene (nifH) expression in the oligotrophic North Pacific Ocean. *Applied and Environmental Microbiology* 71: 5362–5370.

Cleveland, C.C., Townsend, A.R., Schimel, D.S. et al. (1999). Global patterns of terrestrial BNF in natural ecosystems. *Global Biogeochemical Cycles* 13 (2): 623–645.

Curatti, L. and Rubio, L.M. (2014). Challenges to develop nitrogen-fixing cereals by direct *nif*-gene transfer. *Plant Science* 225: 130–137.

Curtis, A.D. and Waller, D.A. (1995). Changes in nitrogen fixation rates in termites (Isoptera: Rhinotermitidae) maintained in the laboratory. *Annals of the Entomological Society of America* 88: 764–767.

Curtis, A.D. and Waller, D.A. (1998). Seasonal patterns of nitrogen fixation in termites. *Functional Ecology* 12: 803–807.

Cusack, D.F., Silver, W., and McDowell, W.H. (2009). BNF in two tropical forests: ecosystem level patterns and effects of nitrogen fertilization. *Ecosystems* 12: 1299–1315.

Darnajoux, R., Zhang, X., McRose, D.L. et al. (2017). BNF by alternative nitrogenases in boreal cyanolichens: importance of molybdenum availability and implications for current BNF estimates. *New Phytologist* 213 (2): 680–689.

Desai, M.S., Assig, K., and Dattagupta, S. (2013). Nitrogen fixation in distinct microbial niches within a chemoautotrophy-driven cave ecosystem. *ISME Journal* 7: 2411–2423.

Dixon, R. and Kahn, D. (2004). Genetic regulation of BNF. *Nature Reviews Microbiology* 2: 621–631.

Forbes, M.S., Broos, K., Baldock, J.A. et al. (2009). Environmental and edaphic factors drive bacterial communities involved in soil N-cycling. *Australian Journal of Soil Research* 47: 380–388.

Fowler, D., Coyle, M., Skiba, U. et al. (2013). The global nitrogen cycle in the twenty-first century. *Philosophical Transactions of the Royal Society, B: Biological Sciences* 368 (1621): 20130164.

Gaby, J.C. and Buckley, D.H. (2011). A global census of nitrogenase diversity. *Environmental Microbiology* 13: 1790–1799.

Gaby, J.C. and Buckley, D.H. (2012). A comprehensive evaluation of PCR primers to amplify the nifH gene of nitrogenase. *PLoS One* 7 (7): e42149. https://doi.org/10.1371/journal.pone.0042149.

Galloway, J.N., Townsend, A.R., Erisman, J.W. et al. (2008). Transformation of the nitrogen cycle: recent trends, questions, and potential solutions. *Science* 320: 889–892.

Geddes, B.A., Ryu, M.H., Mus, F. et al. (2015). Use of plant colonizing bacteria as chassis for transfer of N_2-fixation to cereals. *Current Opinion in Biotechnology* 32: 216–222.

Green, P.A., Vorosmarty, C.J., Meybeck, M. et al. (2004). Pre-industrial and contemporary fluxes of nitrogen through rivers: a global assessment based on typology. *Biogeochemistry* 68: 71–105.

Grote, U., Craswell, E., and Vlek, P. (2005). Nutrient flows in international trade: ecology and policy issues. *Environmental Science & Policy* 8: 439–451.

Gyaneshwar, P., Hirsch, A.M., Moulin, L. et al. (2011). Legume-nodulating betaproteobacteria: diversity, host range, and future prospects. *Molecular Plant-Microbe Interactions* 24: 1276–1288.

Halbleib, C.M. and Ludden, P.W. (2000). Regulation of BNF. *The Journal of Nutrition* 130 (5): 1081–1084.

Hedin, L.O., Brookshire, E.N.J., Menge, D.N.L., and Barron, A.R. (2009). The nitrogen paradox in tropical forest ecosystems. *Annual Review of Ecology, Evolution, and Systematics* 40: 613–635.

Herridge, D.F., Peoples, M.B., and Boddey, R. (2008). Global inputs of biological nitrogen fixation in agricultural systems. *Plant and Soil* 311 (1): 1–18.

Hill, S. (1992). Physiology of nitrogen fixation in free-living heterotrophs. In: *Biological Nitrogen Fixation* (eds. G. Stacey, R.H. Burris and H.J. Evans), 87–134. New York: Chapman & Hall.

Hodkinson, B.P., Allen, J.L., Forrest, L.L. et al. (2014). Lichen-symbiotic cyanobacteria associated with *Peltigera* have an alternative vanadium-dependent nitrogen fixation system. *European Journal of Phycology* 49: 11–19.

Hou, E., Chen, C., McGroddy, M.E., and Wen, D. (2012). Nutrient limitation on ecosystem productivity and processes of mature and old-growth subtropical forests in China. *PLoS One* 7 (12): e52071.

Houlton, B.Z., Wang, Y.P., Vitousek, P.M., and Field, C.B. (2008). A unifying framework for dinitrogen fixation in the terrestrial biosphere. *Nature* 454: 327–330.

Howarth, R.W., Marina, R., and Lane, J. (1988). Nitrogen fixation in freshwater, estuarine, and marine ecosystems. 1. Rates and importance. *Limnology and Oceanography* 33 (4): 669–687.

Hu, Y., Lee, C.C., and Ribbe, M.W. (2012). Vanadium nitrogenase: a two-hit wonder? *Dalton Transactions* 41 (4): 1118–1127.

Huhe, Borjigin, S., Cheng, Y. et al. (2014). Effect of abandonment on diversity and abundance of free-Living nitrogen-fixing bacteria and total bacteria in the cropland soils of Hulun Buir, Inner Mongolia. *PLoS One* 9 (9): e106714.

Ivleva, N.B., Groat, J., Staub, J.M., and Stephens, M. (2016). Expression of active subunit of nitrogenase via integration into plant organelle genome. *PLoS One* 11 (8): e0160951.

Kahindi, J.H.P., Woomer, P., George, T. et al. (1997). Agricultural intensification, soil biodiversity and ecosystem function in the tropics: the role of nitrogen-fixing bacteria. *Applied Soil Ecology* 6: 55–76.

Karl, D., Michaels, A., Bergman, B. et al. (2002). Dinitrogen fixation in the world's oceans. In: *The Nitrogen Cycle at Regional to Global Scales* (eds. E.W. Boyer and R.W. Howarth), 47–98. Dordrecht: Springer.

Kustka, A., Carpenter, E.J., and Sañudo-Wilhelmy, S.A. (2002). Iron and marine nitrogen fixation: progress and future directions. *Research in Microbiology* 153 (5): 255–262.

Laranjo, M., Alexandre, A., and Oliveira, S. (2014). Legume growth-promoting rhizobia: an overview on the *Mesorhizobium* genus? *Microbiological Research* 169 (1): 2–17.

LaRoche, J. and Breitbarth, E. (2005). Importance of the diazotrophs as a source of new nitrogen in the ocean. *Journal of Sea Research* 53: 67–91.

Larson, C.A., Mirza, B., Rodrigues, J.L.M., and Passy, S.I. (2018). Iron limitation effects on nitrogen-fixing organisms with possible implications for cyanobacterial blooms. *FEMS Microbiology Ecology* 94 (5):fiy046. https://doi.org/10.1093/femsec/fiy046.

LeBauer, D.S. and Treseder, K.K. (2008). Nitrogen limitation of net primary productivity in terrestrial ecosystems is globally distributed. *Ecology* 89: 371–379.

Lindsay, E.A., Colloff, M.J., Gibb, N.L., and Wakelin, S.A. (2010). The abundance of microbial functional genes in grassy woodlands is influenced more by soil nutrient enrichment than by recent weed invasion or livestock exclusion. *Applied and Environmental Microbiology* 76 (16): 5547–5555.

Ling, J., Wang, H., Wu, P. et al. (2016). Plant nodulation inducers enhance horizontal gene transfer of *Azorhizobium caulinodans* symbiosis island. *Proceedings of the National Academy of Sciences of the United States of America* 113 (48): 13875–13880.

Liu, J., You, L., Amini, M. et al. (2010). A high-resolution assessment on global nitrogen flows in cropland. *Proceedings of the National Academy of Sciences of the United States of America* 107 (17): 8035–8040.

Ludden, P.W. (2001). *Nitrogenase Complex. Encyclopedia of Life Sciences*. Wiley-Blackwell.

MacKellar, D., Lieber, L., Norman, J.S. et al. (2016). *Streptomyces* thermoautotrophicus does not fix nitrogen. *Scientific Reports* 6: 20086.

Mahaffey, C., Michaels, A.F., and Capone, D.G. (2005). The conundrum of marine N_2 fixation. *American Journal of Science* 305 (6–8): 546–595.

McRose, D.L., Zhang, X., Kraepiel, A.M., and Morel, F.M. (2017). Diversity and activity of alternative nitrogenases in sequenced genomes and coastal environments. *Frontiers in Microbiology* 8: 267.

Mills, M.M., Ridame, C., Davey, M. et al. (2004). Iron and phosphorus co-limit nitrogen fixation in the eastern tropical North Atlantic. *Nature* 429 (6989): 292.

Mirza, B.S., Potisap, C., Nüsslein, K. et al. (2014). Response of free-living nitrogen-fixing microorganisms to land use change in the Amazon rainforest. *Applied and Environmental Microbiology* 80 (1): 281–288.

Mus, F., Crook, M.B., Garcia, K. et al. (2016). Symbiotic nitrogen fixation and the challenges to its extension to nonlegumes. *Applied and Environmental Microbiology* 82 (13): 3698–3710.

Mus, F., Seefeldt, L.C., Alleman, A.B. et al. (2018). Exploring the alternatives of BNF. *Metallomics* 10: 523–538.

Oelze, J. (2000). Respiratory protection of nitrogenase in *Azotobacter* species: is a widely held hypothesis unequivocally supported by experimental evidence? *FEMS Microbiology Reviews* 24 (4): 321–333.

Ohkuma, M., Noda, S., and Kudo, T. (1999). Phylogenetic diversity of nitrogen fixation genes in the symbiotic microbial community in the gut of diverse termites. *Applied and Environmental Microbiology* 65 (11): 4926–4934.

Oldroyd, G.E. and Dixon, R. (2014). Biotechnological solutions to the nitrogen problem. *Current Opinion in Biotechnology* 26: 19–24.

Olivares, J., Bedmar, E.J., and Sanjuán, J. (2013). Biological nitrogen fixation in the context of global change. *Molecular Plant-Microbe Interactions* 26 (5): 486–494.

Paerl, H.W. and Zehr, J.P. (2000). Marine nitrogen fixation. In: *Microbial Ecology of the Oceans* (ed. D.L. Kirchman), 387–426. Wiley-Liss.

Pandey, S., Waller, D.A., and Gordon, A.S. (1992). Variation in acetylene-reduction (nitrogen-fixation) rates in *Reticulitermes* spp. (Isoptera; Rhinotermitidae). *Virginia Journal of Science* 43 (3): 333–338.

Penton, C.R., Yang, C., Wu, L. et al. (2016). NifH-harboring bacterial community composition across an Alaskan permafrost thaw gradient. *Frontiers in Microbiology* 7: 1894.

Peoples, M.B., Herridge, D.F., and Ladha, J.K. (1995). Biological nitrogen fixation: an efficient source of nitrogen for sustainable agricultural production? In: *Management of Biological Nitrogen Fixation for the Development of More Productive and Sustainable Agricultural Systems*, 3–28. Dordrecht: Springer.

Pereira E Silva, M.C., Schloter-Hai, B., Schloter, M. et al. (2013). Temporal dynamics of abundance and composition of nitrogen-fixing communities across agricultural soils. *PLoS One* 8 (9): e74500.

Poza-Carrión, C., Jiménez-Vicente, E., Navarro-Rodríguez, M. et al. (2014). Kinetics of *nif* gene expression in a nitrogen-fixing bacterium. *Journal of Bacteriology* 196 (3): 595–603.

Raymond, J., Siefert, J.L., Staples, C.R., and Blankenship, R.E. (2004). The natural history of nitrogen fixation. *Molecular Biology and Evolution* 21 (3): 541–554.

Reed, S.C., Cleveland, C.C., and Townsend, A.R. (2007). Controls over leaf litter and soil nitrogen fixation in two lowland tropical rain forests. *Biotropica* 39 (5): 585–592.

Reed, S.C., Townsend, A.R., Cleveland, C.C., and Nemergut, D.R. (2010). Microbial community shifts influence patterns in tropical forest nitrogen fixation. *Oecologia* 164 (2): 521–531.

Reed, S.C., Cleveland, C.C., and Townsend, A.R. (2011). Functional ecology of free-living nitrogen fixation: a contemporary perspective. *Annual Review of Ecology, Evolution, and Systematics* 42: 489–512.

Reed, S.C., Cleveland, C.C., and Townsend, A.R. (2013). Relationships among phosphorus, molybdenum and free-living nitrogen fixation in tropical rain forests: results from observational and experimental analyses. *Biogeochemistry* 114 (1–3): 135–147.

Rejmánková, E., Sirová, D., Castle, S.T. et al. (2018). Heterotrophic N_2-fixation contributes to nitrogen economy of a common wetland sedge, *Schoenoplectus californicus*. *PLoS One* 13 (4): e0195570.

Rogers, C. and Oldroyd, G.E. (2014). Synthetic biology approaches to engineering the nitrogen symbiosis in cereals. *Journal of Experimental Botany* 65 (8): 1939–1946.

Sañudo-Wilhelmy, S.A., Kustka, A.B., Gobler, C.J. et al. (2001). Phosphorus limitation of nitrogen fixation by *Trichodesmium* in the Central Atlantic Ocean. *Nature* 411 (6833): 66.

Sapountzis, P., De Verges, J., Rousk, K. et al. (2016). Potential for nitrogen fixation in the fungus-growing termite symbiosis. *Frontiers in Microbiology* 7: 1993.

Schlesier, J., Rohde, M., Gerhardt, S., and Einsle, O. (2015). A conformational switch triggers nitrogenase protection from oxygen damage by Shethna protein II (FeSII). *Journal of the American Chemical Society* 138 (1): 239–247.

Son, Y. (2001). Non-symbiotic nitrogen fixation in forest ecosystems. *Ecological Research* 16 (2): 183–196.

Soto-Urzúa, L. and Baca, B.E. (2001). Mechanisms for protecting nitrogenase from inactivation by oxygen. *Revista Latinoamericana de Microbiología* 43 (1): 37–49.

Stone, M.M., Kan, J., and Plante, A.F. (2015). Parent material and vegetation influence bacterial community structure and nitrogen functional genes along deep tropical soil profiles at the Luquillo Critical Zone Observatory. *Soil Biology and Biochemistry* 80: 273–282.

Temme, K., Zhao, D., and Voigt, C.A. (2012). Refactoring the nitrogen fixation gene cluster from *Klebsiella oxytoca*. *Proceedings of the National Academy of Sciences of the United States of America* 109 (18): 7085–7090.

Teng, Q., Sun, B., Fu, X. et al. (2009). Analysis of nifH gene diversity in red soil amended with manure in Jiangxi, South China. *Journal of Microbiology* 47 (2): 135–141.

Thanganathan, S. and Hasan, K. (2018). Diversity of nitrogen fixing bacteria associated with various termite species. *Pertanika Journal of Tropical Agricultural Science* 41 (3): 925–940.

Tilman, D., Fargione, J., Wolff, B. et al. (2001). Forecasting agriculturally driven global environmental change. *Science* 292 (5515): 281–284.

Townsend, A.R. and Howarth, R.W. (2010). Fixing the global nitrogen problem. *Scientific American* 302 (2): 64–71.

Tu, Q., Zhou, X., He, Z. et al. (2016). The diversity and co-occurrence patterns of N_2-fixing communities in a CO_2-enriched grassland ecosystem. *Microbial Ecology* 71 (3): 604–615.

Turk, K.A., Rees, A.P., Zehr, J.P. et al. (2011). Nitrogen fixation and nitrogenase (*nifH*) expression in tropical waters of the eastern North Atlantic. *The ISME Journal* 5 (7): 1201.

Uddin, M.B., Abu, M., Khan, S.A. et al. (2008). Effects of inorganic fertilizers on BNF and seedling growth of some agroforestry trees in Bangladesh. *Journal of Forestry Research* 19 (4): 303–306.

Vecherskii, M.V., Kostina, N.V., Gorlenko, M.V. et al. (2008). Specific features of nitrogen fixation in the termite *Reticulitermes lucifugus*. *Biology Bulletin* 35 (5): 446–451.

Vitousek, P.M. and Field, C.B. (1999). Ecosystem constraints to symbiotic nitrogen fixers: a simple model and its implications. In: *New Perspectives on Nitrogen Cycling in the Temperate and Tropical Americas* (ed. V.M. Vitousek), 179–202. Dordrecht: Springer.

Vitousek, P.M., Cassman, K.E.N., Cleveland, C. et al. (2002). Towards an ecological understanding of biological nitrogen fixation. In: *The Nitrogen Cycle at Regional to Global Scales* (ed. V.M. Vitousek), 1–45. Dordrecht: Springer.

Vitousek, P.M., Menge, D.N., Reed, S.C., and Cleveland, C.C. (2013). Biological nitrogen fixation: rates, patterns and ecological controls in terrestrial ecosystems. *Philosophical Transactions of the Royal Society, B: Biological Sciences* 368 (1621): 20130119.

Voss, M., Bange, H.W., Dippner, J.W. et al. (2013). The marine nitrogen cycle: recent discoveries, uncertainties and the potential relevance of climate change. *Philosophical Transactions of the Royal Society, B: Biological Sciences* 368 (1621): 20130121.

Wagner, S.C. (2011). Biological nitrogen fixation. *Nature Education Knowledge* 3: 15.

Wakelin, S.A., Gupta, V.V.S.R., and Forrester, S.T. (2010). Regional and local factors affecting diversity, abundance and activity of free-living, N_2-fixing bacteria in Australian agricultural soils. *Pedobiologia* 53 (6): 391–399.

Wang, L., Zhang, L., Liu, Z. et al. (2013). A minimal nitrogen fixation gene cluster from *Paenibacillus* sp. WLY78 enables expression of active nitrogenase in Escherichia coli. *PLoS Genetics* 9 (10): e1003865.

Wang, J., Bao, J.T., Li, X.R., and Liu, Y.B. (2016). Molecular ecology of nifH genes and transcripts along a chronosequence in revegetated areas of the Tengger Desert. *Microbial Ecology* 71 (1): 150–163.

Wurzburger, N., Bellenger, J.P., Kraepiel, A.M., and Hedin, L.O. (2012). Molybdenum and phosphorus interact to constrain asymbiotic nitrogen fixation in tropical forests. *PLoS One* 7 (3): e33710.

Yamada, A., Inoue, T., Wiwatwitaya, D. et al. (2006). Nitrogen fixation by termites in tropical forests, Thailand. *Ecosystems* 9 (1): 75–83.

Yang, J., Xie, X., Wang, X. et al. (2014). Reconstruction and minimal gene requirements for the alternative iron-only nitrogenase in *Escherichia coli*. *Proceedings of the National Academy of Sciences of the United States of America* 111 (35): E3718–E3725.

Yeager, C.M., Kornosky, J.L., Housman, D.C. et al. (2004). Diazotrophic community structure and function in two successional stages of biological soil crusts from the Colorado Plateau and Chihuahuan Desert. *Applied and Environmental Microbiology* 70 (2): 973–983.

Yuan, S. and Peng, S. (2017). Exploring the trends in nitrogen input and nitrogen use efficiency for agricultural sustainability. *Sustainability* 9 (10): 1905.

Zehr, J.P. (2011). Nitrogen fixation by marine cyanobacteria. *Trends in Microbiology* 19 (4): 162–173.

6

Functional Microbiome for Crop Improvement Under a Changing Environment

Abbaci Hocine[1], Bensidhoum Leila[1], Houali Karim[2], and Nabti Elhafid[1]

[1] *Laboratoire de Maitrise des Énergies Renouvelables, Faculté des Sciences de la Nature et de la Vie, Université de Bejaia, Bejaia, Algeria*
[2] *Department of Microbiology and Biochemistry, Université Mouloud Mammeri de Tizi Ouzou, Tizi Ouzou, Algeria*

6.1 Introduction

Feeding a growing world population is among the greatest challenges that humanity is facing in the twenty-first century. Crop productivity needs to increase by 70–100 percent to meet the energy needs of a world population of over nine billion by 2050 (Tilman et al. 2011). A new alternative in the development of an improved and advanced sustainable crop production is the use of beneficial plant–microbe services (Trivedi et al. 2017).

In a context of a changing environment, it is crucial to maximize our knowledge on all the mechanisms involved in plant microbiome interactions at the genetic, physiological, and ecological levels. A gram of soil may harbor thousands to millions of bacterial species and diversity of microbial communities that cover the plant microbiome are mostly unknown. Here, we will not develop topics to gain a solid understanding of the plant microbiome diversity, but rather discuss its potential in improving the quality and quantity of crops. Even though the subject is well-documented, we make no claim for discussing all the aspects related to crop improvement, i.e. reporting most of the advances and applications of genetic engineering in crop farming.

Currently, much effort and enthusiasm are directed to understanding how plants can shape the microbiome and how plants respond to those microbiomes. Consequently, this will contribute in achieving, in the near future, more robust and effective tools to predict outcomes of plant-microbiome interactions either negative or positive.

The soil microbiome affects directly and indirectly the nutrient cycle such as nitrogen, phosphorus, and carbon. Its impact on plant growth protection and improvement under conditions of biotic and abiotic stress must be highlighted. Microbial diversity in soil is controlled by various factors such as changes in soil texture, root exudates, pH, nutrients, or water retention etc. (Fierer and Jackson 2006). The plant–microbiome association is a

The Plant Microbiome in Sustainable Agriculture, First Edition. Edited by Alok Kumar Srivastava, Prem Lal Kashyap and Madhumita Srivastava.

key element in improving plant health and productivity, community balance and ecosystem functioning (Berg et al. 2014).

In this work, we review some recent advances in plant microbiome investigations and describe potential applications of such associations for the mitigation of both abiotic and biotic stresses to improve crop health and productivity.

6.2 Current Climate Change and Its Impact on Agriculture

Climate change negatively impacts biological and human systems in different ways (Haines et al. 2006; Thornton et al. 2014) where precipitation amount, sunshine duration, humidity, temperature, cloud cover, flood, drought, soil moisture and high CO_2 level affect annual crop yields (Siwar et al. 2009). As a result, global warming, a continued and long-term process, is considered as one of the major threats for agriculture, with noticeable eternalized effects (Arndt et al. 2012) such as:

- A decrease in water availability for irrigation.
- Salinization of agricultural land by sea level rising.
- Different pests, weeds, etc., can appear or disappear, and new diseases can emerge in response to changes in temperature and precipitation patterns (Bonfils et al. 2008; Al-Amin et al. 2015).

6.3 Crop Production Under Stressful Environment

Plants live in complex environments. To survive, they react to various environmental challenges by the incorporation of mutual signals associated with biotic and abiotic stresses. Environmental stressors include drought, low/high temperature, salinity and acidic conditions, light intensity, submergence, anaerobiosis, and nutrient starvation (Meena et al. 2017), all causing adverse effects on plant growth and crop productivity. Plants can generally adapt to environmental changes by varying their intrinsic characteristics while implicating different components of their microbiome (Vandenkoornhuyse et al. 2015). There are three major compartments where microbial cells can establish and develop: rhizosphere, endosphere, and phyllosphere (Hardoim et al. 2008; Hirsch and Mauchline 2012). These compartments are populated by a wide array of microorganisms.

On the basis of the nature of their interactions with plants, they can be categorized into three distinct groups: deleterious, neutral, and beneficial. This latter comprises several fungi and bacteria.

Many bacteria such as *Azospirillum* (Nabti et al. 2010), *Rhizobium, Pseudomonas* (Ahmad et al. 2013) and *Bacillus* enhance plants' tolerance to various abiotic stresses where different mechanisms are involved (Figure 6.1). This may be done through nutrient acquisition improvement, nitrogen fixation (Carvalho et al. 2014), and via plant growth regulation substances like cytokinins (Rolli et al. 2015) and indole acetic acid (IAA) (Nabti et al. 2007; Gamalero and Glick 2015). Interactions between these microorganisms and plants are an integral part of the living ecosystem. These microorganisms are the natural

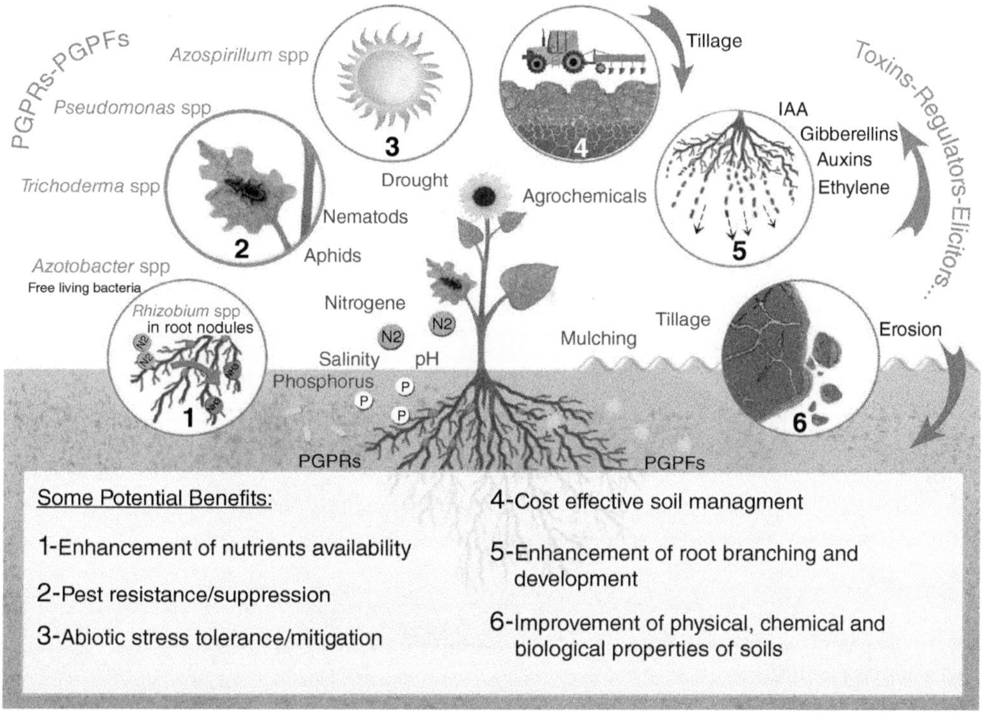

Figure 6.1 Microbiome for crop and soil management.

partners that modulate local and systemic mechanisms in plants to offer defense under adverse external conditions (Meena et al. 2017).

6.3.1 Mitigation of Saline Stress

Salinity is considered as a major threat for agriculture under irrigation. In arid and semi-arid regions, soils are frequently saline with low suitability for any agricultural use (Shrivastava and Kumar 2015). Plant crops differ in their tolerance to salinity. When a stress stimulus was detected, plants responded immediately and effectively to initiate a succession of stress-specific signals. Response usually entails variations in the levels of important categories of phytohormones such as cytokinins, gibberellic acid, ethylene, brassinosteroids, jasmonates, and strigolactones (Wani et al. 2007). In parallel, an accumulation of phenolic acids and flavonoids will take place and the development of numerous antioxidants such as osmolytes and the activation of transcription factors (TF) are initiated at the same-time as the expression of specific genes-stress to support the defense system (Austin and Ballaré 2014).

Plants also manage to mitigate environmental stresses by interacting with the microbiome (Marasco et al. 2012). Some bacterial strains affect the metabolic balance to reduce water stress in wheat plants under reduced irrigation conditions (Nabti et al. 2007; Miliute et al. 2015). *Azospirillum brasilense* (strain NH) restored plant wheat growth by accumulation of osmoprotectants, i.e. glycine betaine, proline, and sugars decreasing salt stress (Nabti et al. 2010). Similarly, Jha et al. (2011) confirmed the effect of *Azophirillum* spp. on the alleviation of water stress in maize by elevation of abscisic acid amount (ABA). This latter is shown as an essential phytohormone for plant growth development increasing proportionally with stress strength (Gamalero and Glick 2015). In their study concerning the functionality of root-associated bacteria along a saltmarsh, Wang et al. (2017) reported that some strains of *Pseudomonas* sp. and *Serratia* sp. exhibited high resistance towards abiotic stress and developed more plant growth promoting traits such as siderophore and IAA-production and biofilm formation.

1-Aminocyclopropane-1-carboxylase (ACC) deaminase produced similarly by many bacteria is considered as a plant growth promoting (PGP) trait. This enzyme plays a crucial role in the modulation of ethylene level in plants under various stresses. It also protects plants from growth inhibition by flooding and anoxia, drought, salinity and pathogens (Gamalero and Glick 2015). ACC deaminase structural and regulatory genes (*acdS* and *acdR*) were found in many rhizospheric bacteria such as the genera: *Pseudomonas*, *Azispirillum*, *Rhizobium*, *Achromobacter* (Blaha et al. 2006). When the ACC amount increases, it is followed by a rise of ethylene in plants, inducing a plant stress (Glick et al. 2007). ACC deaminase converts ACC to ammonia and α-ketobutyrate, thus, ethylene concentration will decrease leading to a prevention of the augmentation of ethylene level by root colonization by bacteria (Glick et al. 2007). The inoculation of canola by *Pseudomonas syringae* ACC deaminase improved significantly the yield and growth of this plant under salinity (Nadeem et al. 2006).

Plant growth promoting 1-aminocyclopropane-1-carboxylic acid (PGPRACC) deaminase – positively enhances the chlorophyll amount in canola, lettuce, and maize (Glick et al. 1998; Han and Lee 2005). Additionally, ACC deaminase produced by *Bradyrhizobium* promoted nodulation by indigenous competitiveness in legumes (Shaharoona et al. 2006).

Ethylene is a gaseous hormone with various biological effects on plant growth at different stages, i.e. germination, tissue differentiation, root and shoot development, flowering and fruit ripening. (Kieber 1997; Glick et al. 1998). Ethylene elevation may occur under symbiosis between plants and rhizobia as well as mycorrhizae (Gamalero and Glick 2015).

Decline in ethylene levels increase plant defenses against a wide variety of adverse environmental stressors (Glick 2005).

Mayak et al. (2004) showed that ethylene production by tomato seedlings is reduced by *Achromobacter piechaudii*, where fresh and dry weights of grown tomatoes are augmented under high saltiness.

6.3.2 Enhancement of Drought Stress Tolerance

Drought restrains a microorganism's aptitude to survive and colonize soil spaces (Carson et al. 2010; Castro et al. 2010). Microbial communities analyzed by bacterial phospholipid profiles in soils revealed that the kinetics of microbial respiration and the structure of the community varied with temperature. The decomposition of organic matter is intrinsically related to changes in the composition of the microbial community that is itself affected by seasonal or annual variations in soil temperature, thus limiting carbon bioavailability (Reich and Oleksyn 2004).

Rolli et al. (2015) demonstrated that the isolates *Acinetobacter* and *Pseudomonas* are capable of adhering and colonizing both the *Arabidopsis* and grapevine rhizoplane under drought stress.

They also concluded that bacteria restored vegetable growth under drought conditions by increasing shoot and leaf biomass, shoot length, and photosynthetic activity, with an enhanced effect in drought-sensitive rootstock. Mommer et al. (2016) and Tripathi et al. (2016) revealed that coumestrol stimulates mycorrhizal colonization leading to improved tolerance to drought.

The study of Kunert et al. (2016) on soybean under drought tolerance showed the importance of direct field experiments on crops and the relevance of the identification of promoters (metabolites, proteins and genes) involved in mechanisms that regulate root architecture as well as the development of bacteroids and the plant lifespan.

To conclude, several polyglycerol polyricinoleate (PGPR) were successfully tested in the restoration of crops under drought conditions, i.e. date palm (Cherif et al. 2015), rice (Gusain et al. 2015), wheat (Kanwal et al. 2017) and maize (Vardharajula et al. 2011) etc.

6.3.3 Microbiome in Extreme Temperatures

Bacteria stimulate plant acclimation by decreasing the effects of cooling temperatures on cells, increasing photosynthetic activity and accumulating cold-related metabolites such as starch, proline, mannitol (Grover et al. 2011; Marasco et al. 2012) and phenolic compounds (Tang and Thomson 1996; Hartmann et al. 2014).

It is revealed that endophytic bacteria improve cold tolerance of grapevine inoculated with *Burkholderia phytofirmans* strain PsJN by modifying the activity patterns of photosynthesis and carbohydrate metabolism of the crop (Grover et al. 2011; Su et al. 2015).

High temperatures may affect symbiotic relationships, nitrogen content, plant production (Pinto et al. 1998); a delay and a fall in rate of nodulation, rhizobial growth and colonization can expose plants to various infectious events (Gopalakrishnan et al. 2015).

Thermotolerant plant growth rhizobacteria could be an effective approach to mitigate the negative impacts of high temperatures on crop yield, i.e. inoculation of wheat with *Pseudomonas putida AKMP7* and *Pseudomonas aeruginosa 2CpS1* (Ali et al. 2011; Meena et al. 2015).

6.3.4 Microbiome and Soil Fertility

Physico-chemical parameters of soils have a direct effect on the sorting of microorganisms by supporting some and eliminating others through root exudate availability affecting the plant–microbiome relationship (Hartmann et al. 2009; Marschner et al. 2011; Lareen et al. 2016). The appropriate pH range for optimum plant growth differs within crop species and varieties. It is well-known that beneficial rhizospheric microorganisms and plants usually prefer neutral pH. Increased acidity in soils frequently influences the microorganism's diversity and alters their activity. This trend in pH generates significant changes in decomposition and leads to the immobilization of basic nutrients, reducing their availability to plants (Andrew et al. 2012; Zhalnina et al. 2015; Santoyo et al. 2017). At low pH, the microbial community is disrupted by cell membrane rupture, enzymatic system alteration and deceleration of cell proliferation. Consequently, this directly affects soil health and productivity (Birgander et al. 2014). Moreover, acidic conditions favor the dominance of fungi compared to bacteria (Rousk et al. 2009) which cause an imbalance in the soil microbial community and enhance the invasion of pathogenic fungal roots. The roles assigned to soil fungi by bacteria in the decomposition and release of nutrients to plants will be disordered (Rousk et al. 2009). Nowadays, it is highly approved that PGPF (plant growth promoting fungi) and PGPR (plant growth promoting rhizobacteria) play important roles in enhancement of soil fertility. PGPR increase nutrient uptake (Ca, K, Fe, Cu, Mn and Zn) and their availability for plants through soil acidification as a result of organic acid production or proton pump ATPase stimulation (Mantelin and Touraine 2004). In the last decades, considerable attention has been paid to PGPR as alternatives to agrochemicals for the improvement of crops. Effective inoculates for wheat, maize, rice, and bean have been intensively tested (Grimes and Mount 1984; Boddey et al. 1986; Garcia de Salamone et al. 1996; Malik et al. 1997).

6.3.5 Bioremediation

All over the world, industrialization and urbanization are considered as major drivers of pollutants such as heavy metals and organics compounds, i.e. polychlorinated biphenyls (PCBs), polycyclic aromatic hydrocarbons (PAHs), fuels, nitroaromatic compounds (NACs), phthalates, and polybrominated biphenyls (PBBs) (Das et al. 2014).

Bioremediation exploits the remarkable ability of the microbe to degrade and transform or accumulate pollutant compounds. Bioremediation is regarded as one of the most effective strategies to manage the polluted environment and recover contaminated soil (Singh 2014). Interestingly, various resistant bacteria isolated from heavy metal-polluted sites showed the ability to reduce these pollutants by bioremediation processes, partly in combination with plants (Nies 1999).

Dell'Amico et al. (2008) showed that *Brassica napus* inoculated with bacterial strains *Pseudomonas tolassi* and *Pseudomonas fluorescens* promoted plant growth under cadmium stress.

This latter is mainly mitigated by production of IAA, siderophores and ACC-deaminase. Similarly, Bensidhoum et al. (2016) reported that heavy metal tolerant *Pseudomonas protegens* strain S5LiBe restored barley germination rate and growth parameters in heavy metal contaminated soil. Equally, fungi have largely proved their potential in recycling or elimination of hazardous wastes from the environment due to their important degradative capabilities (Fritsche and Hofrichter 2005). The role of fungi in bioremediation of soil has been reported by many authors (D'Annibale et al. 2005, 2006; Federici et al. 2007; Leonardi et al. 2008).

6.3.6 Microbiome as Biocontrol Agents (BCAs)

Plant pathogens including bacteria, fungi, insects, and nematodes, are the main cause of the plant diseases. Crop losses worldwide are estimated to one third of the potential food production due to various pests (Borriss 2011).

Synthetic pesticides are used to control plant pathogens, but their excessive use in agricultural practices adversely affects plant growth, soil microbiome and their physiological activities (Lavelle et al. 2004). Biocontrol products have been adopted worldwide and applied in more than 30 million ha as a natural alternative to the intensive application of chemicals in order to reduce the risks associated to these pesticides (van Lenteren et al. 2018). Many studies demonstrated the potential abilities of beneficial microbes like mycorrhizal fungi and PGPR in plant disease control, such as *Bacillus*, *Pseudomonas*, *Burkholderia*, *Trichoderma*, *Verticillium*, *Paecilomyces*, and *Ulocladium* (Elad et al. 1980; Chet 1987; Inglis et al. 2001; Kerry 2001; El-Hamshary et al. 2008; Quan et al. 2010; Nabti et al. 2013; Ronseaux et al. 2013; Bensidhoum et al. 2016; Tabli et al. 2018).

The effectiveness of *Trichoderma harzianum* as a biocontrol agent (BCA) against *Sclerotium rolfsii* and *Rhizoctonia solani* for the reduction of bean cotton and tomato diseases have been reported by many researchers (Elad et al. 1980; Biswas and Sen 2000).

The potential antagonism of several species of *Pseudomonas*, i.e. *P. putida* (Bojanowski 2011), *P. fluorescens* (Duffy and Défago 1999), *Pseudomonas stutzeri* (Lim et al. 1991) and *P. protegens* (Bensidhoum et al. 2016) is well documented.

6.3.6.1 Mechanisms of Biocontrol

Biocontrol activity is based on various mechanisms, like competition, antibiotics and enzyme production, biofilm formation, pathogen parasitism and plant resistance induction (Jamali et al. 2009; Calvo et al. 2010). Biological control also involves interaction between plant, pathogen, BCAs, and the physical environment (Handelsman and Stabb 1996).

6.3.6.1.1 Biocontrol Agent–Plant Pathogen Interaction Most interactions between antagonists and plant pathogens consist of the inhibition of the latter by antimicrobial substances and cell wall-degrading enzymes (Raaijmakers et al. 2002); competition for nutrients and/or space (Antoun and Prévost 2005); and mitigation or degradation of pathogenicity factors such as toxins (Whipps 2001). BCAs can suppress some plant pathogen growth through competition for nutrients like nitrogen, carbon, and microelements in the environment (Elad and Stewart 2007). Additionally, competition for nutrients inhibits spore germination in the soil (Alabouvette et al. 2006). Iron-binding ligands named siderophores secreted by BCAs sequester iron from the micro-environment, depriving phytopathogens from one of their growth factors (Viswanathan and Samiyappan 2004; Pal and Gardener 2006).

Antibiosis consists of the production of substances with antifungal, insecticidal and/or antibiotic properties (Johansson 2003; Adam 2008). BCAs are considered as potential producers of bioactive compounds such aschitinases, antimicrobial peptides, polyketides, phenolic compounds and bio-surfactants (Fernando et al. 2005). Many bacterial strains may affect pathogen growth and eliminate disease by phenazines (Weller 2007), hydrogen cyanide, 2,4-diacetylphloroglucinol (DAPG) and pyoluteorin production, for instance, *P. fluorescens* strain CHA0 (Voisard et al. 1989; Haas and Keel 2003). *Bacillus* and *Pseudomonas* are widely used to control aflatoxigenic fungi by antibiotic production: fengycin A and B, iturin A, mycosubtilin, bacillomycin D and pyochelin (Shams-Ghahfarokhi et al. 2013).

Volatile molecules as hydrogen cyanide, ammonia, 2,3-butanediol and acetoin produced by microorganisms have been identified as antibiotics (Ryu et al. 2003; Charest et al. 2005; Wani et al. 2007; Ahmad et al. 2008). A cyanide-negative mutant of *P. fluorescens* CHA0 was less effective in controlling the take-all of wheat caused by *Gaeumannomyces graminis* var. *tritici*. Biocontrol ability of this strain was restored after its complementation by hcn^+ gene cloned from the wild type strain (Voisard et al. 1989).

A novel *cry* gene encoding for Cry I-protein group with toxic activity toward insects: *Noctuidae*, *Tortricidae*, *Plutellidae*, and *Chrysomelidae*, was already characterized (Ruiz de Escudero et al. 2006). The ability of BCAs to control plant pests is related in some cases to the presence of genes encoding for insecticidal toxins such as: *Mcf* (Makes caterpillars floppy), *Tc* (Toxin complex), *fit*D (fluorescens insect toxin) and *Cry* (Crystal toxins) (Ruiz de Escudero et al. 2006; Péchy-Tarr et al. 2008; Olcott et al. 2010; Garrido-Sanz et al. 2016). Positively, the inoculation of *P. fluorescens* strains on seeds and roots or by foliar spray showed a significant reduction of leaf folder incidence, both under greenhouse and field conditions. These strains also increased larval mortality and incidence of malformed adults under *in vitro* conditions (Kaymak 2010). Similarly, growth of white clover and Luzerne plants inoculated with *Pseudomonas* strains showed a significant resistance to blue-green aphids (Kempster et al. 2002).

Bacillus thuringiensis is a well proven BCA through the action of δ-endotoxins: Cry and Cyt proteins (Crickmore et al. 1998). Ambrose et al. (2014) showed that *Epichloë* fungal endophytes produce a number of toxins similar to the bacterial "makes caterpillars floppy" genes (*mcf1* and *mcf2*). Some *Epichloë* species may improve the abiotic and biotic resistance to the associated plants while others species may be harmful to livestock by producing toxic alkaloids (Song et al. 2016).

The insecticidal activity of this fungus is attributed to its ability to produce anti-herbivore alkaloids (lolines, indole diterpenes, ergot alkaloids, and peramine) (Schardl et al. 2012).

6.3.6.1.2 Biocontrol Agents–Plant Interaction BCAs can protect plants indirectly by activation of defense mechanisms to microbial attack.

An improved ISR (induced systemic resistance) can be triggered by inducing BACs (Bakker et al. 2007) and supports plants under further stress and future attacks by pathogens (Ramamoorthy et al. 2001; Bent 2006), mediated by three endogenous plant signaling molecules such as salicylic acid (SA), jasmonic acid (JA), and ethylene (ET) (Jourdan et al. 2008).

Induction of ISR by PGPRs and PGPFs has been reviewed by several authors (Ryu et al. 2004; Vallad and Goodman 2004; Bent 2006; Rudrappa et al. 2010; Shoresh et al. 2010).

Pseudomonas sp. increased the durability of ISR in rice against disease caused by *R. solani* when applied on seeds, roots, soil, and leaves (Nandakumar et al. 2001). In parallel, volatile organic compounds (VOCs) produced by *Bacillus subtilis* and *Bacillus amyloliquefaciens* reduced disease severity of *Erwinia carotovora* subsp. *carotovora* in *Arabidopsis* seedlings (Szilagyi-Zecchin et al. 2016). VOCs are sufficient to activate ISR in the aforementioned studied plant model (Ryu et al. 2004). Some strains of *Bacillus subtilis* produce acetoin (3-hydroxy-2-butanone) eliciting ISR and protect *Arabidopsis* against *P. syringae* (Rudrappa et al. 2010).

In addition, antibiotic DAPG protects *Arabidopsis thaliana* against *Peronospora parasitica* (Iavicoli et al. 2003; Weller et al. 2012) and tomato against *P. syringae* and the nematode *Meloidogyne javanica* (Siddiqui and Shaukat 2003).

Fungal strains like *Trichoderma* spp. and non-pathogenic *Fusarium* can also induce systemic resistance by MAMP (microbe-associated molecular patterns) production. Arbuscular mycorrhiza (AM) can also trigger plant defense by boosting its ability to respond to pathogen attacks. In this case, the SIR is named mycorrhizal induced resistance (MIR) (Barea 2015).

The use of biocontrol via the modification of environmental characteristics (pH, plant surfaces, etc.) are other interesting alternatives that may also serve as indirect positive actions to suppress plant pathogens. Striking studies on *Trichoderma* strains showed promising results (Manteau et al. 2003; Benítez et al. 2004; Ajouz 2009).

6.4 Challenges in Microbiome Engineering

Understanding the full potential of microbes in the ecosystem functioning in general and their complex beneficial interactions in improving agriculture productivity in particular requires the development and improvement of compatible tools that can be verified in biological assays (Sergaki et al. 2018), always bearing in mind their reproducibility *in situ* on different scales (Knight et al. 2018). There is an urgent need to fill the field-lab gap, i.e. by going beyond fundamental research on interactions within the microbiome. In this respect, recent progress in the engineering microbiome have made it possible to show how meta-omics (metataxonomic, metagenomics, metatranscriptomics, and metaproteomics) can be potentially powerful tools to gain deeper knowledge of the functional capabilities of the microbiome and how they can shape ecosystems (Bulgarelli et al. 2015). While the technical limitations of meta-omics approaches have been discussed in detail by several researchers (Zhang et al. 2010; Segata et al. 2013) and despite the fact that these meta-omics or sequencing techniques share practically common drawbacks such as sampling errors, lack of adequate analytical algorithms and computational power (Sergaki et al. 2018), there is an encouraging and significant leap made in the mitigation of these limitations by developing and exploiting more and more accurate and elaborated meta-omics tools and datasets. Indeed, there is a real potential for meta-omics investigations. New efforts toward the characterization of microbial communities are needed.

6.5 Conclusion

The microbiome has a tremendous potential to confer resistance and adaptation of plants to severe environmental stresses. It is now well-documented that several mechanisms elicited by PGPFs and PGPRs mitigate biotic and abiotic stresses. The spectacular results in studies are leading to changes in agro-system management practices. Consequently, this will reinforce credibility and acceptance of these biocontrol products as alternative methods to chemical inputs. Further intensive efforts in research are needed to optimize soil microbial technology. With the rise in environmental concerns, enhancement of society's awareness, and positive efforts in the legislative area towards the use of biofertilizers, these ecofriendly bioinputs are expected to replace agrochemicals for achieving more sustainable food production.

References

Adam, A. 2008. *Elicitation de la résistance systémique induite chez la tomate et le concombre et activation de la voie de la lipoxygénase par des rhizobactéries non-pathogènes.* [Elicitation of induced systemic resistance in tomato and cucumber and activation of the lipoxygenase pathway by non-pathogenic rhizobacteria]. Doctoral thesis, Université de Liege.

Ahmad, F., Ahmad, I., and Khan, M.S. (2008). Screening of free-living rhizospheric bacteria for their multiple plant growth promoting activities. *Microbiological Research* 163 (2): 173–181.

Ahmad, M., Zahir, Z.A., Nazli, F. et al. (2013). Effectiveness of halo-tolerant, auxin producing *Pseudomonas* and *Rhizobium* strains to improve osmotic stress tolerance in mung bean (*Vigna radiata* L.). *Brazilian Journal of Microbiology* 44 (4): 1341–1348.

Ajouz, S. 2009. December. *Estimation du potentiel de résistance de Botrytis cinerea à des biofongicides*. Avignon. Ciences Agricoles, Université d'Avignon, Français. NNT: 2009AVIG0622.

Alabouvette, C., Olivain, C., and Steinberg, C. (2006). Biological control of plant diseases: the European situation. *European Journal of Plant Pathology* 114 (3): 329–341.

Al-Amin, A.Q., Rasiah, R., and Chenayah, S. (2015). Prioritizing climate change mitigation: an assessment using Malaysia to reduce carbon emissions in future. *Environmental Science & Policy* 50: 24–33.

Ali, S.Z., Sandhya, V., Grover, M. et al. (2011). Effect of inoculation with a thermotolerant plant growth promoting *Pseudomonas putida* strain AKMP7 on growth of wheat (*Triticum* spp.) under heat stress. *Journal of Plant Interactions* 6 (4): 239–246.

Ambrose, K.V., Koppenhöfer, A.M., and Belanger, F.C. (2014). Horizontal gene transfer of a bacterial insect toxin gene into the Epichloë fungal symbionts of grasses. *Scientific Reports* 4: 5562.

Andrew, D.R., Fitak, R.R., Munguia-Vega, A. et al. (2012). Abiotic factors shape microbial diversity in Sonoran Desert soils. *Applied and Environmental Microbiology* 78 (21): 7527–7537.

Antoun, H. and Prévost, D. (2005). Ecology of plant growth promoting rhizobacteria. In: *PGPR: Biocontrol and Biofertilization* (ed. Z.A. Siddiqui), 1–38. Dordrecht: Springer.

Arndt, C., Chinowsky, P., Robinson, S. et al. (2012). Economic development under climate change rode. *Review of Development Economics* 16: 369–377.

Austin, A.T. and Ballaré, C.L. (2014). Plant interactions with other organisms: molecules, ecology and evolution. *New Phytologist* 204 (2): 257–260.

Bakker, P.A., Pieterse, C.M., and Van Loon, L.C. (2007). Induced systemic resistance by fluorescent *Pseudomonas* spp. *Phytopathology* 97 (2): 239–243.

Barea, J.M. (2015). Future challenges and perspectives for applying microbial biotechnology in sustainable agriculture based on a better understanding of plant-microbiome interactions. *Journal of Soil Science and Plant Nutrition* 15 (2): 261–282.

Benítez, T., Rincón, A.M., Limón, M.C., and Codon, A.C. (2004). Biocontrol mechanisms of *Trichoderma* strains. *International Microbiology* 7 (4): 249–260.

Bensidhoum, L., Nabti, E., Tabli, N. et al. (2016). Heavy metal tolerant *Pseudomonas protegens* isolates from agricultural well water in northeastern Algeria with plant growth promoting, insecticidal and antifungal activities. *European Journal of Soil Biology* 75: 38–46.

Bent, E. (2006). Induced systemic resistance mediated by plant growth-promoting rhizobacteria (PGPR) and fungi (PGPF). In: *(eds. S. Tuzun and E. Bent) Multigenic and Induced Systemic Resistance in Plants*, 225–258. Boston, MA: Springer.

Berg, G., Grube, M., Schloter, M., and Smalla, K. (2014). The plant microbiome and its importance for plant and human health. *Frontiers in Microbiology* 5: Article 491. https://doi.org/10.3389/fmicb.2014.00491.

Birgander, J., Rousk, J., and Olsson, P.A. (2014). Comparison of fertility and seasonal effects on grassland microbial communities. *Soil Biology and Biochemistry* 76: 80–89.

Biswas, K.K. and Sen, C. (2000). Management of stem rot of groundnut caused by *Sclerotium rolfsii* through *Trichoderma harzianum*. *Indian Phytopathology* 53 (3): 290–295.

Blaha, D., Prigent-Combaret, C., Mirza, M.S., and Moënne-Loccoz, Y. (2006). Phylogeny of the 1-aminocyclopropane-1-carboxylic acid deaminase-encoding gene acdS in phytobeneficial and pathogenic Proteobacteria and relation with strain biogeography. *FEMS Microbiology Ecology* 56 (3): 455–470.

Boddey, R.M., Baldani, V.L., Baldani, J.I., and Döbereiner, J. (1986). Effect of inoculation of *Azospirillum* spp. on nitrogen accumulation by field-grown wheat. *Plant and Soil* 95 (1): 109–121.

Bojanowski, A. 2011. *Molécules antifongiques et activité antagoniste de deux souches de Pseudomonas envers Helminthosporium solani, agent responsable de la tache argentée de la pomme de terre*. Thesis, Université Laval.

Bonfils, C., Duffy, P.B., Santer, B.D. et al. (2008). Identification of external influences on temperatures in California. *Climatic Change* 87 (1): 43–55.

Borriss, R. (2011). Use of plant-associated *Bacillus* strains as biofertilizers and biocontrol agents in agriculture. In: *Bacteria in Agrobiology: Plant Growth Responses*, 41–76. Berlin, Heidelberg: Springer.

Bulgarelli, D., Garrido-Oter, R., Münch, P.C. et al. (2015). Structure and function of the bacterial root microbiota in wild and domesticated barley. *Cell Host & Microbe* 17 (3): 392–403.

Calvo, P., Ormeño-Orrillo, E., Martínez-Romero, E., and Zúñiga, D. (2010). Characterization of *Bacillus* isolates of potato rhizosphere from Andean soils of Peru and their potential PGPR characteristics. *Brazilian Journal of Microbiology* 41 (4): 899–906.

Carson, J.K., Gonzalez-Quiñones, V., Murphy, D.V. et al. (2010). Low pore connectivity increases bacterial diversity in soil. *Applied and Environmental Microbiology* 76 (12): 3936–3942.

Carvalho, T.L.G., Balsemão-Pires, E., Saraiva, R.M. et al. (2014). Nitrogen signalling in plant interactions with associative and endophytic diazotrophic bacteria. *Journal of Experimental Botany* 65 (19): 5631–5642.

Castro, H.F., Classen, A.T., Austin, E.E. et al. (2010). Soil microbial community responses to multiple experimental climate change drivers. *Applied and Environmental Microbiology* 76 (4): 999–1007.

Charest, M.H., Beauchamp, C.J., and Antoun, H. (2005). Effects of the humic substances of de-inking paper sludge on the antagonism between two compost bacteria and *Pythium ultimum*. *FEMS Microbiology Ecology* 52 (2): 219–227.

Cherif, H., Marasco, R., Rolli, E. et al. (2015). Oasis desert farming selects environment-specific date palm root endophytic communities and cultivable bacteria that promote resistance to drought. *Environmental Microbiology Reports* 7 (4): 668–678.

Chet, I. (1987). Trichoderma: application, mode of action, and potential as biocontrol agent of soilborne plant pathogenic fungi. In: *Innovative Approaches to Plant Disease Control* (ed. I. Chet), 137–160. Wiley.

Crickmore, N., Zeigler, D.R., Feitelson, J. et al. (1998). Revision of the nomenclature for the *Bacillus thuringiensis* pesticidal crystal proteins. *Microbiology and Molecular Biology Reviews* 62 (3): 807–813.

D'Annibale, A., Ricci, M., Leonardi, V. et al. (2005). Degradation of aromatic hydrocarbons by white-rot fungi in a historically contaminated soil. *Biotechnology and Bioengineering* 90 (6): 723–731.

D'Annibale, A., Rosetto, F., Leonardi, V. et al. (2006). Role of autochthonous filamentous fungi in bioremediation of a soil historically contaminated with aromatic hydrocarbons. *Applied and Environmental Microbiology* 72 (1): 28–36.

Das, S., Raj, R., Mangwani, N. et al. (2014). Heavy metals and hydrocarbons: adverse effects and mechanism of toxicity. In: *Microbial Biodegradation and Bioremediation* (ed. S. Das), 23–54. Elsevier.

Dell'Amico, E., Cavalca, L., and Andreoni, V. (2008). Improvement of *Brassica napus* growth under cadmium stress by cadmium-resistant rhizobacteria. *Soil Biology and Biochemistry* 40 (1): 74–84.

Duffy, B.K. and Défago, G. (1999). Environmental factors modulating antibiotic and siderophore biosynthesis by *Pseudomonas fluorescens* biocontrol strains. *Applied and Environmental Microbiology* 65 (6): 2429–2438.

Elad, Y. and Stewart, A. (2007). Microbial control of *Botrytis* spp. In: *Botrytis: Biology, Pathology and Control* (eds. Y. Elad et al.), 223–241. Dordrecht: Springer.

Elad, Y., Chet, I., and Katan, J. (1980). *Trichoderma harzianum*: a biocontrol agent effective against *Sclerotium rolfsii* and *Rhizoctonia solani*. *Phytopathology* 70 (2): 119–121.

El-Hamshary, O.I.M., Abo-Aba, S.E.M., Awad, N., and Gomaa, A.M. (2008). Genetic profile of *Bacillus cereus* and *Bacillus subtilis* indigenous isolates and their performance as bio-control agent against the plant pathogen *Fusarium oxysporum*. *Research Journal of Cell and Molecular Biology* 2 (2): 30–38.

de Escudero, I.R., Estela, A., Porcar, M. et al. (2006). Molecular and insecticidal characterization of a Cry1I protein toxic to insects of the families Noctuidae, Tortricidae, Plutellidae, and Chrysomelidae. *Applied and Environmental Microbiology* 72 (7): 4796–4804.

Federici, E., Leonardi, V., Giubilei, M.A. et al. (2007). Addition of allochthonous fungi to a historically contaminated soil affects both remediation efficiency and bacterial diversity. *Applied Microbiology and Biotechnology* 77 (1): 203–211.

Fernando, W.D., Nakkeeran, S., and Zhang, Y. (2005). Biosynthesis of antibiotics by PGPR and its relation in biocontrol of plant diseases. In: *PGPR: Biocontrol and Biofertilization* (ed. Z.A. Siddiqui), 67–109. Dordrecht: Springer.

Fierer, N. and Jackson, R.B. (2006). The diversity and biogeography of soil bacterial communities. *Proceedings of the National Academy of Sciences of the United States of America* 103 (3): 626–631.

Fritsche, W. and Hofrichter, M. (2005). Aerobic degradation of recalcitrant organic compounds by microorganisms. In: *Environmental Biotechnology: Concepts and Applications* (eds. H.-J. Jördening and J. Winter). Wiley-VCH.

Gamalero, E. and Glick, B.R. (2015). Bacterial modulation of plant ethylene levels. *Plant Physiology* 169 (1): 13–22.

Garrido-Sanz, D., Meier-Kolthoff, J.P., Göker, M. et al. (2016). Genomic and genetic diversity within the *Pseudomonas fluorescens* complex. *PLoS One* 11 (2): e0150183.

Glick, B.R. (2005). Modulation of plant ethylene levels by the bacterial enzyme ACC deaminase. *FEMS Microbiology Letters* 251 (1): 1–7.

Glick, B.R., Penrose, D.M., and Li, J. (1998). A model for the lowering of plant ethylene concentrations by plant growth-promoting bacteria. *Journal of Theoretical Biology* 190 (1): 63–68.

Glick, B.R., Cheng, Z., Czarny, J., and Duan, J. (2007). Promotion of plant growth by ACC deaminase-producing soil bacteria. In: *New Perspectives and Approaches in Plant Growth-Promoting Rhizobacteria Research* (ed. J.C. Shuttle), 329–339. Dordrecht: Springer.

Gopalakrishnan, S., Sathya, A., Vijayabharathi, R. et al. (2015). Plant growth promoting rhizobia: challenges and opportunities. *3 Biotech* 5 (4): 355–377.

Grimes, H.D. and Mount, M.S. (1984). Influence of *Pseudomonas putida* on nodulation of *Phaseolus vulgaris*. *Soil Biology and Biochemistry* 16 (1): 27–30.

Grover, M., Ali, S.Z., Sandhya, V. et al. (2011). Role of microorganisms in adaptation of agriculture crops to abiotic stresses. *World Journal of Microbiology and Biotechnology* 27 (5): 1231–1240.

Gusain, Y.S., Singh, U.S., and Sharma, A.K. (2015). Bacterial mediated amelioration of drought stress in drought tolerant and susceptible cultivars of rice (*Oryza sativa* L.). *African Journal of Biotechnology* 14 (9): 764–773.

Haas, D. and Keel, C. (2003). Regulation of antibiotic production in root-colonizing *Pseudomonas* spp. and relevance for biological control of plant disease. *Annual Review of Phytopathology* 41 (1): 117–153.

Haines, A., Kovats, R.S., Campbell-Lendrum, D., and Corvalán, C. (2006). Climate change and human health: impacts, vulnerability and public health. *Public Health* 120 (7): 585–596.

Han, H.S. and Lee, K.D. (2005). Plant growth promoting rhizobacteria effect on antioxidant status, photosynthesis, mineral uptake and growth of lettuce under soil salinity. *Research Journal of Agriculture and Biological Sciences* 1 (3): 210–215.

Handelsman, J. and Stabb, E.V. (1996). Biocontrol of soilborne plant pathogens. *The Plant Cell* 8 (10): 1855.

Hardoim, P.R., van Overbeek, L.S., and van Elsas, J.D. (2008). Properties of bacterial endophytes and their proposed role in plant growth. *Trends in Microbiology* 16 (10): 463–471.

Hartmann, A., Schmid, M., Van Tuinen, D., and Berg, G. (2009). Plant-driven selection of microbes. *Plant and Soil* 321 (1–2): 235–257.

Hartmann, A., Rothballer, M., Hense, B.A., and Schröder, P. (2014). Bacterial quorum sensing compounds are important modulators of microbe-plant interactions. *Frontiers in Plant Science* 5: 131.

Hirsch, P.R. and Mauchline, T.H. (2012). Who's who in the plant root microbiome? *Nature Biotechnology* 30 (10): 961.

Iavicoli, A., Boutet, E., Buchala, A., and Métraux, J.P. (2003). Induced systemic resistance in *Arabidopsis thaliana* in response to root inoculation with *Pseudomonas fluorescens* CHA0. *Molecular Plant-Microbe Interactions* 16 (10): 851–858.

Inglis, G.D., Goettel, M.S., Butt, T.M., and Strasser, H.E. (2001). Use of hyphomycetous fungi for managing insect pests. In: *Fungi as Biocontrol Agents* (eds. T.M. Butt et al.), 23–69.

Jamali, F., Sharifi-Tehrani, A., Lutz, M.P., and Maurhofer, M. (2009). Influence of host plant genotype, presence of a pathogen, and coinoculation with *Pseudomonas fluorescens* strains on the rhizosphere expression of hydrogen cyanide-and 2, 4-diacetylphloroglucinol biosynthetic genes in *P. fluorescens* biocontrol strain CHA0. *Microbial Ecology* 57 (2): 267–275.

Jha, Y., Subramanian, R.B., and Patel, S. (2011). Combination of endophytic and rhizospheric plant growth promoting rhizobacteria in *Oryza sativa* shows higher accumulation of osmoprotectant against saline stress. *Acta Physiologiae Plantarum* 33 (3): 797–802.

Johansson, P.M., 2003. *Biocontrol of Fusarium in wheat-introducing bacteria to a system of complex interactions*. Doctoral thesis, Swedish University of Agricultural Sciences Uppsala, 2003.

Jourdan, E., Ongena, M. and Thonart, P. 2008. Caractéristiques moléculaires de l'immunité des plantes induite par les rhizobactéries non pathogènes. *BASE*.

Kanwal, S., Ilyas, N., Batool, N., and Arshad, M. (2017). Amelioration of drought stress in wheat by combined application of PGPR, compost, and mineral fertilizer. *Journal of Plant Nutrition* 40 (9): 1250–1260.

Kaymak, H.C. (2010). Potential of PGPR in agricultural innovations. In: *Plant Growth and Health Promoting Bacteria* (ed. K.D. Maheshwari), 45–79. Berlin, Heidelberg: Springer.

Kempster, V.N., Scott, E.S., and Davies, K.A. (2002). Evidence for systemic, cross-resistance in white clover (*Trifolium repens*) and annual medic (*Medicago truncatula* var *truncatula*) induced by biological and chemical agents. *Biocontrol Science and Technology* 12 (5): 615–623.

Kerry, B.R. (2001). Exploitation of the nematophagous fungal *Verticillium chlamydosporium* Goddard for the biological control of root-knot nematodes (*Meloidogyne* spp.). In: *Fungi as Biocontrol Agents: Progress, Problems and Potential*, 155–167. Wallingford, UK: CAB International.

Kieber, J.J. (1997). The ethylene signal transduction pathway in *Arabidopsis*. *Journal of Experimental Botany* 48 (2): 211–218.

Knight, R., Vrbanac, A., Taylor, B.C. et al. (2018). Best practices for analysing microbiomes. *Nature Reviews Microbiology* 16 (7): 410.

Kunert, K.J., Vorster, B.J., Fenta, B.A. et al. (2016). Drought stress responses in soybean roots and nodules. *Frontiers in Plant Science* 7: 1015.

Lareen, A., Burton, F., and Schäfer, P. (2016). Plant root-microbe communication in shaping root microbiomes. *Plant Molecular Biology* 90 (6): 575–587.

Lavelle, P., Blouin, M., Boyer, J. et al. (2004). Plant parasite control and soil fauna diversity. *Comptes Rendus Biologies* 327 (7): 629–638.

van Lenteren, J.C., Bolckmans, K., Köhl, J. et al. (2018). Biological control using invertebrates and microorganisms: plenty of new opportunities. *BioControl* 63 (1): 39–59.

Leonardi, V., Giubilei, M.A., Federici, E. et al. (2008). Mobilizing agents enhance fungal degradation of polycyclic aromatic hydrocarbons and affect diversity of indigenous bacteria in soil. *Biotechnology and Bioengineering* 101 (2): 273–285.

Lim, H.S., Kim, Y.S., and Kim, S.D. (1991). *Pseudomonas stutzeri* YPL-1 genetic transformation and antifungal mechanism against *Fusarium solani*, an agent of plant root rot. *Applied and Environmental Microbiology* 57 (2): 510–516.

Malik, K.A., Bilal, R., Mehnaz, S. et al. (1997). Association of nitrogen-fixing, plant-growth-promoting rhizobacteria (PGPR) with kallar grass and rice. In: *Opportunities for Biological Nitrogen Fixation in Rice and Other Non-Legumes*, 37–44. Dordrecht: Springer.

Manteau, S., Abouna, S., Lambert, B., and Legendre, L. (2003). Differential regulation by ambient pH of putative virulence factor secretion by the phytopathogenic fungus *Botrytis cinerea*. *FEMS Microbiology Ecology* 43 (3): 359–366.

Mantelin, S. and Touraine, B. (2004). Plant growth-promoting bacteria and nitrate availability: impacts on root development and nitrate uptake. *Journal of Experimental Botany* 55 (394): 27–34.

Marasco, R., Rolli, E., Ettoumi, B. et al. (2012). A drought resistance-promoting microbiome is selected by root system under desert farming. *PLoS One* 7 (10): e48479.

Marschner, P., Crowley, D., and Rengel, Z. (2011). Rhizosphere interactions between microorganisms and plants govern iron and phosphorus acquisition along the root axis–model and research methods. *Soil Biology and Biochemistry* 43 (5): 883–894.

Mayak, S., Tirosh, T., and Glick, B.R. (2004). Plant growth-promoting bacteria that confer resistance to water stress in tomatoes and peppers. *Plant Science* 166 (2): 525–530.

Meena, H., Ahmed, M.A., and Prakash, P. (2015). Amelioration of heat stress in wheat, *Triticum aestivum* by PGPR (*Pseudomonas aeruginosa* strain 2CpS1). *Bioscience Biotechnology Research Communications* 8 (2): 171–174.

Meena, K.K., Sorty, A.M., Bitla, U.M. et al. (2017). Abiotic stress responses and microbe-mediated mitigation in plants: the omics strategies. *Frontiers in Plant Science* 8: 172.

Miliute, I., Buzaite, O., Baniulis, D., and Stanys, V. (2015). Bacterial endophytes in agricultural crops and their role in stress tolerance: a review. *Zemdirbyste-Agriculture* 102 (4): 465–478.

Mommer, L., Kirkegaard, J., and van Ruijven, J. (2016). Root–root interactions: towards a rhizosphere framework. *Trends in Plant Science* 21 (3): 209–217.

Nabti, E., Sahnoune, M., Adjrad, S. et al. (2007). A halophilic and osmotolerant *Azospirillum brasilense* strain from Algerian soil restores wheat growth under saline conditions. *Engineering in Life Sciences* 7 (4): 354–360.

Nabti, E., Sahnoune, M., Ghoul, M. et al. (2010). Restoration of growth of durum wheat (*Triticum durum* var. *waha*) under saline conditions due to inoculation with the rhizosphere bacterium *Azospirillum brasilense* NH and extracts of the marine alga *Ulva lactuca*. *Journal of Plant Growth Regulation* 29 (1): 6–22.

Nabti, E.H., Mokrane, N., Ghoul, M. et al. (2013). Isolation and characterization of two halophilic *Bacillus* (*B. licheniformis* and *Bacillus* sp.) with antifungal activity. *Journal of Ecology of Health & Environment* 1 (1): 13–17.

Nadeem, S.M., Hussain, I., Naveed, M. et al. (2006). Performance of plant growth promoting rhizobacteria containing ACC-deaminase activity for improving growth of maize under salt-stressed conditions. *Pakistan Journal of Agricultural Sciences* 43 (3–4): 114–121.

Nandakumar, R., Babu, S., Viswanathan, R. et al. (2001). Induction of systemic resistance in rice against sheath blight disease by *Pseudomonas fluorescens*. *Soil Biology and Biochemistry* 33 (4–5): 603–612.

Nies, D.H. (1999). Microbial heavy-metal resistance. *Applied Microbiology and Biotechnology* 51 (6): 730–750.

Olcott, M.H., Henkels, M.D., Rosen, K.L. et al. (2010). Lethality and developmental delay in *Drosophila melanogaster* larvae after ingestion of selected *Pseudomonas fluorescens* strains. *PLoS One* 5 (9): e12504.

Pal, K.K. and Gardener, B.M. 2006. Biological control of plant pathogens. The Plant Health Instructor [electronic resource].

Péchy-Tarr, M., Bruck, D.J., Maurhofer, M. et al. (2008). Molecular analysis of a novel gene cluster encoding an insect toxin in plant-associated strains of *Pseudomonas fluorescens*. *Environmental Microbiology* 10 (9): 2368–2386.

Quan, C.S., Wang, X., and Fan, S.D. (2010). Antifungal compounds of plant growth promoting rhizobacteria and its action mode. In: *Plant Growth and Health Promoting Bacteria* (ed. K.D. Maheshwari), 117–156. Berlin, Heidelberg: Springer.

Raaijmakers, J.M., Vlami, M., and De Souza, J.T. (2002). Antibiotic production by bacterial biocontrol agents. *Antonie van Leeuwenhoek* 81 (1–4): 537.

Ramamoorthy, V., Viswanathan, R., Raguchander, T. et al. (2001). Induction of systemic resistance by plant growth promoting rhizobacteria in crop plants against pests and diseases. *Crop Protection* 20 (1): 1–11.

Reich, P.B. and Oleksyn, J. (2004). Global patterns of plant leaf N and P in relation to temperature and latitude. *Proceedings of the National Academy of Sciences of the United States of America* 101 (30): 11001–11006.

Rolli, E., Marasco, R., Vigani, G. et al. (2015). Improved plant resistance to drought is promoted by the root-associated microbiome as a water stress-dependent trait. *Environmental Microbiology* 17 (2): 316–331.

Ronseaux, S., Clément, C., and Barka, E. (2013). Interaction of *Ulocladium atrum*, a potential biological control agent, with *Botrytis cinerea* and grapevine plantlets. *Agronomy* 3 (4): 632–647.

Rousk, J., Brookes, P.C., and Bååth, E. (2009). Contrasting soil pH effects on fungal and bacterial growth suggest functional redundancy in carbon mineralization. *Applied and Environmental Microbiology* 75 (6): 1589–1596.

Rudrappa, T., Biedrzycki, M.L., Kunjeti, S.G. et al. (2010). The rhizobacterial elicitor acetoin induces systemic resistance in *Arabidopsis thaliana*. *Communicative & Integrative Biology* 3 (2): 130–138.

Ryu, C.M., Farag, M.A., Hu, C.H. et al. (2003). Bacterial volatiles promote growth in *Arabidopsis*. *Proceedings of the National Academy of Sciences of the United States of America* 100 (8): 4927–4932.

Ryu, C.M., Farag, M.A., Hu, C.H. et al. (2004). Bacterial volatiles induce systemic resistance in *Arabidopsis*. *Plant Physiology* 134 (3): 1017–1026.

de Salamone, I.G., Döbereiner, J., Urquiaga, S., and Boddey, R.M. (1996). Biological nitrogen fixation in *Azospirillum* strain-maize genotype associations as evaluated by the 15 N isotope dilution technique. *Biology and Fertility of Soils* 23 (3): 249–256.

Santoyo, G., Pacheco, C.H., Salmerón, J.H., and León, R.H. (2017). The role of abiotic factors modulating the plant-microbe-soil interactions: toward sustainable agriculture. A review. *Spanish Journal of Agricultural Research* 15 (1): 13.

Schardl, C.L., Young, C.A., Faulkner, J.R. et al. (2012). Chemotypic diversity of epichloae, fungal symbionts of grasses. *Fungal Ecology* 5 (3): 331–344.

Segata, N., Boernigen, D., Tickle, T.L. et al. (2013). Computational meta'omics for microbial community studies. *Molecular Systems Biology* 9 (1): 666.

Sergaki, C., Lagunas, B., Lidbury, L. et al. (2018). Challenges and approaches in microbiome research: from fundamental to applied. *Frontiers in Plant Science* https://doi.org/10.3389/fpls.2018.01205.

Shaharoona, B., Arshad, M., Zahir, Z.A., and Khalid, A. (2006). Performance of *Pseudomonas* spp. containing ACC-deaminase for improving growth and yield of maize (*Zea mays* L.) in the presence of nitrogenous fertilizer. *Soil Biology and Biochemistry* 38 (9): 2971–2975.

Shams-Ghahfarokhi, M., Kalantari, S., and Razzaghi-Abyaneh, M. (2013). Terrestrial bacteria from agricultural soils: versatile weapons against aflatoxigenic fungi. In: *Aflatoxins: Recent Advances and Future Prospects* (ed. M. Razzaghi-Abyaneh), 23–40. https://doi.org/10.5772/45918.

Shoresh, M., Harman, G.E., and Mastouri, F. (2010). Induced systemic resistance and plant responses to fungal biocontrol agents. *Annual Review of Phytopathology* 48: 21–43.

Shrivastava, P. and Kumar, R. (2015). Soil salinity: a serious environmental issue and plant growth promoting bacteria as one of the tools for its alleviation. *Saudi Journal of Biological Sciences* 22 (2): 123–131.

Siddiqui, I.A. and Shaukat, S.S. (2003). Suppression of root-knot disease by *Pseudomonas fluorescens* CHA0 in tomato: importance of bacterial secondary metabolite, 2, 4-diacetylpholoroglucinol. *Soil Biology and Biochemistry* 35 (12): 1615–1623.

Singh, R. (2014). Microorganism as a tool of bioremediation technology for cleaning environment: a review. *Proceedings of the International Academy of Ecology and Environmental Sciences* 4 (1): 1.

Siwar, C., Alam, M., Murad, M., and Al-Amin, A.Q. (2009). A review of the linkages between climate change, agricultural sustainability and poverty in Malaysia. *International Review of Business Research Papers* (ISSN 1832-9543) 5 (6): 309–321.

Song, H., Nan, Z., Song, Q. et al. (2016). Advances in research on *Epichloë* endophytes in Chinese native grasses. *Frontiers in Microbiology* 7: 1399.

Su, F., Jacquard, C., Villaume, S. et al. (2015). *Burkholderia phytofirmans* PsJN reduces impact of freezing temperatures on photosynthesis in *Arabidopsis thaliana*. *Frontiers in Plant Science* 6: 810.

Szilagyi-Zecchin, V.J., Mógor, Á.F., and Figueiredo, G.G.O. (2016). Strategies for characterization of agriculturally important bacteria. In: *Microbial Inoculants in Sustainable Agricultural Productivity* (ed. D.P. Singh), 1–21. New Delhi: Springer.

Tabli, N., Rai, A., Bensidhoum, L. et al. (2018). Plant growth promoting and inducible antifungal activities of irrigation well water-bacteria. *Biological Control* 117: 78–86.

Tang, C. and Thomson, B.D. (1996). Effects of solution pH and bicarbonate on the growth and nodulation of a range of grain legume species. *Plant and Soil* 186 (2): 321–330.

Thornton, P.K., Ericksen, P.J., Herrero, M., and Challinor, A.J. (2014). Climate variability and vulnerability to climate change: a review. *Global Change Biology* 20 (11): 3313–3328.

Tilman, D., Balzer, C., Hill, J., and Befort, B.L. (2011). Global food demand and the sustainable intensification of agriculture. *Proceedings of the National Academy of Sciences of the United States of America* 108 (50): 20260–20264.

Tripathi, P., Rabara, R.C., Reese, R.N. et al. (2016). A toolbox of genes, proteins, metabolites and promoters for improving drought tolerance in soybean includes the metabolite coumestrol and stomatal development genes. *BMC Genomics* 17 (1): 102.

Trivedi, P., Schenk, P.M., Wallenstein, M.D., and Singh, B.K. (2017). Tiny Microbes, Big Yields: enhancing food crop production with biological solutions. *Microbial Biotechnology* 10 (5): 999–1003.

Vallad, G.E. and Goodman, R.M. (2004). Systemic acquired resistance and induced systemic resistance in conventional agriculture. *Crop Science* 44 (6): 1920–1934.

Vandenkoornhuyse, P., Quaiser, A., Duhamel, M. et al. (2015). The importance of the microbiome of the plant holobiont. *New Phytologist* 206 (4): 1196–1206.

Vardharajula, S., Zulfikar Ali, S., Grover, M. et al. (2011). Drought-tolerant plant growth promoting *Bacillus* spp.: effect on growth, osmolytes, and antioxidant status of maize under drought stress. *Journal of Plant Interactions* 6 (1): 1–14.

Viswanathan, R. and Samiyappan, R. (2004). Secondary metabolites production by *Pseudomonas* spp. strains antagonistic to *Colletotrichum falcatum* causing red rot disease in sugarcane. *Acta Phytopathologica Entomologica et Hungarica* 39: 28–37.

Voisard, C., Keel, C., Haas, D., and Dèfago, G. (1989). Cyanide production by *Pseudomonas fluorescens* helps suppress black root rot of tobacco under gnotobiotic conditions. *The EMBO Journal* 8 (2): 351–358.

Wang, M., Li, E., Liu, C. et al. (2017). Functionality of root-associated bacteria along a salt marsh primary succession. *Frontiers in Microbiology* 8: 2102.

Wani, P.A., Khan, M.S., and Zaidi, A. (2007). Synergistic effects of the inoculation with nitrogen-fixing and phosphate-solubilizing rhizobacteria on the performance of field-grown chickpea. *Journal of Plant Nutrition and Soil Science* 170 (2): 283–287.

Weller, D.M. (2007). *Pseudomonas* biocontrol agents of soilborne pathogens: looking back over 30 years. *Phytopathology* 97 (2): 250–256.

Weller, D.M., Mavrodi, D.V., van Pelt, J.A. et al. (2012). Induced systemic resistance in *Arabidopsis thaliana* against *Pseudomonas syringae* pv. tomato by 2, 4-diacetylphloroglucinol-producing *Pseudomonas fluorescens*. *Phytopathology* 102 (4): 403–412.

Whipps, J.M. (2001). Microbial interactions and biocontrol in the rhizosphere. *Journal of Experimental Botany* 52: 487–511.

Zhalnina, K., Dias, R., de Quadros, P.D. et al. (2015). Soil pH determines microbial diversity and composition in the park grass experiment. *Microbial Ecology* 69 (2): 395–406.

Zhang, W., Li, F., and Nie, L. (2010). Integrating multiple 'omics' analysis for microbial biology: application and methodologies. *Microbiology* 156 (2): 287–301.

7

Agricultural Importance of Phyllosphere Microbiome: Recent Trends and Future Perspectives

Subha Swaraj Pattnaik, Parasuraman Paramanantham, and Siddhardha Busi

Department of Microbiology, School of Life Sciences, Pondicherry University, Puducherry-605 014, India

7.1 Introduction

The interactions between plants and microorganisms have been extensively explored. In earlier times, these studies concentrated mostly on the pathogenic interaction between the plant and the microorganism. The majority of plant-associated microorganisms possess either neutral or beneficial aspects in plant growth and development (Braga, Dourado and Araújo 2016; Rey and Schornack 2013). Recent studies emphasize the influence of beneficial plant microbe interactions and their application towards enhancing plant productivity. It is of interest that plant-associated microorganisms facilitate plants with several beneficial aspects, including providing access to gain metabolite exchange and conversion, physiochemical modification, signalling, chemotaxis and exchange of genetic information. In most environmental conditions, plants fail to escape from pathogenic attacks and stress conditions; and in this scenario it is important for the plants to acquire potent defence mechanisms to overcome pathogenic infections and stress. The association between plants and microorganisms helps plants in adaptation and specialization, and in developing strong antagonistic mechanisms to potentially ward off pathogenic attack. The plant-associated microorganism directly influences plant productivity, for instance by controlling nutrient availability/acquisition, development of improved defence mechanisms against pathogens and facilitating tolerance capability towards different environmental conditions (Andreote, Gumiere and Durrer 2014).

In general all the regions in the plant possess their respective microbial communities, depending on nutrient availability and plant genotype. Recent investigations are highly focused on certain regions of the plant, including the rhizosphere (below the ground), phyllosphere (aerial surfaces above the ground) and endosphere (internal tissues). The rhizosphere region, located below the ground, is enriched with a complex microbial diversity, influenced by intense availability of plant mucilage and root exudates that potentially facilitate microbial growth and development. In contrast, the second region, namely the

The Plant Microbiome in Sustainable Agriculture, First Edition. Edited by Alok Kumar Srivastava, Prem Lal Kashyap and Madhumita Srivastava.

phyllosphere, holds comparatively fewer nutrients and also experiences different environmental stress conditions including temperature, radiation and moisture.

The microorganisms that successfully colonize these regions, specifically the microbes residing near or on the plant are termed epiphytes, whereas the microorganisms residing within the plant tissues are referred to as endophytes. Among the mentioned plant region specific microorganisms, the phyllosphere-associated microorganisms have gained considerable attention owing to the wide range of diversity and ability to show tolerance to different environmental stress conditions (Turner, James and Poole 2013). Leaves are the dominant part of the phyllosphere of the plants, which harbour dense populations of microbial communities. The availability of these dense diverse microorganisms contributes to the microclimatic condition of the leaves, notably the leaves which are covered with waxy cuticle, giving continuous protection of plants from ultraviolet (UV) radiation, water deficiency, nutrient availability and temperature (Remus-Emsermann and Schlechter 2018).

Ongoing research on this topic aims to explore the functional role of the specific microbial community coordinates with the phyllosphere and their notable role in plant growth and development. However, among the dense populations of microorganisms in the phyllosphere, very low percentages of the microbial communities are cultivatable under laboratory conditions; most are non-cultivatable microorganisms. Hence, these are the major limitation to the researcher who wishes to explore more about plant–microbe interaction in the phyllosphere. Recent advances in science and technology have facilitated exploitation of phyllosphere microbiotas; in particular advances in genomic and metagenomic approaches have opened up new avenues for understanding phyllosphere microbiotas in the growth and development of plants.

Extensive investigation is required to acquire detailed knowledge about the impact of the physico-chemical behaviour of the environment, including different climate conditions, concentrations of volatile organic compounds, plant hormones, pollutants and trace gas composition on plant–microbe interaction in the phyllosphere (Bringel and Couee 2015). With the available knowledge about the phyllosphere and their associated microbial community, it was explained that the adaption mechanism of the microorganisms to the phyllospheric environment is dependent on the different properties of physico-chemical and biotic constraints. Moreover, the plant-based microbiotas able to express tolerance towards the variety of antimicrobial compounds produced by the plant and/or the pathogenic microorganisms. Interestingly, recent research put emphasis on the relationship between the signalling system within the microbiotas and towards the plants (Patel et al. 2013).

The microorganisms in the phyllosphere are effective participants in a dual mode of impact creators in the microbial signalling system. Thus in one way they enhance the microbial communities to potentially establish a biofilm by using microbial signalling. However in another way they are actively participating in the production of anti-signalling molecules that selectively block the signalling mechanisms of the pathogen and inhibit biofilm formation. Moreover, research showed their active participation in the synthesis of phytohormones and metabolites which are required for plant growth and development (Vorholt 2012).

Recent research on agricultural sectors concentrated on biological aid to enhance crop productivity and inhibit the crop pests. This is positive, because chemical crop yield promoter and the pesticides hold the toxic profiles with their application to agriculture. It is well-established that microbial communities closely associated by symbiotic relationships with plants control the growth and developmental kinetics of the host plant. This kind of practice is not new; it has been practised from ancient times without scientific information. By manipulating the soil microbiome the crop yield was enhanced. The phyllosphere-associated microbes possess a number of special characteristic features which protect plants from different environmental factors, and potentially promote their product yield. Bio-inoculation technology gains a significant focus as a potential alternative for chemical fertilizers and pesticides (Finkel et al. 2017).

This chapter emphasizes the microbial diversity in the phyllosphere, and the structural and functional importance of the phyllospheric microbiome. Recent technological advances to explore the phyllosphere-based microbiota and different environmental factors that affect the dynamics of the phyllosphere microbiome are also discussed. It is necessary to enlist the application of the phyllosphere microbiome in agriculture when we are discussing phyllosphere–microorganism interaction. The current development and future research on the microbiomes in the phyllosphere are discussed.

7.2 The World of the Phyllosphere

Plants harbour a diverse group of microorganisms inhabiting different parts of plants. The phyllosphere represents the aerial parts of the plants, providing a highly dynamic and ecologically important habitat for establishment of microbial colonization. The phyllosphere is considered to be one of the most vibrant and largest biological surfaces, providing a breeding platform for the establishment of bacteria, archaebacteria, fungi, algae, protozoa and nematodes (Wagi and Ahmed 2017). It is estimated that the phyllosphere accounts for contributing approximately 60 percent of the biomass across all the taxonomic plants on earth, suggesting its candidature as a potential habitat for microbial colonization (Koskella 2020). The term "phyllosphere" was used for the first time by a plant pathologist, F.T. Last, as a microhabitat on the leaf surface for colonization of characteristic microflora analogous to that of the rhizosphere in the roots of plant (Vacher et al. 2016). The phyllosphere represents a total global leaf surface area of approximately 1,017,260,200 km^2 inhabiting an approximate 10^{26} of bacterial cells, thereby providing promising avenues for integrative studies on phyllosphere-associated microbial communities and their role in environmental sustainability (Vorholt 2012).

The phyllosphere can be categorized into different types according to the different surfaces of plants such as the phylloplane (leaf surface), caulosphere (stem surface), anthosphere (flower surface), and carposphere (fruit surface). The above mentioned phyllospheric components are significantly different in terms of the colonization of characteristically different microbial communities (Farre-Armegnol et al. 2016). According to recent trends in understanding the phyllosphere microbiology, the phyllosphere interface can be categorized into two overlapping and closely associated areas: the phylloplane (depicting the leaf surface landscape) and the phyllotelma (depicting the leaf surface waterscape) with a unique dimensional niche for colonization of a diverse range of microbial communities (Thapa and Prasanna 2018). Among the different plant surfaces inhabited by microorganisms, the phyllosphere represents a relatively nutrient poor but mostly dynamic environment, owing to the large influxes of abiotic factors and frequent exposure to different environmental cues (Turner et al. 2013).

7.3 Importance of the Phyllosphere

The plant phyllosphere, though nutritionally poor unlike that of the rhizosphere, provides a dynamic ecosystem in the earth by providing a habitat for a diverse range of microflora to colonize. It also furnishes a multifaceted platform to allow interaction between host plant and colonizing microbial communities in association with the plant-environment interactions (Berg et al. 2016). The phyllospheric bacteria have the inherent property of affecting the plant physiological and developmental process, thereby influencing plant growth promotion and development, and providing resistance against phytopathogens (Figure 7.1). Similarly, fungal communities associated with the phyllosphere play a crucial role in maintaining systemic resistance of the host plant, and enhancing the ability of drought tolerance (Rossmann et al. 2017). Understanding the community composition and dynamics of the phyllospheric microbiota with reference to different abiotic and biotic factors is crucial for the management of plant health, growth and development (Copeland et al. 2015).

Phyllospheric microbial communities are not only crucial for the development of the plant and protection against plant pathogens, but also involve in the process of nitrogen fixation as well as facilitating bioremediation of harmful chemicals and pollutants and playing an important role in carbon sequestration as well as nutrient recycling in the atmosphere (Thapa et al. 2017; Chen et al. 2021). The recent development in high-throughput molecular

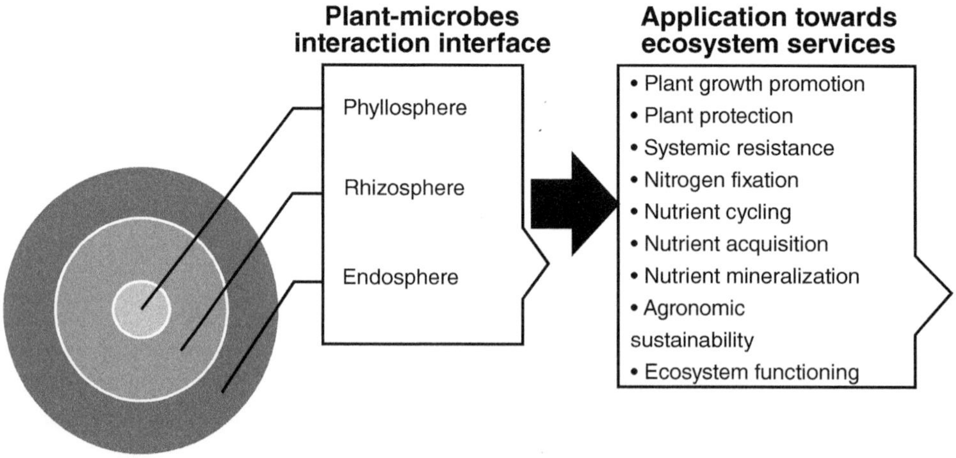

Figure 7.1 A schematic representation of different plant–microbe interaction interfaces and their significant contribution towards ecosystem services.

techniques, and advanced sequencing approaches in the last few decades have provided new avenues for compositional analysis of phyllospheric microbiotas by culture-independent methods which characteristically provide a new dimension to the current knowledge on the phyllosphere and associated microbiota. The advent of culture-independent sequencing analysis has proven to be important for improving scientific knowledge on the phyllosphere microbiotas and their valuable contribution to socio-economic values (Chaudhary et al. 2017).

Understanding the correlations between phyllosphere-associated microbial communities with their host plant will provide an insight into the functional attributes of plant biogeography and the potential factors reciprocating ecosystem functioning (Kembel et al. 2014). Since the leaf microbiotas are responsible for controlling the fate of both agricultural and non-agricultural plants, considerable attention has been focused in understanding the interaction between the inhabiting microbiota and the host plant (Aleklett et al. 2014).

7.4 Microbial Life in the Phyllosphere

The phyllosphere constitutes one of the most important and vibrant niches for the colonization of typical microbial communities, and it provides adequate interface for intricate synergism between the colonizing microbiotas and the host plant for the maintenance of health and the developmental profile of the plant (Yadav et al. 2017). The phyllosphere forms the basis of a heterogeneous interface for the colonization of diverse microbial communities, starting from bacteria to filamentous fungi, algae and protozoa (Sivakumar et al. 2020). Among the different microbial communities inhabiting the phyllosphere, bacteria remain the most commonly colonizing community, and are known as phyllospheric bacteria or phyllobacteria. Though both non-pathogenic as well as pathogenic bacteria co-exist in the phyllosphere, alpha-proteobacteria, including the genera *Methylobacterium* and *Sphingomonas* dominate the microbial communities in the phyllosphere and play a crucial role in phyllosphere dynamics and ecosystem functioning (Tani et al. 2015).

As evidenced from molecular techniques, species of beta-proteobacteria and gamma-proteobacteria are the dominant phyllospheric inhabitants following the dominance of alpha-proteobacterial species (Sivakumar et al. 2020). *Methylobacterium* sp. popularly known as pink pigmented facultative methylotrophs (PPFMs), are the predominant bacterial community to colonize the phyllosphere. PPFMs have the inherent potential to utilize methane, methanol and other volatile organic compounds as carbon and energy sources emitted by the host plants. They have an influential role in determining the plant–microbe interactions in the phyllospheric interface, nutrient recycling, and plant growth promotion by synthesis of phytohormones and they provide resistance to stress conditions (Iguchi et al. 2012; Agafonova et al. 2016). As shown in recent studies, bacterial phyla belonging to Alphaproteobacteria, Gammaproteobacteria (predominantly genus *Pseudomonas*) and Actinobacteria constitute one-third of the total bacterial phyla inhabiting the phyllosphere. In addition, bacterial phyla such as Bacteriodetes (belonging to the genus *Spirosoma*), Cyanobacteria and Acidobacteria also inhabit the phyllosphere and represent an important aspect in maintaining phyllospheric microbial dynamics and environmental sustainability (Vorholt 2012; Rossmann et al. 2017).

Among the filamentous fungi inhabiting the phyllosphere, genera such as *Aureobasidium*, *Clostridium* and *Taphrina* belonging to the group Ascomycota predominate the fungal species and contribute in ecosystem functioning and services (Chaudhary et al. 2017). In addition, filamentous fungal genera belonging to *Cladosporium*, *Aspergillus*, *Penicillium*, *Alternaria* etc. frequently inhabit the phyllosphere (Sivakumar et al. 2020). The diverse microbiotas present in the phyllosphere are closely associated with host plants through different types of interactions such as commensalism, mutualism and pathogenic relationships (Rastogi et al. 2013). The bacterial communities colonizing the phyllosphere region of the host plants are categorized into epiphytes and endophytes based upon their distributional pattern (Costa et al. 2012; Batool, Rehman and Hasnain 2016).

Conventional culture-dependent and advanced culture-independent methods are exploited to determine the structural and functional attributes of phyllospheric microbiota. The intervention of culture-independent approaches has revolutionized present knowledge on microbial communities colonizing the plant phyllosphere, giving insight into the description of phyllospheric microbiota under different physiological and environmental conditions and specific adaptation strategies shown by the phyllospheric microbiota for the development of plant and systemic resistance against phytopathogens (Meyer and Leveau 2012).

7.5 Phyllosphere Microbiome and Its Importance

The phyllosphere represents an integrated regulatory network of different plant-associated traits with a direct or indirect influence on host plant physiological and metabolic activities thereby regulating plant growth, development, reproduction and systemic resistance mechanisms (Rosado et al. 2018). The phyllospheric microbiota utilize the plant exudates and in turn play a crucial role in protecting the host plant from phytopathogenic microorganisms, maintaining the developmental dynamics of the plant by enhancing phytohormone production, maintaining biogeochemical cycles, maintaining nutrient uptake and acquisition, and enhancing agronomic sustainability, ecosystem functioning and productivity (Penuelas and Terradas 2014; Yadav and Saxena 2018). The role of phyllospheric microbial communities as potential biofertilizers, biopesticides and phytostimulators is well-established, depicting their importance in maintaining plant growth and systemic resistance, thereby improving crop productivity and agro-ecosystem sustainability (Muller and Ruppel 2014). Recent development in the field of environmental genomics and metagenomic strategies has also revolutionized the present knowledge on the socio-economic impact of the phyllospheric microbiome on plant–environment interactions and thus has an influence on ecosystem services and environmental sustainability (Bringel and Couee 2015).

The microbial communities, especially the phyllosphere-associated bacterial communities colonizing the phyllosphere, play a crucial role in growth promotion by modulating the production of plant growth regulators such as auxins, gibberellins, cytokinins and abscisic acid. They provide resistance to host plants against phytopathogen invasion by producing bioactive metabolites such as alkaloids, terpene and its derivatives, steroids and antioxidant enzymes (Banik et al. 2016). From ancient times, plants and plant-based microbiota have been inherently associated and characteristically affect each other's metabolic and physiological activities. The microbial communities associated with the phyllosphere also greatly expanded the metabolic profile and physiological pattern, thereby enhancing the ability of the host plants in resource uptake, nutrient acquisition, and stress tolerance against an extending range of phytopathogens and environmental conditions (Andreote and Silva 2017).

7.6 Structure and Function of the Phyllospheric Microbiome

The microbiomes of the phyllosphere are directly dependent upon the physiological nature of the leaf. The microhabitat conditions in the phyllosphere are a difficult environment for the survival of microorganisms. However, microorganisms readily change their physiological behaviour to adapt to microhabitat conditions, and initiate their colonization (Schlechter et al. 2019). Often microorganisms colonize veins and grooves of epidermal cells in the leaf where the availability of the nutrients are in abundance in nature (Sivakumar et al. 2020). Commonly, the leaf surface is a waxy cuticle that poorly allows penetration of nutrients. The water drops on the leaf solubilize the nutrients and diffuse into the leaf, subsequently reaching the phyllosphere. The available nutrients in the phyllosphere then facilitate the microorganisms in establishing their growth. This leads to the production of biosurfactants that potentially act on the surface of the leaf and enhance the availability of water and nutrients (Roman and Wagner 2018). The availability of water and nature of the nutrients in the phyllosphere may alter the physiology of the bacterial cell in the community, which is achieved through cell-to-cell communication among the bacteria, most often through physiological contact (Tecon et al. 2018).

The phyllosphere provides a highly dynamic interface for the association of the plant with the inhabiting microbial communities as well as closely associated insect communities. The role of plant–microbe and insect–microbe interactions in the phyllosphere characteristically influence the structural and functional diversity and performance of the associated host plants. Recently, a new concept of plant–microbe–insect (PMI) interaction has been developed, underlying the concept of both plant–microbe and insect–microbe interaction to provide new and interesting avenues for maintaining functional attributes of the ecosystem and its sustainability (Saleem et al. 2017).

7.6.1 Available Technology for Analyzing Microbial Communities in the Phyllosphere

Science and technology offer a wide range of methodology for exploiting the microbial diversity and intensity of the microbial population residing on the plant phyllosphere, providing an opportunity to evaluate their potential plant growth promotion and its attributes. In earlier times two different approaches were followed to determine the presence of microorganisms on the phyllosphere; the phyllosphere-residing microorganisms were directly or indirectly observed by subjecting the plant part to the microscopic unit. In another method, the microorganisms were separated out from the phyllosphere and evaluated for diversity. Later, these approaches evolved and were categorized as culture-dependent and culture-independent approaches.

7.6.2 Culture-Dependent Approaches

The culture-dependent methods for identification of phyllospheric microbiota have been followed until the present time. Herein, the phyllosphere-associated microorganisms are directly cultured on a suitable medium by imprinting the surface of the leaf on the medium. These methods provide vast information on spatial distributional patterns of colonized microbiota on the surface of the leaf and their coordination among neighbouring microbial communities on the surface of the leaf. Culture-based identification of the microbial diversity on the surface of the phyllosphere is the most followed and reliable method. However, these methods encounter certain limitations in that they fail to provide information about the impact of topography of the phyllosphere on the diversity of the microbial community. These limitations arose due the differences in the cultural requirements of each of the microbial communities residing in the phyllosphere. Hence, the microorganisms that found the required growth supplements in the media would show a dominant growth pattern, and completely overwhelm the slow growing as well as the facultative microorganisms. Recently, this limitation has been overcome with selective and enriched media, which allow the growth of a particular group of microorganisms and enhance the slow growing as well as facultative microorganisms respectively.

Advances in microscopic technology offer a new opportunity to visualize the microbial community on the surface of the phyllosphere in their native state, and to observe their coordination and distribution on the plant surfaces (Thapa and Prasanna 2018).

Classical microscopic observation of the microorganisms associated with the phyllosphere was performed by peeling off the surface layer of the plant leaf, and observing it directly under a brightfield light microscope. This method faced certain difficulties including failure to observe the small and transparent objects on the leaf surface. This complication was overcome with different staining procedures. With growth of science and technology, other novel methodologies have been discovered for observing microbial communities on the surface of the leaf. The microscopic technologies including epifluorescence, scanning electron and confocal laser scanning microscopy were developed for visualizing microbial communities on the leaf (Morris, Monier and Jacques 1997). In a recent study, microscopic evidence was obtained to show the intra-tissue colonization of microorganisms in banana plants. Herein, a micro-section of banana leaf was treated with dual stain such as SYTO 9 and propidium iodide to observe the live and dead bacterial cells on the leaf surface by subjecting the stain-treated sample to epifluorescence or a confocal laser scanning microscope. The study concluded that extensive microbial colonies reside in the phyllosphere of the plants and it also emphasized that changes in the microbial community for each component of the leaf surface were observed. Finally, the study concluded the cell wall–plasma membrane peri-space is an extensive niche for the colonization of microbiota (Thomas and Reddy 2013).

7.6.3 Culture-Independent Approaches

Even though the culture-based methods are widely accepted as important in exploring the microbial communities on the phyllosphere, the results showed a biased determination of a certain group of microorganisms which do not represent the total available microorganisms in the region. These situations were created because certain microorganisms that reside on the phyllosphere are able to be cultivated under laboratory conditions, or when the culture media are provided with special media supplements to facilitate the growth of slow growing/facultative microorganisms. However, certain microbial communities residing on the phyllosphere completely fail to grow under laboratory conditions. These groups of microorganisms are metabolically active, but fail to reproduce in the laboratory, and are called non-cultivatable microorganisms. It may be possible that the abundance of the non-cultivatable microorganisms is comparatively higher than cultivatable microorganisms. However, other conditions may also possible in that non-cultivatable microorganisms fail to develop in the laboratory due to the insufficient numbers to be cultured. Or, it may be due to the selective modulation of the plant, which is required for microbial community growth and development. However, these two conditions have not been explored completely (Batool, Rehman and Hasnain 2016).

It has been realized that the microorganisms that are associated with plants and which show metabolic activity, but fail to be cultured on the common media and culture conditions, have significant importance in exploring the microbiota of plants. To overcome this limitation, a novel approach was adapted called the culture-independent microorganism detection method. This method was designed to determine the microbial population in the leaf sample without culturing them, but based on 16S rRNA gene amplification and amplicon sequencing (Singh et al. 2019). This method facilitates identification of a broader spectrum of microbial populations than the culture-based identification technology. However, this method also holds certain limitations including poor PCR amplification due to availability of the inhibitory compounds, non-specific primer, and non-availability of the quantitative information. Also, certain plant RNAs become amplified in this process and create a negative impact in the interpretation of results with plant–microbe interactions.

In recent times metagenomic shotgun sequencing of DNA has gained significant attention in that it could enable exploration of the complex microbial population from the sample by using high-throughput genome analysis technology (Noble et al. 2020). Herein, the analysis provides detailed information on individual genes or the particular group of the microbial community of the sample. Similarly another method has been developed, namely, metaproteomics which may facilitate determination and isolation of all the protein produced by the microbial population in the complex environmental sample, without culturing the microorganisms residing in the phyllosphere. These two approaches were further evolved by combining both the approaches including metagenomics and metaproteomics, and employing them to evaluate microbial communities on the phyllosphere (Müller and Ruppel 2014). The continuous growth in science and technology undoubtedly enhances the present knowledge about the microbial interaction on the phyllosphere.

7.7 Factors Affecting Dynamics of the Phyllosphere Microbiome

A number of biotic and abiotic factors directly or indirectly affect the structure and function of the phyllosphere microbiome. The abiotic factors such as temperature, elevation gradient, UV-radiation, water availability and moisture content significantly modulate the biochemical, physiological and metabolic activities of phyllospheric microbial communities (Cordier et al. 2012; Gomes et al. 2018; Lopez-Velasco et al. 2018). The structural and functional attributes of the phyllospheric microbiota strictly depend upon the atmospheric level of CO_2 as well as on temperature (Ren et al. 2015). Solar radiation is regarded as an influential parameter in shaping the structure and function of phyllospheric microbiota, colonization efficacy and survival pattern (Truchado et al. 2017). Biotic factors such as leaf surface topology, nutrient availability conditions, leaf physiology and plant species also contribute in regulating the phyllospheric microbiome (Turner et al. 2013; Wagi and Ahmed, 2017).

Apart from these biotic factors, plant genotype, the developmental stage of the host plant, occurrence of pollination, interference of phytopathogens, and competitive and cooperational interplay between inhabiting microbial species characteristically influence the phyllosphere dynamics. The anthropogenic events, particularly agricultural management tools, critically influence the diversity of the phyllosphere microbiome by modulating the soil and plant traits (Liu et al. 2020). Sustainable agricultural practices including crop diversification, organic farming and intercropping also significantly influence the composition and functional attributes of the phyllosphere microbiome (Compant et al. 2019). Phytopathogen invasion exhibits a strong influence on the diversity, composition, structural characteristics and functional attributes of phyllosphere microbiota. Thus, this effect could be considered as an opportunity to characterize the competitiveness and co-operation of the phyllospheric microbiota with invading pathogens (Zhang et al. 2019). Interestingly, the increase in the diversity of phyllosphere microbiota is directly proportional to the increased disease control strategies (Cernava et al. 2019). The application of biocontrol agents critically modulates the diversity and complexity of phyllospheric microbiota and in turn the complex microbiota facilitates enhanced biocontrol efficacy (Qin et al. 2019).

Apart from these factors, the frequent exposure to anthropogenic disturbances observed in the phyllosphere results in severe implications in the composition and functional traits of inhabiting microbiota (Xiang et al. 2020). The mineral contents present in the leaf interface also critically determine the community profile of inhabiting phyllosphere microbiota (Darlison et al. 2019). The structure and function of the phyllospheric microbiome strictly depends upon phenotypes and genotypes of the host plant species, plant–microbe interaction in the phyllosphere, and morphological, physiological and environmental factors (Figure 7.2).

Moreover, light also potentially influences the microbiomes of the phyllosphere. Both plants and the microorganisms in the phyllosphere use light as a stimulus to activate their biological processes. The photoreceptor in plants selectively absorbs the specific wavelength and activates the internal system, potentially altering hormonal concentration, production of secondary metabolites, and release of volatile compounds that subsequently

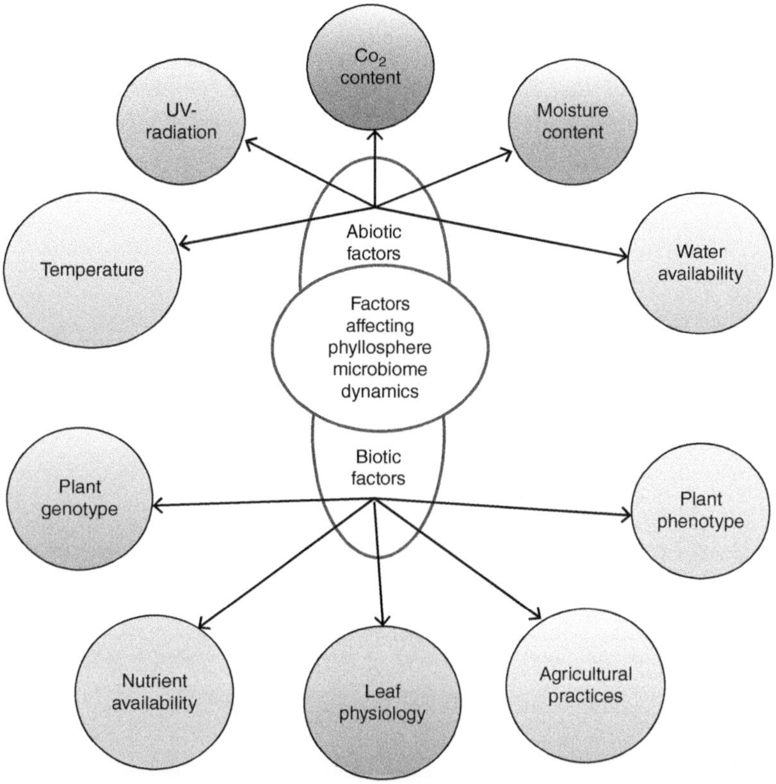

Figure 7.2 Different abiotic and biotic factors responsible for determining plant–microbe interactions in the phyllosphere.

positively impact the plant–microbe interaction. Mutually, microorganisms are involved in making essential elements like nitrogen, phosphorus, and iron as biologically available for plant growth and development (Carvalho and Castillo 2018). The composition and species richness of microbial communities in the phyllospheric interface depends upon the phylogenetic aspect of biogeographical patterns of the host plant (Redford et al. 2010; Chaudhary et al. 2017). The role of different environmental factors and their interaction with the host plant is also an important aspect in the constitution of heterogeneity in structural and functional diversity of colonizing phyllosphere microbiotas (Andreote et al. 2014).

In addition to plant species, site-specificity also matters in understanding the community composition of phyllospheric microbiota, suggesting a range of differential factors in determining the functional attributes of phyllospheric microbiota with special reference to environmental sustainability (Knief et al. 2010). The structure and function of phyllospheric microbial communities not only differs between different species of host plants, but also differs exclusively between different cultivars of the same species (Kim et al. 2012). The indiscriminate exploitation of synthetic and semi-synthetic insecticides in agriculture has a significant impact on both rhizospheric as well as phyllospheric microbial communities (Moulas et al. 2013; Ottesen et al. 2016). In addition to the use of different fertilizers and pesticides as critical factors influencing phyllospheric microbiota, the cultivation method also plays an important role in maintaining the compositional set-up for the phyllospheric microbiota (Thapa et al. 2018). The denaturing gradient gel electrophoresis (DGGE) banding patterns infer colonization of similar bacterial populations isolated from the same plant species, whereas a differential bacterial community was observed when different species of plants were considered for sampling (Jackson et al. 2015).

7.8 Application of the Phyllosphere Microbiome in Agriculture

7.8.1 Agricultural Pest Management

The rapid increase in global population is a human threat in terms of food security and sustainability. Hence significant attention has been given to the agricultural sector to enhance productivity and quality. Biological aids, namely, bio-fertilizers, biological control agents etc., have gained considerable attention as an alternative to the chemical aids for plant growth and development. In an ecosystem, plants encounter numerous pests including weeds, plant pathogens and certain animals. These phenomena create a major impact on the agricultural economy; it has been estimated that they lead to an approximately 30-40 percent reduction in the annual crop yield. Insects remain the most challenging pest to control, since they occur in a vast array of diversity, with special features like quick adaptation to the environment in agricultural sectors (Kupferschmied, Maurhofer and Keel 2013). Utilization of plant-associated microorganisms to control plant pests has been considered a viable alternative technology to maintain sustainable agricultural practices.

The phyllospheric microbiota forms the basis of integrated pest management (IPM) to control the nuisance of pests (Francis et al. 2020). For example, phyllosphere microbiota are actively engaged as biocontrol agents to control the wildfire disease in tobacco plants. Study results confirmed that the phyllosphere microorganisms employed acted as potential biocontrol agents (Qin et al. 2019). The central mechanism of biological control of pests is blocking the growth of one organism by using another organism. It was reported that phyllospheric microbiota are capable of inducing plant resistance against different pests in specific environments. Some pathogenic microbial attacks on plants were successfully controlled by employing biological agents. Some research criteria are still needed to gain a better understanding of biological control on the plant-pest and plant–microbe interaction (Heydari and Pessarakli 2010). Duke and co-workers investigated the biocontrol potential of *Pseudomonas chlororaphis* PA23, a phyllospheric bacterial strain against necrotrophic fungus *Sclerotinia sclerotiorum* of the *Brassica napus* plant. The investigation result revealed that the bacterial strain PA23 could effectively decrease the disease severity of *S. sclerotiorum* on the *B. napus* plant. The molecular identification of the effect of the biocontrol agent on gene regulation in the aerial surface lesion-forming pest were studied. Herein, the host pathogen interface showed that the biocontrol treated sample showed an intense reduction in the number of genes upregulated in response to the pest. The study postulated that awareness of plant–microbe interaction would potentially enhance the development of biocontrol methodology as an alternative mode to chemical pesticides (Duke et al. 2017).

7.8.2 Environmental Stress and Resistance Management

In the earlier sections it was discussed that unlike other microbiomes of the plant, the phyllosphere microbiome encounters a series of stress conditions including poor availability of nutrients, and different environmental stress factors such as fluctuations in temperature, humidity, continuous exposure to UV radiation, desiccation, osmotic pressure, intense environment of reactive oxygen species (ROS) and starvation. Hence, it clear that these phyllosphere microbiota are capable of mitigating different environmental stresses. As the interaction among the plant and the microorganisms are an integral portion of the living ecosystem, it was widely accepted that this interaction could positively modulate local and systemic mechanisms in plants to provide defence mechanisms against environmental conditions (Meena et al. 2017). A study investigated stress resistance involved in plant colonization of the phyllospheric bacterium *Methylosinus* sp. B4S. The test result showed that a particular gene, namely *phyR*, a general stress response gene, effectively enhanced the stress resistance against heat shock and UV radiation (Iguchi et al. 2013). Recently, the foliar application of the phyllospheric bacterium, *Bacillus megaterium* PB50 not only improved growth promotion in rice plants under osmotic stress conditions, but also facilitated the plants to bypass drought conditions by modulating stomatal closure (Arun et al. 2020). The advanced methodology still required to evaluate the plant response against various environmental factors and phyllosphere microbiota-mediated abiotic stress mitigation strategies in plants.

7.8.3 Plant Growth and Promotion

The phyllospheric microbial communities, especially phyllospheric bacterial communities, have the inherent property to intricately associate with host plants, adapting valuable metabolic and physiological activities of their host (Stone et al. 2018). In this context, a number of bacterial species isolated from the phyllosphere are reported for improving the production of phytohormones such as auxins, gibberellins and cytokinins, which facilitate plant growth and development. The phyllobacteria also mimic the pathways involved in amino acid production which enable them to facilitate increased nutritional acquisition to host plants (Pascazio et al. 2015; Chaudhary et al. 2017). Among the different Proteobacteria inhabiting the phyllosphere, the genus *Methylobacterium* is involved in maintaining plant growth kinetics by significantly modulating the nutrient uptake and supplementing essential nutrients to plant hosts (Kumar et al. 2016). The inherent properties of different species of *Methylobacterium* in growth promotion attributes could be due to their close association with plant hosts which may vary from symbiotic to epiphytic or endophytic associations. The majority of *Methylobacterium* spp. are intriguingly associated with host plant metabolism and physiological activities such as nitrogen fixation, nodule formation, phytohormone production and enhanced photosynthetic activity (Dourado et al. 2015).

Different species of *Methylobacterium*, isolated from the phyllosphere of different crop species are involved in enhancing seed vigour, seedling germination and growth parameters in wheat (*Triticum aestivum*) by producing phytohormones such as cytokinin and other plant hormones (Meena et al. 2012). In this regard, the foliar expression of *M. extorquens* MP_1 and *M. zatmanii* MS_4 isolated from the phyllosphere of peach (*Prunus persica* L.) and strawberry (*Fragaria ananassa* L.) exhibited significant plant growth promoting attributes on tomato (*Solanum lycopersicum* L.) by mimicking IAA production and improving seed vigour and germination of tomato seeds (Pattnaik et al. 2017). In continuation of the work, *M. extorquens* MM_2 sampled from the phyllosphere of mustard leaves (*Brassica niagra*) also showed promising growth promoting properties suggesting its application towards achieving sustainability in agriculture and crop productivity (Subhaswaraj et al. 2017). One of the important members of the bacterial community colonizing the phyllosphere is *Sphingomonas* sp. which has the ability to produce a plant growth promoting agent, indole acetic acid (IAA), which enables the enhanced growth and development of the associated host plant (Chaudhary et al. 2017).

7.8.4 Crop Productivity and Agricultural Sustainability

Plant-associated microbial communities are the key determinants in efficient nutrient mobilization and transport, thereby promoting plant health and productivity (Berg et al. 2014). The phyllospheric microbial communities also play a crucial role in the nitrogen fixation process, nutrient cycling between host plant and atmosphere, nutrient acquisition and solubilization, indicating a role in improving crop productivity (Chaudhary et al. 2017). The microbial communities associated with the phyllosphere also show 1-aminocyclopropane-1-carboxylate (ACC) deaminase activity, which regulates endogenous ethylene production, thereby improving the growth and productivity of economically important crops (de Souza et al. 2015). A recent study evidenced that *Acinetobacter* from the phyllosphere significantly enhanced crop productivity in the tomato plant (Dong et al. 2019). The phyllosphere microbiota is not only involved in growth promoting attributes, but also encompasses the maintenance of the nutrient status of the host plant by influencing nutrient cycling. Thus the influence of the phyllosphere microbiota enables health benefits to the host plant in terms of essential nutrients, thus improving crop productivity (Abadi et al. 2020).

7.9 Current Trends and Future Perspectives

Over the last few decades, microbial communities inhabiting the phyllosphere have provided new insights into the structure and function of the phyllosphere microbiome in relation to environmental and biogenic factors for the proper management of agro-ecosystems, ecosystem functioning and environmental sustainability. The advance of high-throughput technologies such as transcriptomics and metaproteogenomics has revolutionized the comprehensive understanding of the phyllosphere microbiome in maintaining a variety of ecosystem services for mankind and the environment (Remus-Emsermann and Schlechter 2018). The introduction of global molecular approaches such as metagenomics and proteomics in the field of phyllosphere microbiology will not only provide the physiological potential of phyllospheric microbial communities, but also give an insight into the impact of plant–microbe interactions towards ecosystem functioning and agricultural sustainability (Knief et al. 2012).

The intervention of metagenomic and metaproteomic approaches in phyllosphere-based microbiology determines the presence of species-specific microbial proteins which are responsible in providing differential adaptation strategies to the colonizing phyllospheric microbiota (Bulgarelli et al. 2013). The intervention of next-generation molecular technologies and high-throughput strategies in the area of phyllosphere microbiology provide further insights into the structure and function of the phyllospheric microbial community and their composition, factors and drivers of plant–microbe interaction and plant–environment interaction in ecosystem functioning, and interactions with biotic and abiotic components for environmental sustainability (Rastogi et al. 2013). The high-throughput next generation sequencing strategies critically empower understanding the composition, microbial diversity and functional attributes of phyllosphere microbiota, and thus support the agrotechnical practices for biotechnological and agricultural applications (Arya and Harel 2019).

The introduction of high-throughput metabarcoding strategies provides a new dimension for the identification of phyllospheric microbiota and its interaction with the host plant species in a culture-independent method for the proper management of phyllospheric microbiota in achieving sustainability of the environment (Abdelfattah et al. 2015). The objective of the present scientific knowledge on phyllosphere microbiology is to regulate the high-throughput molecular approaches, to understand the key drivers of microbial diversity which will be influential in the development of environmental strategies for the management of phyllospheric microbiota favouring plant/crop health and widespread ecosystem services (Laforest-Lapointe et al. 2016). Thus, the key factors such as metagenomic approaches, temporal variation, microbe–microbe interaction, spatial variation and microbial cultures may be taken into consideration for understanding the community profile of phyllosphere microbiota and defining functional attributes in biotechnological and agricultural fields (Laforest-Lapointe and Whitaker 2019). The complexity and diversity shown by the phyllosphere microbiota, and their widespread biotechnological and agricultural implications have influenced the paradigm shift in minimizing chemical products, and have emerged as a viable alternative for agronomical applications in the future (Barrera et al. 2019).

7.10 Conclusion

The phyllosphere is the region in plants which is observed above the ground. This region encounters different biotic and abiotic stress factors. Plants are continuing to be established in the earth, and insights into plant biology reveal several adaptations and relationships with neighbouring organisms. The relationship between plants and microorganisms has solved several questions regarding plant management strategies against different biotic and abiotic stress factors. Microorganisms coordinating with the phyllosphere region are potentially promoting plant growth, increasing product yield, developing resistance to different environmental stress factors and reducing pathogen infection. Recent advances in environmental research have introduced an alternative technology such as biofertilizers and biopesticides, as opposed to chemical entities. Research on the phyllosphere remains an effervescent topic, where researchers have been engaged in the investigation of phyllosphere microbiota to determine how the microbial community may support plant growth, development and productivity for successful and sustained agricultural practice.

References

Abdelfattah, A., Nicosia, M.G.L.D., Cacciola, S.O., Droby, S., and Schena, L. (2015). Metabarcoding analysis of fungal diversity in the phyllosphere and carposphere of olive (*Olea europaea*). *PLoS ONE* 10(7): e0131069.

Abadi, V.A.J.M., Sepehri, M., Rahmani, H.A., Zarei, M., Ronaghi, A., Taghavi, S.M., and Shamshiripour, M. (2020). Role of dominant phyllosphere bacteria with plant growth-promoting characteristics on growth and nutrition of Maize (*Zea mays* L.). *J. Soil. Sci. Plant. Nutr.* 20: 2348–2363.

Agafonova, N.V., Doronina, N.V., and Trotsenko, Y.A. (2016). Enhanced resistance of pea plants to oxidative stress caused by paraquat during colonization by aerobic methylobacteria. *Appl. Biochem. Microbiol.* 52(2): 199–204.

Aleklett, K., Hart, M., and Shade. A. (2014). The microbial ecology of flowers: an emerging frontier in phyllosphere research. *Botany* 92: 253–266.

Andreote, F.D., Gumiere, T., and Durrer, A. (2014). Exploring interactions of plant microbiomes. *Scientia Agricola* 71(6): 528–539.

Andreote, F.D., and Silva, M.C.P. (2017). Microbial communities associated with plants: learning from nature to apply it in agriculture. *Curr. Opin. Microbiol.* 37: 39–34.

Arun K.D., Sabarinathan, K.G., Gomathy, M., Kannan, R., and Balachandar, D. (2020). Mitigation of drought stress in rice crop with plant growth-promoting abiotic stress-tolerant rice phyllosphere bacteria. *J. Basic. Microbiol.* 60: 768–786.

Arya, G.C. and Harel, A. (2019). Phyllosphere and its potential role in sustainable agriculture. In: *Microbial Genomics in Sustainable Agroecosystems*. Tripathi, V., Kumar, P., Tripathi, P., and Kishore, A., (Eds.). pp. 39-65.

Banik, A., Mukhopadhaya, S.K., and Dangar, T.K. (2016). Characterization of N_2-fixing plant growth promoting endophytic and epiphytic bacterial community of Indian cultivated and wild rice (*Oryza* spp.) genotypes. *Planta* 243: 799–812.

Barrera, S.E., Sarango-Flores, S.W., and Montenegro-Gomez, S.P. (2019). The phyllosphere microbiome and its potential application in horticultural crops: A review. *Rev. Colomb. Cienc. Hortic.* 13(3): 384–396.

Batool, F., Rehman, Y., and Hasnain, S. (2016). Phylloplane associated plant bacteria of commercially superior wheat varieties exhibit superior plant growth promoting abilities. *Front. Life. Sci.* 9(4): 313–322.

Berg, G., Grube, M., Schloter, M., and Smalla, K. (2014). Unraveling the plant microbiome: looking back and future perspectives. *Front. Microbiol.* 5: 148.

Berg, G., Rybakova, D., Grube, M., Koberl, M., and Price, A. (2016). The plant microbiome explored: implications for experimental botany. *J. Exp. Bot.* 67(4): 995–1002.

Braga, R.M., Dourado, M.N., and Araújo, W.L. (2016). Microbial interactions: ecology in a molecular perspective. *Brazilian J. Microbiol.* 47: 86–98.

Bringel, F. and Couee, I. (2015). Pivotal roles of phyllosphere microorganisms at the interface between plant functioning and atmospheric trace gas dynamics. *Front. Microbiol.* 6: 1–14.

Bulgarelli, D., Schlaeppi, K., Spaepen, S., van Themaat, E.V.L., and Schulze-Lefert, P., (2013). Structure and functions of the bacterial microbiota of plants. *Annu. Rev. Plant. Biol.* 64, 807–838.

Carvalho, S.D. and Castillo, J.A., (2018). Influence of light on plant-phyllosphere interaction. *Front. Plant. Sci.* 9: 1482.

Cernava, T., Chen, X., Krug, L., Li, H., Yang, M., and Berg, G. (2019). The tea leaf microbiome shows specific responses to chemical pesticides and biocontrol applications. *Sci. Total. Environ.* 667: 33–40.

Chaudhary, D., Kumar, R., Sihag, K., Rashmi, and Kumari, A. (2017). Phyllospheric microflora and its impact on plant growth: A review. *Agr. Rev.* 38(1): 51–59.

Chen, X., Wicaksono, W.A., Berg, G., and Cernava, T. (2021). Bacterial communities in the plant phyllosphere harbor distinct responders to a broad-spectrum pesticide. *Sci. Total. Environ.* 751: 141799.

Compant, S., Samad, A., Faist, H., and Sessitsch, A. (2019). A review on the plant microbiome: Ecology, Functions, and emerging trends in microbial application. *J. Adv. Res.* 19: 29–37.

Copeland, J.K., Yuan, L., Layeghifard, M., Wang, P.W., and Guttman, D.S. (2015). Seasonal community succession of the phyllosphere microbiome. *Molecular Plant-Microbe Interactions* 28(3): 274–285.

Cordier, T., Robin, C., Capdevielle, X., Fabreguettes, O., Desprez-Loustau, M.L., and Vacher, C. (2012). The composition of phyllosphere fungal assemblages of European beech (*Fagus sylvatica*) varies significantly along an elevation gradient. *New Phyologist* 196: 510–519.

Costa, L.E.O., de Queiroz, M.V., Borges, A.C., de Moraes, C.A., and de Araujo, E.F., (2012). Isolation and characterization of endophytic bacteria isolated from the leaves of the common bean (*Phaseolus vulgaris*). *Braz. J. Microbiol.* 2012: 1562–1575.

Darlison, J., Mogren, L., Rosberg, A.K., Gruden, M., Minet, A., Line, C., Mieli, M., Bengtsson, T., Hakansson, A., Uhlig, E., Becher, P.G., Karlsson, M., and Alsanius, B.W. (2019). Leaf mineral content govern microbial community structure in the phyllosphere of spinach (*Spinacia oleracea*) and rocket (*Diplotaxis tenuifolia*). *Sci. Total. Environ.* 675: 501–512.

de Souza, R., Ambrosini, A., and Passaglia, L.M.P. (2015). Plant growth-promoting bacteria as inoculants in agricultural soils. *Genet. Mol. Biol.* 38(4): 401–419.

Dong, C.J., Wang, L.L., Li, Q., and Shang, Q.M. (2019). Bacterial communities in the rhizosphere, phyllosphere and endosphere of tomato plants. *PLoS ONE* 14(11): e0223847.

Dourado, M.N., Neves, A.A.C, Santos, D.S., and Araujo, W.L. (2015). Biotechnological and agronomic potential of endophytic pink-pigmented methylotrophic *Methylobacterium* spp. *Biomed. Res. Int.* 2015: Article ID 909016.

Duke, K.A., Becker, M.G., Girard, I.J., Millar, J.L., Dilantha, Fernando, W.G., Belmonte, M., and de Kievit, T.R. (2017). The biocontrol agent *Pseudomonas chlororaphis* PA23 primes *Brassica napus* defenses through distinct gene networks. *BMC Genomics* 18(1): 467.

Farre-Armegnol, G., Filella, I., Llusia, J., and Penuelas, J. (2016). Bidirectional interaction between phyllospheric microbiotas and plant volatile emissions. *Trends. Plant. Sci.* 21(10): 854–860.

Finkel, O.M., Castrillo, G., Herrera Paredes, S., Salas González, I., and Dangl, J.L., (2017). Understanding and exploiting plant beneficial microbes. *Curr. Opin. Plant Biol.* 38: 155–163.

Francis, F., Jacquemyn, H., Delvigne, F., and Lievens, B. (2020). From diverse origins to specific targets: Role of microorganisms in indirect pest biological control. *Insects* 11(8): 533.

Gomes, T., Pereira, J.A., Benhadi, J., Lino-Neto, T., and Baptista, P., (2018). Endophytic and epiphytic phyllosphere fungal communities are shaped by different environmental factors in a Mediterranean ecosystem. *Microb. Ecol.* 76(3): 668–679.

Heydari, A. and Pessarakli, M. (2010). A review on biological control of fungal plant pathogens using microbial antagonists. *J. Biol. Sci.* 10(4): 273–290.

Iguchi, H., Sato, I., Sakakibara, M., Yurimoto, H., and Sakai, Y. (2012). Distribution of methanotrophs in the phyllosphere. *Bioscience Biotechnology, and Biochemistry* 76(8): 1580–1583.

Iguchi, H., Sato, I., Yurimoto, H., and Sakai, Y. (2013). Stress resistance and C1 metabolism involved in plant colonization of a methanotroph *Methylosinus* sp. B4S. *Arch. Microbiol.* 195(10–11): 717–726.

Jackson, C.R., Stone, B.W.G., and Tyler, H.L. (2015). Emerging perspectives on the natural microbiome of fresh produce vegetables. *Agriculture* 5: 170–187.

Kembel, S.W., O'Connor, T.K., Arnold, H.K., Hubbell, S.P., Wright, S.J., and Green, J.L. (2014). Relationships between phyllosphere bacterial communities and plant functional traits in a neotropical forest. *Proceedings of National Academy of Sciences USA* 111(38): 13715–13720.

Kim, M., Singh, D., Lai-Hoe, A., Go, R., Rahim, R.A., Ainuddin, A.N., Chun, J., and Adams, J.M., (2012). Distinctive phyllosphere bacterial communities in tropical trees. *Microb. Ecol.* 63: 674–681.

Knief, C., Delmotte, N., Chaffron, S., Stark, M., Innerebner, G., Wassmann, R., von Mering, C., and Vorholt, J.A. (2012). Metaproteogenomic analysis of microbial communities in the phyllosphere and rhizosphere of rice. *ISME J.* 6: 1378–1390.

Knief, C., Ramette, A., Frances, L., Alonso-Blanco, C., and Vorholt, J.A. (2010). Site and plant species are important determinants of the *Methylobacterium* community composition in the plant phyllosphere. *ISME J.* 4: 719–728.

Koskella, B. (2020). The phyllosphere. *Curr. Biol.* 30(19), R1143–R1146.

Kumar, M., Tomar, R.S., Lade, H., and Paul, D. (2016). Methylotrophic bacteria in sustainable agriculture. *World J. Microbiol. Biotechnol.* 32: 120.

Kupferschmied, P., Maurhofer, M., and Keel, C. (2013). Promise for plant pest control: root-associated pseudomonads with insecticidal activities. *Front. Plant Sci.* 4: 1–17.

Laforest-Lapointe, I., Messier, C., and Kembel, S.W. (2016). Host species identity, site and time drive temperate tree phyllosphere bacterial community structure. *Microbiome* 4: 27.

Laforest-Lapointe, I. and Whitaker, B.K. (2019). Decrypting the phyllosphere microbiota: progress and challenges. *Am. J. Bot.* 106(2): 171–173.

Liu, H., Brettell, L.E., and Singh, B. (2020). Linking the phyllosphere microbiome to plant health. *Trend. Plant. Sci.* 25(9): 841–844.

Lopez-Velasco, G., Carder, P.A., Welbaum, G.E., and Ponder, M.A. (2018). Diversity of the spinach (*Spinacia oleracea*) spermosphere and phyllosphere bacterial communities. *FEMS Microbiol. Lett.* 346: 146–154.

Meena, K.K., Kumar, M., Kalyuzhnaya, M.G., Yandigeri, M.S., Singh, D.P., Saxena, A.K., and Arora, D.K. (2012). Epiphytic pink-pigmented methylotrophic bacteria enhance germination and seedling growth of wheat (*Triticum aestivum*) by producing phytohormone. *Antonie Leeuwenhoek* 101: 777–786.

Meena, K.K., Sorty, A.M., Bitla, U.M., Choudhary, K., Gupta, P., Pareek, A., and Minhas, P.S. (2017). Abiotic stress responses and microbe-mediated mitigation in plants: the omics strategies. *Front. Plant Sci.* 8: 1–25.

Meyer, K.M. and Leveau, J.H.J. (2012). Microbiology of the phyllosphere: a playground for testing ecological concepts. *Oecologia* 168: 621–629.

Morris, C.E., Monier, J.M., and Jacques, M.A. (1997). Methods for observing microbial biofilms directly on leaf surfaces and recovering them for isolation of culturable microorganisms. *Appl. Environ. Microbiol.* 63(4): 1570–1576.

Moulas, C., Petsoulas, C., Rousidou, K., Perruchon, C., Karas, P., and Karpouzas, D.G. (2013). Effects of systemic pesticides imidacloprid and metalaxyl on the phyllosphere of pepper plants. *Biomed. Res. Int.* 2013: Article ID 969750.

Müller, T. and Ruppel, S. (2014). Progress in cultivation-independent phyllosphere microbiology. *FEMS Microbiol. Ecol.* 87(1): 2–17.

Noble, A.S., Noe, S., Clearwater, M.J., and Lee, C.K. (2020). A core phyllosphere microbiome exists across distant populations of a tree species indigenous to New Zealand. *PLoS ONE* 15(8): e0237079.

Ottesen, A.R., Gorham, S., Reed, E., Newell, M.J., Ramachandran, P., Canida, T., Allard, M., Evans, P., Brown, E., and White, J.R. (2016). Using a control to better understand phyllosphere microbiota. *PLoS ONE* 11(9): e0163482.

Pascazio, S., Crecchio, C., Ricciuti, P., Palese, A.M., Xiloyannis, C., and Sofo, A. (2015). Phyllosphere and carposphere bacterial communities in olive plants subjected to different cultural practices. *Int. J. Plant. Biol.* 6: 6011.

Patel, H.K., Suárez-Moreno, Z.R., Degrassi, G., Subramoni, S., González, J.F., and Venturi, V. (2013). Bacterial LuxR solos have evolved to respond to different molecules including signals from plants. *Front. Plant Sci.* 4: 1–5.

Pattnaik, S., Rajkumari, J., Paramanandham, P., and Busi, S. (2017). Indole acetic acid production and growth-promoting activity of *Methylobacterium extorquens* MP_1 and *Methylobacterium zatmanii* MS_4 in tomato. *Int. J. Veg. Sci.* 23(4): 321–330.

Penuelas, J. and Terradas, J. (2014). The foliar microbiome. *Trends. Plant. Sci.* 19(5): 278–280.

Qin, C., Tao, J., Liu, T., Liu, Y., Xiao, N., Li, T., Gu, Y., and Yin, H., and Meng, D. (2019). Responses of phyllosphere microbiota and plant health to application of two different biocontrol agents. *AMB Exp.* 9: 42.

Rastogi, G., Coaker, G.L., and Leveau, J.H.J. (2013). New insights into the structure and function of phyllosphere microbiota through high-throughput molecular approaches. *FEMS Microbiol. Lett.* 348: 1–10.

Redford, A.J., Bowers, R.M., Knight, R., Linhart, Y., and Flerer, N. (2010). The ecology of the phyllosphere: geographic and phylogenetic variability in the distribution of bacteria on tree leaves. *Environ. Microbiol.* 12(11): 2885–2893.

Remus-Emsermann, M.N.P. and Schlechter, R.O. (2018). Phyllosphere microbiology: at the interface between microbial individuals and the plant host. *New Phytol.* 218(4): 1327–1333.

Ren, G., Zhu, C., Alam, M.S., Tokida, T., Sakai, H., Nakamura, H., Usui, Y., Zhu, J., Hasegawa, T., and Jia, Z. (2015). Response of soil, leaf endosphere and phyllosphere bacterial communities to elevated CO2 and soil temperature in a rice paddy. *Plant Soil.* 392: 27–44.

Rey, T. and Schornack, S. (2013). Interactions of beneficial and detrimental root-colonizing filamentous microbes with plant hosts. *Genome Biol.* 14(6): 121.

Roman, M.S. and Wagner, A. (2018). An enormous potential for niche construction through bacterial cross-feeding in a homogenous environment. *PLoS. Comput. Biol.* 14(7): e1006340.

Rosado, B.H.P., Almeida, L.C., Alves, L.F., Lambais, M.R., and Oliveira, R.S. (2018). The importance of phyllosphere on plant functional ecology: a phyllo trait manifesto. *New Phytologist* 219: 1145–1149.

Rossmann, M., Sarango-Flores, S.W., Chiaramonte, J.B., Kmit, M.C.P, and Mendes, R. (2017). Plant microbiome: composition and functions in plant compartments. In: Pylro, V. and Roesch, L. (Eds.) *The Brazilian Microbiome*, pp. 7-20.

Saleem, M., Meckes, N., Pervaiz, Z.H., and Traw, M.B. (2017). Microbial interactions in the phyllosphere increase plant performance under herbivore biotic stress. *Front. Microbiol.* 8: 41.

Schlechter, R.O., Miebach, M., and Remus-Emsermann, M.N.P. (2019). Driving factors of epiphytic bacterial communities: A review. *J. Adv. Res.* 19: 57–65.

Singh, P., Santoni, S., Weber, A., This, P., and Peros, J.P. (2019). Understanding the phyllosphere microbiome assemblage in grape species (*Vitaceae*) with amplicon sequence data structures. *Sci. Rep.* 9: 14294.

Sivakumar, N., Sathishkumar, R., Selvakumar, G., Shyamkumar, R., and Arjunekumar, K. (2020). Phyllospheric microbiomes: diversity, ecological significance, and biotechnological applications. In: *Plant Microbiomes for Sustainable Agriculture, Sustainable Development and Biodiversity*. Yadav, A.N., (Eds.). pp. 113-172.

Stone, B.W.G., Weingarten, E.A., and Jackson, C.R. (2018). The role of the phyllosphere microbiome in plant health and function. *Ann. Plant. Rev.* 1(2): 533–556.

Subhaswaraj, P., Jobina, R., Parasuraman, P., and Siddhardha, B. (2017). Plant growth promoting activity of pink pigmented facultative Methylotroph – *Methylobacterium extorquens* MM_2 on *Lycopersicon esculentum* L. *J. Appl. Biol. Biotechnol.* 5(1): 42–46.

Tani, A., Sahin, N., Fujitani, Y., Kato, A., Sato, K., and Kimbara, K. (2015). *Methylobacterium* species promoting rice and barley growth and interaction specificity revealed with Whole-Cell Matrix-Assisted Laser Desorption/Ionization-Time of Flight Mass Spectrometry (MALDI-TOF/MS) Analysis. *PLoS ONE* 10(6): e0129509.

Tecon, R., Ebrahimi, A., Kleyer, H., Erev, S.L., and Or, D. (2018). Cell-to-cell bacterial interactions promoted by drier conditions on soil surfaces. *Proc. Nat. Acad. Sci. USA* 115(39): 9791–9796.

Thapa, S. and Prasanna, R., (2018). Prospecting the characteristics and significance of the phyllosphere microbiome. *Ann. Microbiol.* 68(5): 229–245.

Thapa, S., Prasanna, R., Ranjan, K., Velmourougane, K., and Ramakrishnan, B., (2017). Nutrients and host attributes modulate the abundance and functional traits of phyllosphere microbiome in rice. *Microb. Res.* 204: 55–64.

Thapa, S., Ranjan, K., Ramakrishnan, B., Velmourougane, K., and Prasanna, R. (2018). Influence of fertilizers and rice cultivation methods on the abundance and diversity of phyllosphere microbiome. *J. Basic. Microbiol.* 58: 172–186.

Thomas, P. and Reddy, K.M. (2013). Microscopic elucidation of abundant endophytic bacteria colonizing the cell wall–plasma membrane peri-space in the shoot-tip tissue of banana. *AoB Plants* 5: 1–12.

Truchado, P., Gil, M.I., Reboleiro, P., Rodelas, B., and Allende, A. (2017). Impact of solar radiation exposure on phyllosphere bacterial community of red-pigmented baby leaf lettuce. *Food. Microbiol.* 66: 77–85.

Turner, T.R., James, E.K., and Poole, P.S. (2013). The plant microbiome. *Genome Biol.* 14(6): 209.

Vacher, C., Hampe, A., Porte, A.J., Sauer, U., Compant, S., and Morris, C.E. (2016). The Phyllosphere: Microbial jungle at the plant–climate interface. *Annu. Rev. Ecol. Evol. Syst.* 47: 1–24.

Vorholt, J.A. (2012). Microbial life in the phyllosphere. *Nat. Rev. Microbiol.* 10(12): 828–840.

Wagi, S. and Ahmed, A. (2017). Phyllospheric plant growth promoting bacteria. *J. Bacteriol. Mycol.* 5(1): 00119.

Xiang, Q., Chen, Q.L., Zhu, D., Yang, X.R., Qiao, M., Hu, H.W., and Zhu, Y.G., (2020). Microbial functional traits in phyllosphere are more sensitive to anthropogenic disturbance than in soil. *Environ. Pollut.* 265: 114954.

Yadav, A.N. and Saxena, A.K. (2018). Biodiversity and biotechnological applications of halophilic microbes for sustainable agriculture. *J. Appl. Biol. Biotechnol.* 6(1), 48–55.

Yadav, A.N., Verma, P., Kour, D., Rana, K.L., Kumar, V., Singh, B., Chauahan, V.S., Sugitha, T.C.K., Saxena, A.K., and Dhaliwal, H.S. (2017). Plant microbiomes and its beneficial multifunctional plant growth promoting attributes. *Int. J. Environ. Sci. Nat. Res.* 3(1): 555601.

Zhang, Z., Kong, X., Jin, D., Yu, H., Zhu, X., Su, X., Wang, P., Zhang, R., Jia, M., and Deng, Y. (2019). *Euonymus japonicas* phyllosphere microbiome significantly changed by powdery mildew. *Arch. Microbiol.* 201: 1099–1109.

8

Microbial Consortia

Emerging Conglomerate for Better and Superior Sustainable Agricultural Practices

Rishi Kumar Verma, Manisha Sachan, and Shivesh Sharma

Department of Biotechnology, Motilal Nehru National Institute of Technology, Allahabad, Uttar Pradesh, India

8.1 Introduction

Plant-associated microbiomes and their agglomerates have remarkable potential to improve plant pliability and revenues in agriculture systems. There is enough evidence that agricultural practices that use microbes or their metabolic products can augment nutrient uptake, resist pests, suppress diseases and lessen plant stress responses.

By seeing real world conditions, utilization of microbes up to their full efficiency, potential, and consistency is the current need. The global population is estimated to be increased by nine billion by 2050. To serve this population, we need to increase productivity many fold to a maximum of double the present capacity. Traditional agricultural practices are providing a good yield, but at the same time degrading and polluting the soil, as well as badly affecting the ecology of the natural microbiome (Tilman et al. 2002). Efforts should be made towards microbiome engineering which could lead the development and exploration of formulations of microbial consortia which should be plant specific. Thus, there would be a specific remedy for a particular requirement. These formulations can provide better plant health, improved crop yield and greater protection to plants against diseases (Trivedi et al. 2017). Major agricultural countries are facing substantial problems: either they are not getting adequate yields after hefty inputs of inorganic constituents, or they are unable to increase the farming costs. Facing both the challenges, they are expected to improve quality as well as quantity of crops and foods.

8.2 Microbial Consortia: Significant Explication for This Dilemma

Plants carefully design their specific rhizosphere milieu by oozing specific exudates such as carbohydrates, vitamins, organic acids, siderophores, amino acids, mucilage, phenolic compounds and various organic constituents to improve nutrient availability, interaction

The Plant Microbiome in Sustainable Agriculture, First Edition. Edited by Alok Kumar Srivastava, Prem Lal Kashyap and Madhumita Srivastava.

with specific beneficial microbes and to attract other similar ones (Sarma et al. 2015; Vishwakarma et al. 2017a). This rhizosphere expedites growth, development, and functioning of plant growth promoting rhizobacteria (PGPR). This community of microbes colonizes the root surfaces, promotes plant growth (Vacheron et al. 2013), provides safeguards to plants from pathogens (Lugtenberg and Kamilova 2009), and also supports plants to endure abiotic stresses (He and Yang 2007).

In several studies, it has been demonstrated that microbes are helpful in increasing the crop yield and they are the best alternative to chemically synthesized fertilizers which are harmful for the soil and its ecology. Unfortunately, it is also evident that greenhouse trials with microbes as biofertilizers often fail or are unremarkable once applied in field trials, or their effects are inconsistent with different climatic conditions. This is the prime obstacle behind substantial ratification of this particular technology. For large-scale enactment, it is necessary to advance and revamp the present selection process of microbes. Its application technology has to be reformed after detailed study of the interaction between available native microbiome of the crop in its original field and inoculating microbes (Trivedi et al. 2017).

Microbial consortia, are an excellent microbial conglomerate, which is compatible in itself and can form beneficial associations with the already present microbial networks in natural rhizospheric soils. In conglomerate form they have a better chance to survive and can provide more benefits to the host compared with single strain. As we are experiencing a population boom all over the world, it will become a great challenge to provide food for everyone; thus we need to increase agricultural yields. These microbial consortia have all the qualities to play a vital role in overcoming these future challenges.

8.3 Microbial Consortia in Plant Health

Kloepper et al. (1989), named PGPR (plant growth promoting rhizobacteria) as root inhabiting microbes which were a hopeful option for sustainable agriculture. This definition of PGPR was further advanced to include advantageous fungi also, now a new term has emerged, PGPMs, i.e. plant growth promoting microbes which include both bacteria and fungi with plant growth promoting properties. PGPMs tend to improve agricultural soil fertility and help in plant growth promotion via nutrient mineralization, organic matter cycling and maintaining ecosystem health. Several bacteria and fungi are known to support plant growth. Some known bacteria are *Azotobacter*, *Azospirillum*, *Pseudomonas*, *Acetobacter*, *Burkholderia*, and *Bacillus*, etc., and fungi like *Trichoderma*, *Ampelomyces*, *Coniothyrium*, *Dactylella*, *Gliocladium*, and *Paecilomyces*, which can be considered as PGPMs (Kloepper et al. 1980a,b; Bashan and Levanony 1990).

Several researchers presented their studies claiming that microbial consortia are apparently the improved option for better plant growth, health, and their survival in challenging environment as well as in natural conditions. These microbial combinations stimulate plant roots to exude more amino acids, growth regulators and sugars as well as to make plant roots better at utilizing minerals and other constituents present in rhizosphere in an efficient way. This helps in better nitrogen fixation, which likewise supports plants to anticipate any change in environmental conditions also (Raja et al. 2006; Zlotnikova et al. 2007; Nadeem et al. 2013; Vishwakarma et al. 2017b).

Roots of *Withania somnifera* are rich in withanolides, which have medicinal importance in India. Withanolides are the effective ingredient of this medicinal plant, having a noted therapeutic potential. There is better plant growth, dry biomass, plant nitrogen and phosphorus, spore count and root colonization once plants are inoculated with microbial consortia made up of *Bacillus licheniformis* and *Aculospora laevis* (Anuroopa and Bagyaraj 2017). Pigeon pea is an important pulse crop in India. An experiment reported that microbial consortia made up of *Pseudomonas fluorescens* + *Bacillus subtilis* have recorded 23.7 cm root length and *P. fluorescens* + *Rhizobium* recorded 21.4 cm after 30 DAS. Both combinations also showed better shoot length 24.2 cm and 22.6 cm simultaneously after 30 DAS (Rajasekhar et al. 2016). *Vaccinium corymbosum* is a highly valued fruit crop which is commonly known as blueberry. A field experiment carried out with a commercially available microbial consortium named Oiko bac 174 (Oikos Chile Ltd.) which is a combination of 10 soil microorganisms: *B. subtilis*, *B. licheniformis*, *Bacillus megaterium*, *Bacillus polymyxa*, *Bacillus macerans*, *P. fluorescens*, *Pseudomonas putida*, *Nocardiac orallina*, *Saccharomyces cerevisiae*, and *Trichoderma viride*. This microbial consortia treatment showed increased shoot and root dry weights, with values 32 and 31 per cent higher than the control plants (Schoebitz et al. 2016). *Phaseolus vulgaris*, commonly known as French bean, is a far-reaching legume crop, farmed for its nutritive and succulent pods and soft seeds. An experiment was performed in greenhouse conditions to investigate the outcomes of microbial consortium with *Paenibacillus polymyxa*, *Pantoea agglomerans*, and *Funneliformis mosseae* on the various parameters of French bean. As a result it was found that growth of the plant, nutritive value, per plant yield, mycorrhizal root colonization and activity of alkaline phosphatase in rhizospheric soil were found better in comparison with the un-inoculated control. Once this experiment was carried out in a field with different combinations of inorganic nitrogen, phosphorus, and potassium (NPK) fertilizer, it was found that height of plant, stem thickness, shoot dry weight and outturn fruits are higher in the treatment of microbial consortia with 50 per cent NPK and these results were on a par with the 100 per cent NPK trial. This showed that microbial consortia could be an alternative for inorganic fertilizers in future (Chauhan and Bagyaraj 2015). Some undefined consortia which were obtained from continuous enrichment were evaluated in greenhouse conditions with rice plants. They showed improved nitrogenase activity, increased root dry weight, stem dry weight and foliar dry weight. Such types of random enriched consortia should be explored more (Vanegas and Uribe-Vélez 2014). A comparative study of a commercial microbial consortia named Omega have been carried out against nine different microbial strains (*Bacillus* spp., *Streptomyces* spp., *Azotobacter* spp., *and Frauteria* spp.), over *Vigna mungo* in pots. Seeds treated with Omega showed significant increases in height of shoots, roots, and total dry weight (11.21 cm, 0.07 g and 0.23 g) against un-inoculated controls (8.51 cm, 0.04 g and 0.14 g). An on-farm trial was also proposed for sugarcane after seeing its promising results (E Leo Daniel et al. 2011). Plant growth of *W. somnifera* was tested with microbial consortia containing *Azospirillum, Azotobacter, Pseudomonas*, and *Bacillus* against control over two consecutive years. The amalgamation of quoted PGPR strains resulted in significantly increased plant height, root length, and alkaloid content (Rajasekar and Elango 2011).

Figure 8.1 shows the effect of microbial consortia on plant growth.

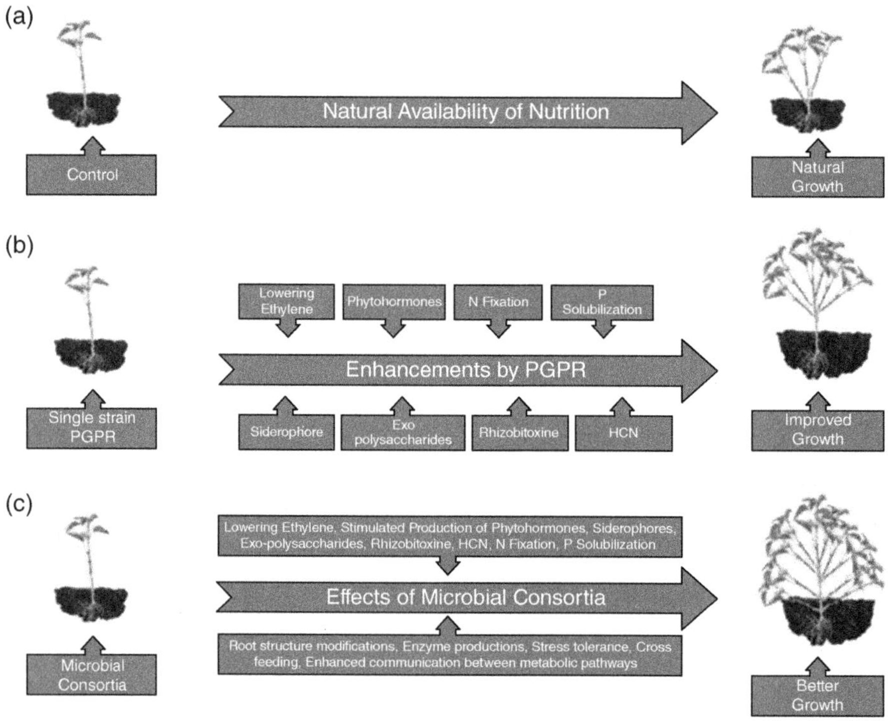

Figure 8.1 Effect of microbial consortia on plant growth. (a) Control: No additional supplement and grown naturally. (b) Single strain of PGPR supplied and it showed improved growth comparatively. (c) Effect of microbial consortia, plant showed better growth in comparison with others.

8.4 Microbial Consortia in Disease Suppression

Scientists started using microbial consortia for better plant health and yield, but their use in disease suppression has recently attracted attention. Isolation and identification of rhizospheric microbes that can play an important role to save plants from disease in environment friendly manner was commenced. Jain et al. (2012) reported that these microbial consortia can arbitrate induced systemic resistance (ISR). This sort of plant defense relies on jasmonic acid, ethylene signaling. In this defense mechanism, plants raise their innate defense to confront biotic challenges (Lugtenberg and Kamilova 2009). Apart from this, other known plant hormones, such as auxins, abscisic acid (ABA), cytokinins, gibberellins, and brassinosteroids, which are well known as regulators for plant progressions, have recently come to light as important players in plant immunity (Denancé et al. 2013; Vishwakarma et al. 2016).

It has been shown that plants exclusively prefer some of the microbes over others, to obtain tailor-made benefits. If we could understand the criteria on which basis plants recruit these microbes, we could provide similar benefits to other plants by designing microbial consortia with efficient microbial strains. In such a way this technique could be more efficient, reliable, consistent, and will get wider acceptance across the globe under different and varied environmental circumstances (Stockwell et al. 2011; Bakker et al. 2013; Yadav et al. 2017). The only concern with this combination is that participating microbes must be compatible and should not be producing any suppressive or deleterious components for others in the consortia (Jain et al. 2012; Singh et al. 2013). It would be best for participating microbes if they secrete constituents which can be utilizable by others, or which would help them in proliferation and establishing connections among all. Different microbes tend to occupy different parts of roots which is advantageous, as there is least chance of competition among them. This diversified availability of microbes with non-competitive nature allows better biocontrol and disease management in plants. Several former reports presented that various microorganisms like *Trichoderma*, *Bacillus*, *Pseudomonas*, *Rhizobium*, *Glomus*, etc., were used to develop microbial consortia (Duffy et al. 1996; Rudresh et al. 2005; Jetiyanon 2007; Kannan and Sureendar 2009; Srivastava et al. 2010) As per these reports, these microbes are adequate enough to provide plant health individually. Since they are compatible, their microbial consortia are possible; upon application together they not only allowed better nitrogen and phosphorus uptake, but also facilitated disease free plant growth. These results were helpful in designing better, more prominent and effective strategies for disease free plants and also led agriculture towards the use of minimum chemicals. Table 8.1 contains an exhaustive list of microbial consortia which are reported helpful in suppressing diseases caused by plant pathogens, and Figure 8.2 shows the effect of microbial consortia on disease resistance.

8.5 Engineering of Microbial Consortia

To develop a more effective and customized solution for betterment of crops, we can engineer microbial consortia. We should go for complete characterization of the whole microbiota of the plant. It should include a whole plant and all its parts: leaf, root, stem, etc. The microbial community is helping plants in different ways. Root microbes are known for helping in nutrient uptake, whereas microbes of other plants also contribute in plant defense and other functions.

In microbial consortia the fitness of microbes and their regulation mechanisms play an important role in its design and construction, because they need to survive and work together. While designing microbial consortia a few mechanisms have to be taken into consideration such as interaction modes among microbes, exchange of metabolites and energy, etc. In the natural microbial consortia, there are numerous means of communication according to the ways of metabolic exchange, it including commensalism, synergism, mutualism, and competition. Once we are able to understand the way of inter-talk among the microbes, designing and engineering of microbial consortia will be an easier task (Song et al. 2014; Ding et al. 2016). Another point about this combination which has to be taken into consideration is the optimization of key factors like metabolic balance and energy consumption. We already know that information among microbes is communicated through signal molecules and these interactions play an important role in building up microbial ecosystems. We also need to focus on their production, their diffusion or absorption pattern with nearby cells and the talk-back response of microbial cells. These patterns are important in engineering, designing, and advancement of microbial consortia.

Table 8.1 Microbial consortia against plant pathogens.

S·N	Microbial Consortia	Plant pathogen(s)	Plants	References
1	*Pseudomonas aeruginosa* PJHU15 + *Trichoderma harzianum* TNHU27 + *Bacillus subtilis BHHU100*	*Sclerotinia sclerotiorum*	*Pisum sativum*	Jain et al. (2015)
2	*Bacillus cereus* AR156 + *B. subtilis* SM21 + *Serratia* sp. XY21	*Verticillium dahliae*	Cotton	Yang et al. (2014)
3	*Paenibacillus xylanilyticus* YUPP-1 + *Paenibacillus polymyxa* YUPP-8 + *Bacillus subtilis* YUPP-2	*Verticillium dahliae*	Cotton	Yang et al. (2013)
4	*Pseudomonas flourescens* Pf1, TDK1, TV5 + *Bacillus subtilis isolate* TH10	*Sarocladium oryzae*	*Oryza sativa*	Sundaramoorthy et al. (2013)
5	*Trichoderma harzianum* Tr6 + *Pseudomonas* sp. Ps14	*Fusarium oxysporum* and *Radicis cucumerinum*	*Cucumis sativus*	Alizadeh et al. (2013)
6	*P. aeruginosa* + *T. harzianum* + *Mesorhizobium* sp.	*Sclerotium rolfsii*	*Cicer arietinum*	Singh et al. (2013)
7	*Azospirillum* sp. AZ204 + *Pseudomonas fluorescens* Pf1	*Rhizoctonia bataticola*	*Gossypium hirsutum*	Marimuthu et al. (2013)
8	*Ochrobactrum lupini* KUDC1013 + *Novosphingobium pentaromativorans* KUDC1065 + *Paenibacillus polymyxa* E681	*Xanthomonas axonopodis*	*Capsicum annuum*	Hahm et al. (2012)
9	*B. subtilis* (EPCO16) + *P. fluorescens* (Pf1)	*F. solani*	*Capsicum annum*	Sundaramoorthy et al. (2012)
10	*B. megaterium* + *B. cereus* + *P. fluorescens*	*Macrophomina phaseolina*	Soybean	Attia et al. (2011)
11	*Glomus sinuosum* + *Gigaspora albida* + *P. fluorescens*	*Rhizoctonia solani*	*Phaseolus vulgaris*	Neeraj and Singh (2011)
12	*Trichoderma viride* + *Pseudomonas fluorescens* + *Bacillus subtilis*	*Lasiodiplodia theobromae*	*Jatropha curcas*	Latha et al. (2011)
13	*Pseudomonas fluorescens* + *Bacillus subtilis*	*Fusarium oxysporum*	Banana	Akila et al. (2011)
14	*B. licheniformis* + *Bacillus* sp. + *P. aeruginosa* + *S. fradiae*	*Sunflower necrosis virus*	Sunflower	Srinivasan and Mathivanan (2011)

15	*P. fluorescens* EBC5 + *P. fluorescens* EBC6	*Pythium aphanidermatum*	*Capsicum annum*	Muthukumar et al. (2010)
16	*Pseudomonas sp.* (UPMP3) + *Burkholderia* sp. (UPMB3)	*Fusarium oxysporum*	Banana	Fishal et al. (2010)
17	*Bacillus pumilus* + *Bacillus subtilis* + *Bacillus safensis* + *Bacillus sphaericus*	*Phytophthora capsici*	Squash	Zhang et al. (2010)
18	*P. fluorescens* (Pf1) + *Endophytic Bacillus sp.* (EPB22)	*Banana bunchy top virus* (BBTV)	Banana	Harish et al. (2009)
19	*P. chlororaphis* PCL1391 + *P. fluorescens* WCS365	*Colletotrichum lindemuthianum*	*Phaseolus vulgaris*	Bardas et al. (2009)
20	*Pseudomonas aeruginosa* + *Burkholderia cepacia* + *Serratia marcescens*	*Sclerotium rolfsii*	Chili	Siddiqui and Meon (2009)
21	*P. fluorescens* Aur 6 + *Chryseobacterium balustinum* Aur 9	*Pyricularia oryzae*	*Oryza sativa*	Lucas et al. (2009)
22	*B. licheniformis* MML2501 + *Bacillus sp.* MML 2551 + *P. aeruginosa* MML 2212 + *Streptomyces fradiae* MML1042	Sunflower necrosis virus	*Helianthus annuus*	Srinivasan and Mathivanan (2009)
23	*G. intraradices* + *Rhizobium* sp. + *P. straita*	*Meloidogyne incognita*, *Macrophomina phaseolina*	*C. arietinum*	Akhtar and Siddiqui (2008)
24	*Rhizobia* + *B. cereus strain* BS03 + *P. aeruginosa* RRLJ04	*Fusarium udum*	*Cajanus cajan*	Dutta et al. (2008)
25	*Pseudomonas fluorescens* Pf1 + *Bacillus subtilis* EPB22	*Banana bunchy top virus*	Banana	Kavino et al. (2007)
26	*Bacillus amyloliquefaciens* IN937a + *B. pumilus* IN937b	*Sclerotium rolfsii*, *Ralstonia solanacearum*	*Capsicum annum*, *Solanum lycopersicum*	Jetiyanon (2007)
27	Rhizobium + *P. straita*	*Meloidogyne incognita*	*Pisum sativum*	Siddiqui and Singh (2005)

(*Continued*)

Table 8.1 (Continued)

S·N	Microbial Consortia	Plant pathogen(s)	Plants	References
28	*P. fluorescens* MTCC 1749 + *T. viride* NCC 34	*Rhizoctonia solani*	*Oryza sativa*	Mathivanan et al. (2005)
29	*Trichoderma harzianum* DB11 + *Gliocladium catenulatum Gliomix*	*Phytophthora cactorum* and *P. fragariae*	*Fragaria ananassa*	Vestberg et al. (2004)
30	*P. putida* WCS358 + *P. putida* strain RE8	*Fusarium oxysporum*	*Raphanus sativus*	de Boer et al. (2003)
31	*P. fluorescens* NBRI-N6 + *P. fluorescens* NRI-N	*Sclerotium rolfsii*	*Piper betel*	Singh et al. (2003)
32	*B. subtilis* MBI600 + *Rhizobium tropici* UMR 1899	*Fusarium solani*	*Phaseolus vulgaris*	Estevez De Jensen et al. (2002)
33	*Pichia guilliermondii* Y2 + *B. mycoides* B16	*B. cinerea*	*Fragaria ananassa*	Guetsky et al. (2002)

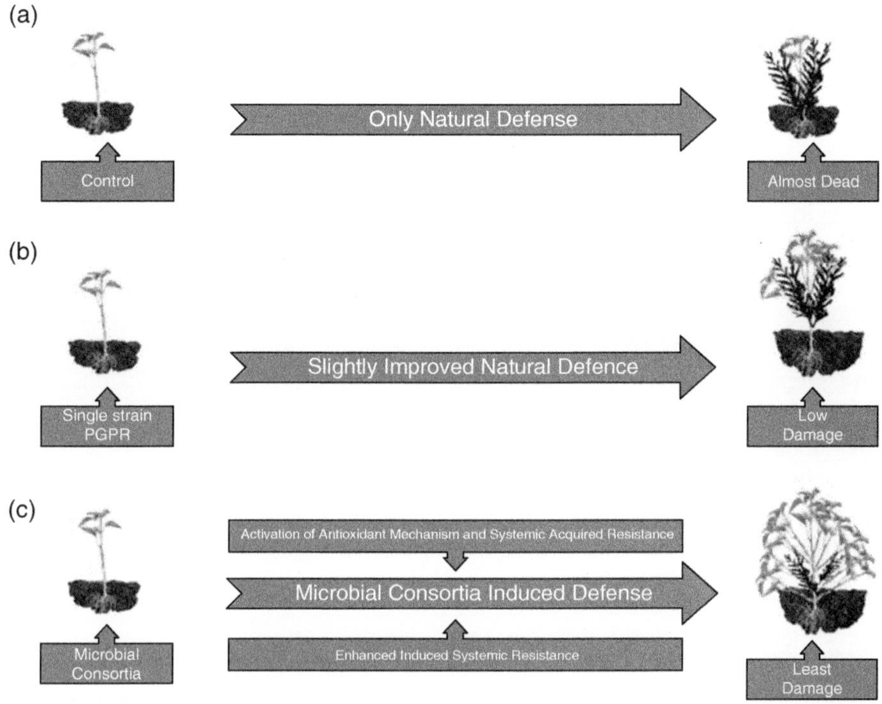

Figure 8.2 Effect of microbial consortia on disease resistance. (a) Control: Almost dead once exposed with plant pathogen. (b) Upon single strain of PGPR application, plant less damaged compared with control. (c) Effect of microbial consortia, plant showed better growth and least damage in comparison with the others.

It is recommended that the total microbiota of plant should be isolated, characterized and their role must be studied in the context of plant health. This much elaborated data will be helpful to identify or design any tool which can select the best combination of beneficial microbes which can work and perform in any climatic condition or soil. Such selected combinations may be useful in enhanced production of crops, their improved quality and are also helpful in environment protection. Once this technique is fully developed and functional, it will have the capacity to revolutionize agricultural production (Mueller and Sachs 2015; Sheth et al. 2016).

8.6 Research Based Initiatives to Explore Microbial Consortia

Two powerful associations have shown great interest in exploiting the microbiome to increase crop and food yield. In 2016, two very important initiatives were launched, which clearly identified the prospects for the microbiome approach. (i) The US microbiome initiative was launched by The White House in May 2016, with an investment of $450 million to encourage inventions as well as developing new approaches which could be easily adapted by industries. This initiative was specially designed to obtain results in the form of good yielding crops and better performing soil microbiomes. (ii) The International Bioeconomy Forum (IBF) was launched by the EU Commission on October 2016 for utilizing microbiomes to obtain better food security as their primary and vital component, along with economic growth and regional employment creation (Singh and Trivedi 2017).

Likewise the year 2014 was also very crucial from the research point of view on microbial consortia. Two initiatives were taken at the time to encourage research and progress of microbial consortia. (i) The Defense Advanced Research Projects Agency (DARPA) USA, announced a program "Biological Robustness in Complex Settings (BRICS)" in August 2014. This program was aimed to design robust microbial consortia which should be stable and safe for biological systems. (ii) The National Basic Research Program (NBRP) China, in January 2014 proposed project 973 – "Design and Construction of Microbial Consortia" which was aimed to eliminate glitches for concocting natural microbial consortia to meet industrial demands (Ding et al. 2016).

8.7 Conclusions

The challenge of codifying a multifunctional microbial inoculum by adding suitable microbes to improve plant growth and health requires corresponding effort, yet, such a multipurpose consortium can be tailored to help plants to establish, grow well, and survive in any environmental conditions. Synthetic microbial consortia are attracting more and more attention with the rapid progress of synthetic biology. Academic authorities including Ron Weiss, Frances Arnold, and Pam Silver are all working on the system design and construction, robust research, and dynamic stability analysis. As one of the important aspects in synthetic microbial consortia, signal delivery and metabolite exchange are the basic principles and guidelines. Ecosystems not only contain interaction among strains, but also between microbial consortia and the external environment, and are widely used in natural relations simulation. Hence there is a phenomenon that with the increase of species, principles for design and analysis of microbial consortia are becoming more specific and novel, but the mechanisms are less understood, and the strains are more inclined to wild type rather than genetically engineered in application. Several challenges need to be overcome for revealing the details of multispecies consortia; the foremost of these is to optimize media compositions that satisfy all the multiple species, which greatly influence population dynamics and metabolism. Moreover, we need to take advantage of design and computational tools that take biological variability, uncertainty and evolution into account to provide rational guidance. Microbial consortia are recognized as an immense treasure trove. Plentiful key crop species and their natural relatives have not yet been studied for their related bacterial populations and their expending or impeding effects. With an approximate number of 500,000 plant species, a lot of work lies ahead for plant microbiome research to explore new aspects of plant growth and disease resistance.

References

Akhtar, M.S. and Siddiqui, Z.A. (2008). Biocontrol of a root-rot disease complex of chickpea by *Glomus intraradices*, *Rhizobium* sp. and *Pseudomonas straita*. *Crop Protection* 27 (3–5): 410–417.

Akila, R., Rajendran, L., Harish, S. et al. (2011). Combined application of botanical formulations and biocontrol agents for the management of *Fusarium oxysporum* f. sp. *cubense* (Foc) causing *Fusarium* wilt in banana. *Biological Control* 57 (3): 175–183.

Alizadeh, H., Behboudi, K., Ahmadzadeh, M. et al. (2013). Induced systemic resistance in cucumber and *Arabidopsis* thaliana by the combination of *Trichoderma harzianum* Tr6 and *Pseudomonas* sp. Ps14. *Biological Control* 65 (1): 14–23.

Anuroopa, N. and Bagyaraj, D.J. (2017). Inoculation with selected microbial consortia enhanced the growth and yield of *Withania somnifera* under polyhouse conditions. *Imperial Journal of Interdisciplinary Research* 3 (2): 127–133.

Attia, M., Awad, N.M., Turky, A.S., and Hamed, H.A. (2011). Induction of defense responses in soybean plants against *Macrophomina phaseolina* by some strains of plant growth promoting rhizobacteria. *Journal of Applied Sciences Research* 7 (11): 1507–1517.

Bakker, P.A., Berendsen, R.L., Doornbos, R.F. et al. (2013). The rhizosphere revisited: root microbiomics. *Frontiers in Plant Science* 4: 165.

Bardas, G.A., Lagopodi, A.L., Kadoglidou, K., and Tzavella-Klonari, K. (2009). Biological control of three *Colletotrichum lindemuthianum* races using Pseudomonas chlororaphis PCL1391 and *Pseudomonas fluorescens* WCS365. *Biological Control* 49 (2): 139–145.

Bashan, Y. and Levanony, H. (1990). Current status of *Azospirillum* inoculation technology: Azospirillum as a challenge for agriculture. *Canadian Journal of Microbiology* 36 (9): 591–608.

de Boer, M., Bom, P., Kindt, F. et al. (2003). Control of *Fusarium* wilt of radish by combining *Pseudomonas putida* strains that have different disease-suppressive mechanisms. *Phytopathology* 93 (5): 626–632.

Chauhan, H. and Bagyaraj, D.J. (2015). Inoculation with selected microbial consortia not only enhances growth and yield of French bean but also reduces fertilizer application under field condition. *Scientia Horticulturae* 197: 441–446.

Denancé, N., Sánchez-Vallet, A., Goffner, D., and Molina, A. (2013). Disease resistance or growth: the role of plant hormones in balancing immune responses and fitness costs. *Frontiers in Plant Science* 4: 155.

Ding, M.Z., Song, H., Wang, E.X. et al. (2016). Design and construction of synthetic microbial consortia in China. *Synthetic and Systems Biotechnology* 1 (4): 230–235.

Duffy, B.K., Simon, A., and Weller, D.M. (1996). Combination of *Trichoderma koningii* with fluorescent pseudomonads for control of talk-all on wheat. *Phytopathology* 86 (2): 188–194.

Dutta, S., Mishra, A.K., and Kumar, B.D. (2008). Induction of systemic resistance against fusarial wilt in pigeon pea through interaction of plant growth promoting rhizobacteria and rhizobia. *Soil Biology and Biochemistry* 40 (2): 452–461.

Fishal, E.M.M., Meon, S., and Yun, W.M. (2010). Induction of tolerance to Fusarium wilt and defense-related mechanisms in the plantlets of susceptible berangan banana pre-inoculated with *Pseudomonas* sp. (UPMP3) and *Burkholderia* sp.(UPMB3). *Agricultural Sciences in China* 9 (8): 1140–1149.

Guetsky, R., Shtienberg, D., Elad, Y. et al. (2002). Improving biological control by combining biocontrol agents each with several mechanisms of disease suppression. *Phytopathology* 92 (9): 976–985.

Hahm, M.S., Sumayo, M., Hwang, Y.J. et al. (2012). Biological control and plant growth promoting capacity of rhizobacteria on pepper under greenhouse and field conditions. *Journal of Microbiology* 50 (3): 380–385.

Harish, S., Kavino, M., Kumar, N. et al. (2009). Induction of defense-related proteins by mixtures of plant growth promoting endophytic bacteria against Banana bunchy top virus. *Biological Control* 51 (1): 16–25.

He, Z.L. and Yang, X.E. (2007). Role of soil rhizobacteria in phytoremediation of heavy metal contaminated soils. *Journal of Zhejiang University. Science. B* 8 (3): 192–207.

Jain, A., Singh, S., Kumar Sarma, B., and Bahadur Singh, H. (2012). Microbial consortium–mediated reprogramming of defence network in pea to enhance tolerance against *Sclerotinia sclerotiorum*. *Journal of Applied Microbiology* 112 (3): 537–550.

Jain, A., Singh, A., Singh, S., and Singh, H.B. (2015). Biological management of *Sclerotinia sclerotiorum* in pea using plant growth promoting microbial consortium. *Journal of Basic Microbiology* 55 (8): 961–972.

de Jensen, C.E., Percich, J.A., and Graham, P.H. (2002). Integrated management strategies of bean root rot with *Bacillus subtilis* and *Rhizobium* in Minnesota. *Field Crops Research* 74 (2–3): 107–115.

Jetiyanon, K. (2007). Defensive-related enzyme response in plants treated with a mixture of *Bacillus* strains (IN937a and IN937b) against different pathogens. *Biological Control* 42 (2): 178–185.

Kannan, V. and Sureendar, R. (2009). Synergistic effect of beneficial rhizosphere microflora in biocontrol and plant growth promotion. *Journal of Basic Microbiology* 49 (2): 158–164.

Kavino, M., Harish, S., Kumar, N. et al. (2007). Rhizosphere and endophytic bacteria for induction of systemic resistance of banana plantlets against bunchy top virus. *Soil Biology and Biochemistry* 39 (5): 1087–1098.

Kloepper, J.W., Schroth, M.N., and Miller, T.D. (1980a). Effects of rhizosphere colonization by plant growth-promoting rhizobacteria on potato plant development and yield. *Phytopathology* 70 (11): 1078–1082.

Kloepper, J.W., Leong, J., Teintze, M., and Schroth, M.N. (1980b). Enhanced plant growth by siderophores produced by plant growth-promoting rhizobacteria. *Nature* 286 (5776): 885.

Kloepper, J.W., Lifshitz, R., and Zablotowicz, R.M. (1989). Free-living bacterial inocula for enhancing crop productivity. *Trends in Biotechnology* 7 (2): 39–44.

Latha, P., Anand, T., Prakasam, V. et al. (2011). Combining Pseudomonas, Bacillus and Trichoderma strains with organic amendments and micronutrient to enhance suppression of collar and root rot disease in physic nut. *Applied Soil Ecology* 49: 215–223.

Lucas, J.A., Solano, B.R., Montes, F. et al. (2009). Use of two PGPR strains in the integrated management of blast disease in rice (*Oryza sativa*) in Southern Spain. *Field Crops Research* 114 (3): 404–410.

Lugtenberg, B. and Kamilova, F. (2009). Plant-growth-promoting rhizobacteria. *Annual Review of Microbiology* 63: 541–556.

Maiyappan, S., Amalraj, E.L.D., Santhosh, A., and Peter, A.J. (2010). Isolation, evaluation and formulation of selected microbial consortia for sustainable agriculture. *Journal of Biofertilizers & Biopesticides* 2 (109): 2.

Marimuthu, S., Ramamoorthy, V., Samiyappan, R., and Subbian, P. (2013). Intercropping system with combined application of *Azospirillum* and *Pseudomonas fluorescens* reduces root rot incidence caused by *Rhizoctonia bataticola* and increases seed cotton yield. *Journal of Phytopathology* 161 (6): 405–411.

Mathivanan, N., Prabavathy, V.R., and Vijayanandraj, V.R. (2005). Application of talc formulations of *Pseudomonas fluorescens* Migula and *Trichoderma viride* Pers. ex SF Gray decrease the sheath blight disease and enhance the plant growth and yield in rice. *Journal of Phytopathology* 153 (11–12): 697–701.

Mueller, U.G. and Sachs, J.L. (2015). Engineering microbiomes to improve plant and animal health. *Trends in Microbiology* 23 (10): 606–617.

Muthukumar, A., Bhaskaran, R., and Sanjeevkumar, K. (2010). Efficacy of endophytic *Pseudomonas fluorescens* (Trevisan) migula against chilli damping-off. *Journal of Biopesticides* 3 (Special Issue): 105.

Nadeem, S.M., Naveed, M., Zahir, Z.A., and Asghar, H.N. (2013). Plant–microbe interactions for sustainable agriculture: fundamentals and recent advances. In: *Plant Microbe Symbiosis: Fundamentals and Advances* (ed. N.K. Arora), 51–103. New Delhi: Springer.

Raja, P., Uma, S., Gopal, H., and Govindarajan, K. (2006). Impact of bio inoculants consortium on rice root exudates, biological nitrogen fixation and plant growth. *Journal of Biological Sciences* 6: 815–823.

Rajasekar, S. and Elango, R. (2011). Effect of microbial consortium on plant growth and improvement of alkaloid content in *Withania somnifera* (Ashwagandha). *Current Botany*.

Rajasekhar, L., Satish, K.S., and Divya, J. (2016). Evaluation of microbial consortium for 'plant health management' pigeon pea. *International Journal of Plant, Animal and Environmental Sciences* 6 (2): 107–113.

Rudresh, D.L., Shivaprakash, M.K., and Prasad, R.D. (2005). Effect of combined application of Rhizobium, phosphate solubilizing bacterium and *Trichoderma* spp. on growth, nutrient uptake and yield of chickpea (*Cicerariteniun* L.). *Applied Soil Ecology* 28 (2): 139–146.

Sarma, B.K., Yadav, S.K., Singh, S., and Singh, H.B. (2015). Microbial consortium-mediated plant defense against phytopathogens: readdressing for enhancing efficacy. *Soil Biology and Biochemistry* 87: 25–33.

Schoebitz, M., López, M.D., Serrí, H. et al. (2016). Combined application of microbial consortium and humic substances to improve the growth performance of blueberry seedlings. *Journal of Soil Science and Plant Nutrition* 16 (4): 1010–1023.

Sheth, R.U., Cabral, V., Chen, S.P., and Wang, H.H. (2016). Manipulating bacterial communities by in situ microbiome engineering. *Trends in Genetics* 32 (4): 189–200.

Siddiqui, Y. and Meon, S. (2009). Effect of seed bacterization on plant growth response and induction of disease resistance in chilli. *Agricultural Sciences in China* 8 (8): 963–971.

Siddiqui, Z.A. and Singh, L.P. (2005). Effects of fly ash, *Pseudomonas striata* and *Rhizobium* on the reproduction of nematode Meloidogyne incognita and on the growth and transpiration of pea. *Journal of Environmental Biology* 26 (1): 117–122.

Singh, K. (2011). Organic amendments to soil inoculated arbuscular mycorrhizal fungi and *Pseudomonas fluorescens* treatments reduce the development of root-rot disease and enhance the yield of *Phaseolus vulgaris* L. *European Journal of Soil Biology* 47 (5): 288–295.

Singh, B.K. and Trivedi, P. (2017). Microbiome and the future for food and nutrient security. *Microbial Biotechnology* 10 (1): 50–53.

Singh, A., Mehta, S., Singh, H.B., and Nautiyal, C.S. (2003). Biocontrol of collar rot disease of betelvine (*Piper betel* L.) caused by *Sclerotium rolfsii* by using rhizosphere-competent *Pseudomonas fluorescens* NBRI-N6 and *P. fluorescens* NBRI-N. *Current Microbiology* 47 (2): 0153–0158.

Singh, A., Sarma, B.K., Upadhyay, R.S., and Singh, H.B. (2013). Compatible rhizosphere microbes mediated alleviation of biotic stress in chickpea through enhanced antioxidant and phenylpropanoid activities. *Microbiological Research* 168 (1): 33–40.

Song, H., Ding, M.Z., Jia, X.Q. et al. (2014). Synthetic microbial consortia: from systematic analysis to construction and applications. *Chemical Society Reviews* 43 (20): 6954–6981.

Srinivasan, K. and Mathivanan, N. (2009). Biological control of sunflower necrosis virus disease with powder and liquid formulations of plant growth promoting microbial consortia under field conditions. *Biological Control* 51 (3): 395–402.

Srinivasan, K. and Mathivanan, N. (2011). Plant growth promoting microbial consortia mediated classical biocontrol of sunflower necrosis virus disease. *Journal of Biopesticides* 4 (1): 65.

Srivastava, R., Khalid, A., Singh, U.S., and Sharma, A.K. (2010). Evaluation of arbuscular mycorrhizal fungus, fluorescent *Pseudomonas* and *Trichoderma harzianum* formulation against *Fusarium oxysporum* f. sp. *lycopersici* for the management of tomato wilt. *Biological Control* 53 (1): 24–31.

Stockwell, V.O., Johnson, K.B., Sugar, D., and Loper, J.E. (2011). Mechanistically compatible mixtures of bacterial antagonists improve biological control of fire blight of pear. *Phytopathology* 101 (1): 113–123.

Sundaramoorthy, S., Raguchander, T., Ragupathi, N., and Samiyappan, R. (2012). Combinatorial effect of endophytic and plant growth promoting rhizobacteria against wilt disease of *Capsicum annum* L. caused by *Fusarium solani*. *Biological Control* 60 (1): 59–67.

Sundaramoorthy, S., Karthiba, L., Raguchander, T., and Samiyappan, R. (2013). Ecofriendly approaches of potential microbial bioagents in management of sheath rot disease in rice caused by *Sarocladium oryzae* (Sawada). *Plant Pathology Journal* 12: 98–103.

Tilman, D., Cassman, K.G., Matson, P.A. et al. (2002). Agricultural sustainability and intensive production practices. *Nature* 418 (6898): 671.

Trivedi, P., Schenk, P.M., Wallenstein, M.D., and Singh, B.K. (2017). Tiny microbes, big yields: enhancing food crop production with biological solutions. *Microbial Biotechnology* 10 (5): 999–1003.

Vacheron, J., Desbrosses, G., Bouffaud, M.L. et al. (2013). Plant growth-promoting rhizobacteria and root system functioning. *Frontiers in Plant Science* 4: 356.

Vanegas, J. and Uribe-Vélez, D. (2014). Selection of mixed inoculants exhibiting growth-promoting activity in rice plants from undefined consortia obtained by continuous enrichment. *Plant and Soil* 375 (1-2): 215–227.

Vestberg, M., Kukkonen, S., Saari, K. et al. (2004). Microbial inoculation for improving the growth and health of micropropagated strawberry. *Applied Soil Ecology* 27 (3): 243–258.

Vishwakarma, K., Sharma, S., Kumar, N. et al. (2016). Contribution of microbial inoculants to soil carbon sequestration and sustainable agriculture. In: *Microbial Inoculants in Sustainable Agricultural Productivity* (eds. D.P. Singh et al.), 101–113. New Delhi: Springer.

Vishwakarma, K., Mishra, M., Jain, S. et al. (2017a). Exploring the role of plant-microbe interactions in improving soil structure and function through root exudation: a key to sustainable agriculture. In: *Plant-Microbe Interactions in Agro-Ecological Perspectives* (eds. D.P. Singh et al.), 467–487. Singapore: Springer.

Vishwakarma, K., Sharma, S., Kumar, V. et al. (2017b). Current scenario of root exudate–mediated plant-microbe interaction and promotion of plant growth. In: *Probiotics in Agroecosystem* (eds. D.P. Singh et al.), 349–369. Singapore: Springer.

Yadav, G., Vishwakarma, K., Sharma, S. et al. (2017). Emerging significance of rhizospheric probiotics and its impact on plant health: current perspective towards sustainable agriculture. In: *Probiotics and Plant Health* (eds. V. Kumar et al.), 233–251. Singapore: Springer.

Yang, P., Sun, Z.X., Liu, S.Y. et al. (2013). Combining antagonistic endophytic bacteria in different growth stages of cotton for control of *Verticillium* wilt. *Crop Protection* 47: 17–23.

Yang, W., Zheng, L., Liu, H.X. et al. (2014). Evaluation of the effectiveness of a consortium of three plant-growth promoting rhizobacteria for biocontrol of cotton *Verticillium* wilt. *Biocontrol Science and Technology* 24 (5): 489–502.

Zhang, S., White, T.L., Martinez, M.C. et al. (2010). Evaluation of plant growth-promoting rhizobacteria for control of *Phytophthora* blight on squash under greenhouse conditions. *Biological Control* 53 (1): 129–135.

Zlotnikova, A.K., Kazakova, M.L., Zlotnikov, K.M. et al. (2007). Physiological and biochemical properties of the bacterial association of *Klebsiella terrigena* E6 and *Bacillus firmus* E3. *Applied Biochemistry and Microbiology* 43 (3): 304.

9

Rhizomicrobiome for Sustainable Crop Growth and Health Management

Tualar Simarmata, Mieke R. Setiawati, Betty N. Fitriatin, and Diyan Herdiyantoro

Department of Soil Sciences and Land Resources Management, Faculty of Agriculture of Padjadjaran University, Jatinangor, West Java, Indonesia

9.1 Introduction

Soil as the most magnificent living system, an excellent natural medium and major source of nutrients for plant growth. It harbors an amazingly diverse community of tiniest bacteria, algae, fungi, and protozoa, to the more complex nematodes and micro-arthropods, to the visible earthworms, insects, small vertebrates, and plants (Ingham 2001; NCRS 2004; Moravec et al. 2009). They form a unique or specific relationship and make up a community of organisms living all or part of their lives in the soil, known as the soil food web or energy flow in the soil ecosystem (Ingham 2001). The soil food web is the community of organisms living all or part of their lives in the soil and the structure of relations among the organisms within an ecosystem based on what each consumes. Primary producers consume water, minerals, carbon dioxide, and a few other things to produce organic matter, which is consumed by most of the rest of the creatures that are, in turn, consumed by others (Ingham 2001; Powell and Kalironomos 2007). In addition, soil shelters many animals from invertebrates such as worms and insects to mammals like rabbits, rodents, and badgers (Sullivan 2001; NCRS 2004).

The healthy and fertile soil ecosystem contains about 10^8–10^9 of bacteria, 10^7–10^8 of Actinomycetes, 10^5–10^6 of fungi, 10^4–10^5 of algae, 10^3–$10^4$of protozoa and 10^2–10^3 of nematode per gram of soil (Ingham 2001; Fitter et al. 2005; Hoorman 2010; Bhattarai et al. 2015).

The total fresh biomass in a hectare of living topsoil contains approximately 1008 kg of earthworms, 2688 kg of fungi, 1680 kg of bacteria, 149 kg of protozoa, 997 kg of arthropods and algae, and even small mammals in some cases (Bollen 1959; Sullivan 2001). Another study revealed that the highest percentage of population are belong to bacteria, actinomycetes, and molds (Table 9.1). Agricultural soil that has about 4 per cent of organic matter contain about 8–5 tons biomass of bacteria, fungi, protozoa, nematodes, earthworms, and arthropods has the following count of soil organisms (Sullivan 2001). In contrast, the

The Plant Microbiome in Sustainable Agriculture, First Edition. Edited by Alok Kumar Srivastava, Prem Lal Kashyap and Madhumita Srivastava.

Table 9.1 Typical numbers of soil organisms in different ecosystems and total fresh biomass of organisms in upper topsoil of fertile soils (Bollen 1959; Ingham 2001; Bhattarai et al. 2015).

	Agriculture Land	Prairie	Forest	Fresh biomass	
Organism	**Organisms per gram (teaspoon) of soil**			**kg ha^{-1}**	**%**
Bacteria	10^8–10^9	10^8–10^9	10^8–10^9	1120	13,8
Actinomycetes	10^7–10^8	10^7–10^8	10^7–10^9	1120	13,8
Fungi	Several yards	10s–100's of yards	1–40 miles (in conifers)	2240	27,6
Algae	10^4–10^5			112	1,4
Protozoa	1000's	1000's	100 000's	224	2,8
Nematodes	10–20	10's–100's	100's	56	0,7
	Organisms per square foot				
Arthropods	<100	500–2000	10 000–25 000	112	1,4
Earthworms	5–30	10–50	10–50	1120	13,8
Plant roots				2000	24,7

intensive agricultural soils has a relative low population of beneficially organisms and relative higher the population of a harm microbes, particularly soil-borne diseases (Simarmata et al. 2016).

The healthy soils are tilled, fertilized, and mixed or bioturbated by soil organisms. Native forest soils or soil ecosystems that are covered with a layer of organic material of plant litter and/or growing plants are rich in biodiversity and complexity of soil organisms, and decompose plant residue and dead roots, releasing their stored nutrients slowly over time, known as mineralization. The fertile soils may contain up to a hundred earthworms per shovel, while the degraded soil contain only none to a few per shovel (NCRS 2004). Therefore, the soil can be viewed as a living community and complex system or sophisticated ecosystems (Ingham 2001; Sullivan 2001; Moravec et al. 2009; Simarmata 2013).

The highest population and biodiversity of organisms are found in the area of the root system, known as rhizosphere and the bacteria that occupy the rhizosphere are called the rhizobacteria, the community of organisms being called the rhizomicrobiome (Mendes et al. 2013; Bashir et al. 2016). The term rhizosphere is from the Greek words "*rhiza*" which means root and "*sphere*" which means field or area of influence (Hiltner 1904; Hartmann et al. 2008).The rhizosphere as the key habitat of soil organisms is a rich zone of highly available nutrients, acting as a junction for nutrient exchange between plants, soil, and microbes (Hinsinger et al. 2009; Mendes et al. 2011; Bashir et al. 2016) and as the hot spot and harbor for the beneficially microbial community. The activity and biodiversity of the microbial community composition of the rhizosphere are highly dependent on root exudates (Dias et al. 2012; Aranda et al. 2013).The rhizosphere as an effective root zone may contain up to 10^{11} microbes cells per gram of root and more than 30,000 prokaryotic species (Manoharachary and Mukerji 2006; Mendes et al. 2011; Parmar and Dufresne 2011). The secreted or exudated organic substances have a great influences on microorganism activity and biodiversity or community in the rhizosphere. Generally, the root exudates are affected by plant species or cultivars, and with plant age and development. The exudates consist of proteins, carbohydrates, organic acids, vitamins, nucleotides, flavonoids, enzymes, hormones, volatile compounds phenolics and secondary metabolites (Chaparro et al. 2012; Huang et al. 2014; Pii et al. 2015).

Briefly, the rhizosphere can be divided into the zones of endorhizosphere (the internal root parts), rhizoplane (the root surface), and ectorhizosphere (thin layer of soil adhering to the roots) (Clark 1949; Lynch 1989; Pinton et al. 2000). The zone of endorhizosphere consists of root tissues including cortical cells and the endodermis and the zone rhizoplane is the area of root surface where soil microbes and soil particles interact. The rhizoplane comprises the cortex, epidermis, and mucilage. The zone of ectorhizosphere is formed from soil particles adjacent to the roots (Figure 9.1). Bulk soil is the portion of soil which is not the component of the rhizosphere (Lambers et al. 2008; Brundrett 2009; Bashir et al. 2016) (Figure 9.1). In addition, the rhizoplane is referred to as the surface of the plant tissues in contact with the soil (i.e. roots and rhizomes) and microbes that can exist in an adherent form to the plant tissues are termed epiphytes, while the microbial genomes located inside plant tissues are called endophytes (Bulgarelli et al. 2013).

The rhizosphere is an excellent habitat of an enormous microbial community and in which soil as a medium and complex signaling occurs among microbes and plants, accomplished by exudates, creating zones of interaction between roots and the surrounding soil (Egamberdieva et al. 2008; Mendes et al. 2011); this is known as the rhizomicrobiome. The rhizomicrobiome is composed of a diversified microbial community with specific functions apart from promoting plant growth, and is highly influenced by the plant type, soil type and environmental conditions. The function of the rhizomicrobiome can be compared to microbial communities in the human gut (Turnbaugh et al. 2007; Bernedsen and Pieterse 2012) (Table 9.2). Similar to human, the intestinal microbial communities have a highly significant influence on soil and plant health. Consequently, a comprehensive understanding of the characteristics, importance, and role of the rhizomicrobiome are the key factors for sustainable agriculture.

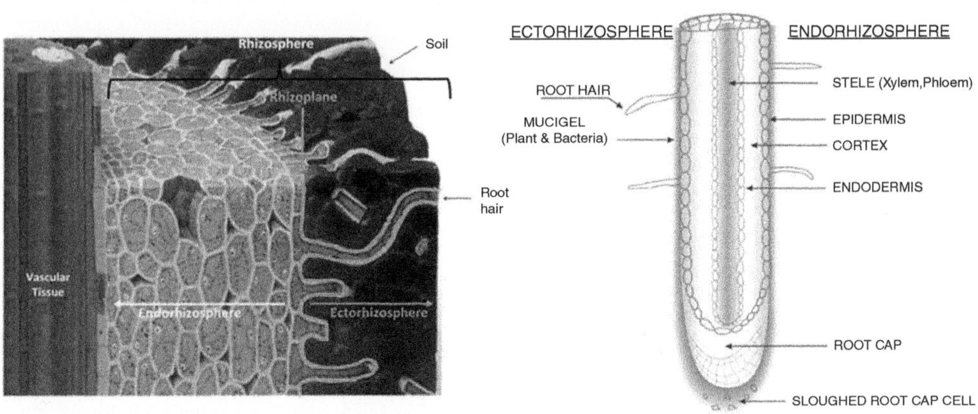

Figure 9.1 The rhizosphere area consists of endorhizosphere, rhizoplane and ectorhizosphere (Fageria and Stone 2006; Bashir et al. 2016).

Table 9.2 Similarities of the microbiomes of the human gut and plant roots (Bernedsen and Pieterse 2012).

Characteristic	The human gut microbiome	The rhizosphere microbiome
Important for nutrient uptake	Microbiota assist in the breakdown of dietary products and produce essential nutrients, such as vitamins B and D. In return, the microbiome is provided with a steady supply of carbon in the form of mucins (Derrien et al. 2010; Fagundes et al. 2012).	Mycorrhiza and rhizobia assist plants with the uptake of phosphorus and nitrogen, respectively. Furthermore, the microbiome assists in the weathering of minerals and the degradation of recalcitrant organic matter. In return, the microbiome is provided with carbon in root exudates and other rhizodeposits (Bais et al. 2006; Van Der Heijden et al. 2008).
Prevent colonization by pathogens	Mechanisms include competition for nutrients or for adhesion receptors, production of inhibitory metabolites and modulation of toxin production or action (Fagundes et al. 2012).	Mechanisms include competition for (micro) nutrients, production of antibiotic compounds or lytic enzymes and consumption of pathogen stimulatory compounds (Lugtenberg and Kamilova 2009; Doornbos et al. 2012).
Modulate host immunity	Host innate immune systems adapt to colonization by microflora and shift to a primed state that not only affects the intestinal mucosa, but can also regulate immune response in respiratory mucosa. Furthermore, development of the gut microflora in the first year of life is of crucial importance in the development of the immune system and for susceptibility to the development of disease later in life (Ichinohe et al. 2011; Bron et al. 2012; Fagundes et al. 2012).	Many beneficial soil-borne microorganisms have been found to boost systemically the defensive capacity of the plant. This ISR is a state in which the plant immune system is primed for accelerated activation of defense (Van der Ent et al. 2009).
Distinguish friend from foe	As symbionts and pathogens express similar molecular patterns that are recognized by the innate immune system, it is not entirely known how the innate immune system distinguishes friend from foe. Multiple mechanisms are present to avoid aberrant activation of the immune system (e.g. physical barrier provided by mucus, induced desensitization of epithelial cells to bacterial lipopolysaccharide and low levels of pathogen receptors on apical surface of epithelial cells (Lathrop et al. 2011; Chinen and Rudensky 2012).	Symbionts and pathogens express similar molecular patterns that are recognized by the innate immune system, and it is largely unknown how plants distinguish friend from foe. Both pathogens and beneficials are also known to suppress plant immune response to promote their own colonization through secretion of effector molecules (Zamioudis and Pieterse 2012).
	Many commensal microbes are potentially pathogenic but are controlled by the host immune system; commensals can become pathogenic in immunity-impaired mice.	Both pathogens and beneficials are also known to suppress plant immune response to promote their own colonization through secretion of effector molecules.

Table 9.2 (Continued)

Characteristic	The human gut microbiome	The rhizosphere microbiome
	The adaptive immune system is trained to be tolerant of commensals. Regulatory T-cells not only suppress immune response to self, but are also educated to suppress immune responses to the commensal gut microbiota.	
Microbiome density and diversity	Although microbial density is high, with typically 10^{11}–10^{12} microbial cells per ml of intestinal fluid, the phylogenetic diversity is relatively low. Only seven of the 55 described bacterial phyla are found in the human gut, which is dominated by Firmicutes and Bacteroidetes. It is estimated that some 500–1000 species of bacteria exist in the human gut (Ley et al. 2006; Arumugam et al. 2011).	In the rhizosphere, the microbial density is typically higher than in bulk soil and ranges from 10^8 to 10^9 bacteria per gram. However, soil microbial communities are considered to hold the most diverse microbial communities in the world, with up to 10^4 bacterial species per gram (Roesch et al. 2007; Weinert et al. 2011).
	There is an indication that intestinal microbial variation between individuals is stratified rather than continuous and that there is a limited amount of classifiable communities that can exist in the gut, coined "enterotypes."	Root microbiomes of plants grown in the same soil differ between plant species and between genotypes within a species. However, the existence of classifiable "rhizotypes" has not yet been reported.

9.2 Characteristics of the Rhizomicrobiome

9.2.1 Microbial Community and Soil Food Web

The microbial community residing in the rhizomicrobiome includes bacteria, fungi, archaea, algae, nematodes, protozoa, virus, oomycetes, and microarthropods (Lynch 1990; Buee et al. 2009; Raaijmakers et al. 2009; Mendes et al. 2013; Chaparro et al. 2014). Based on their functions in the rhizobiome, they can be grouped as photosynthesizers, decomposers, mutualists, grazers and predators, pathogens, parasites, root feeders, bacterial feeders, fungal feeders, shredders, and high level predators (Figure 9.2; Table 9.3) (Ingham 2004). Briefly, Ingham (2004) presented the group or community that plays a great role in the rhizomicrobiome of the soil ecosystem system as follows:

- **Photosynthesizers** consists of green plants, algae, and bacteria that have the capability to capture and convert the solar energy to chemical energy through the photosynthetic process to organic substance.
- **Decomposers** consist of bacteria, actinomycetes (filamentous bacteria) and saprophytic fungi which degrade plant and animal residue, organic compounds and some pesticides.
- **Grazers and predators** are protozoa, mites, nematodes and other organisms which "graze" on bacteria or fungi; prey on other species of protozoa and nematodes; or both graze and prey. Grazers and predators release plant-available nutrients as they consume microbes. Often organisms specialize in one type of prey, such as either bacteria or fungi. Certain collembolans (springtails) even specialize on specific species of fungi. Other organisms are generalists and will feed on any microbial species they encounter.

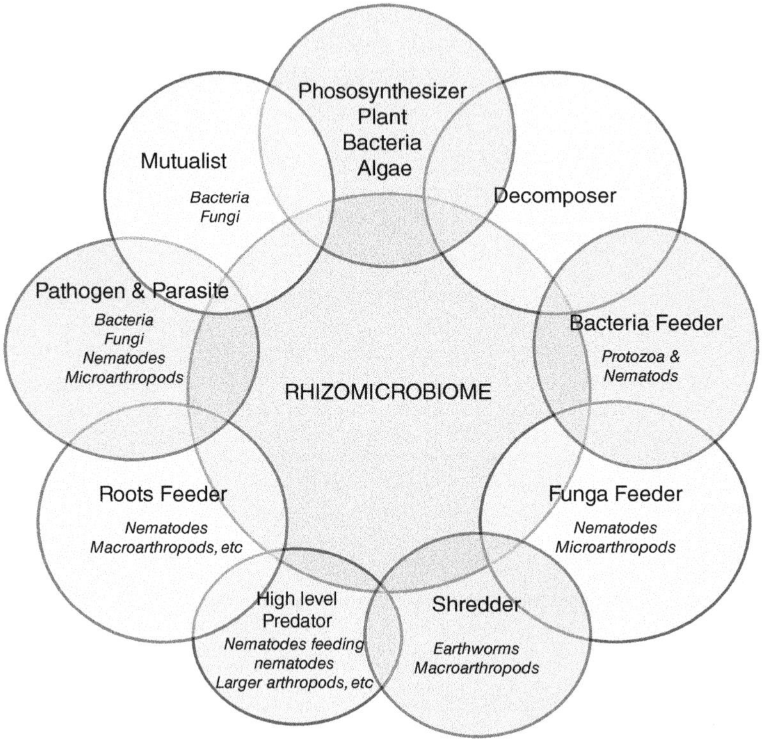

Figure 9.2 The diversity of the microbial community in the rhizomicrobiome.

Table 9.3 Major community of organisms and functions in the rhizomicrobiome.

Type of soil organisms		**Major functions**
Photosynthesizers	• Plants • Algae • Bacteria	Capture energy • Use solar to fix CO_2 • Add organic matter to soil (biomass such as dead cells, plant litter, and secondary metabolites
Decomposers	• Bacteria • Fungi	Break down residue • Immobilize (retain) nutrients in their biomass • Create new organic compounds (cell constituent, waste product) that are sources of energy and nutrients for other organisms • Produce compound that help to bind soil into aggregates • Bind soil aggregates with fungal hyphae • Nitrifying and denitrifying bacteria convert forms of nitrogen • Compete with disease causing organism

Table 9.3 (Continued)

Type of soil organisms		Major functions
Mutualists	• Bacteria • Fungi	Enhance plant growth • Protect plant roots from disease causing organisms • Some bacteria fix N_2 • Some fungi from mycorrhiza associations with roots and deliver nutrients (such as P) and water to plant
Pathogens Parasites	• Bacteria • Fungi • Nematodes • Microarthropods	Promote disease • Consume roots and other plants part, causing diseases • Parasite nematodes or insect, including diseases causing organisms
Root feeders	• Nematodes • Macroarthropods (e.g. cutworm, weevil larvae and symphylans)	Consume plants roots • Potentially cause significant crop yield losses
Bacterial feeders	• Protozoa • Nematodes	Graze • Release plant available nitrogen (NH_4^+) and other nutrients when feeding on bacteria • Control many roots feeding or disease causing pests • Stimulate and control the activity of bacterial population
Fungal feeders	• Nematodes • Microarthropods	Graze • Release plant available nitrogen (NH_4^+) and other nutrients when feeding on fungi • Control many roots feeding or disease causing pests • Stimulate and control the activity of fungal population
Shredders	• Earthworms • Macroarthropods	Break down residue and enhance soil structure • Shred plant litter as they feed on bacteria and fungi • Provide habitat for bacterial in their guts and fecal pellets • Enhance soil structure as they produce fecal pellets and burrow through soil
Higher-level predators	Nematodes feeding nematodes, larger arthropods, mice, voles, shrews, birds, other above ground animals	Control populations • Control the population of lower trophic level predators • Larger organisms improve soil structure by burrowing and by passing soil through guts • Larger organisms carry smaller organisms long distance

Source: adopted from Ingham (2001), NCRS (2004).

- **Mutualists** belong to mycorrhizal fungi, nitrogen-fixing bacteria, and some free-living microbes which have co-evolved together with plants to form mutually beneficial associations with plants. Mycorrhizae are associations between fungi and plant roots in which the fungus supplies nutrients and perhaps water to the plant, and the plant supplies food to the fungi. These fungi can exist inside (endomycorrhizae) or outside (ectomycorrhizae) the plant root cell wall. The common arbuscular mycorrhizae (AM or vesicular-arbuscular mycorrhiza (VAM) fungi) are endomycorrhizae.
- **Pathogens, parasites, and root feeders** are organisms that caused disease and make up a tiny fraction of the organisms in the soil, but have been most studied by researchers. Disease-causing organisms include certain species of bacteria, fungi, protozoa, nematodes, insects, and mites.
- **Litter transformers** are arthropods (invertebrates) with jointed legs, including insects, spiders, mites, springtails, centipedes, and millipedes. Many soil arthropods shred and consume plant litter and other organic matter, increasing the surface area accessible to decomposers. The organic matter in their fecal pellets is frequently more physically and chemically accessible to microbes than was the original litter. Some litter transformers, especially ants, termites, scarab beetles, and earthworms are "ecosystem engineers" that physically change the soil habitat for other organisms by chewing and burrowing through the soil. Microbes (decomposers) living within their guts break down the plant residue, dung, and fecal pellets consumed along with the soil.

Rhizomicrobiome as the community of organisms living all or part of their lives in the soil form a food web of energy and nutrients flow as one organism eats another (Figure 9.3).

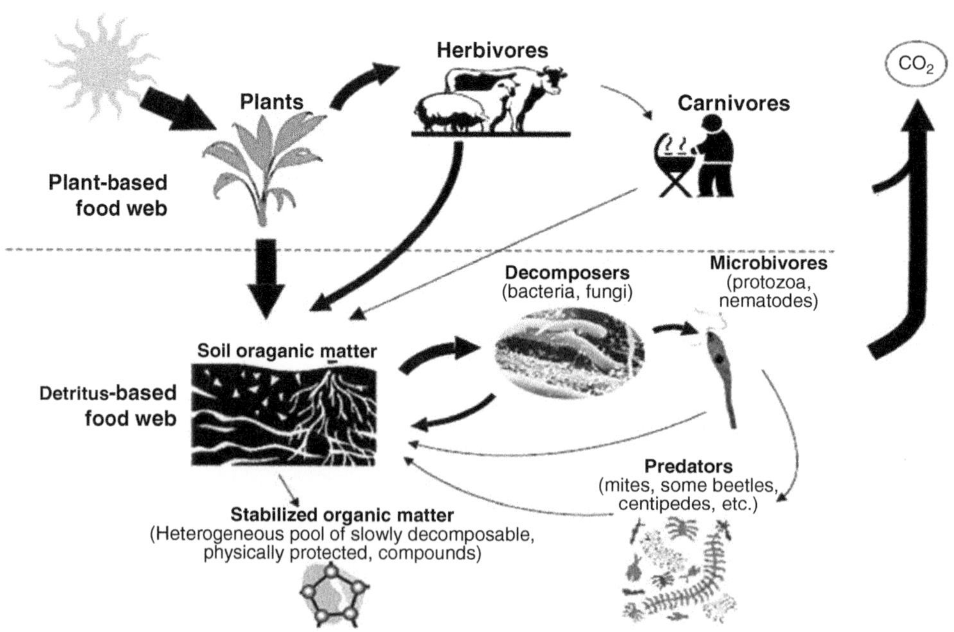

Figure 9.3 The food web and energy flow (arrows represent energy flow) of the plant-based (above-ground) and detritus-based (below-ground) in the soil ecosystem. (*Source:* adapted from NCRS (2004) based on Chapin et al. (2011)).

All food webs are fueled by the primary producers (the plants, lichens, moss, photosynthetic bacteria, and algae) that use the sun's energy to fix carbon dioxide from the atmosphere. Most other soil organisms get energy and carbon by consuming the organic compounds found in plants, other organisms, and waste by-products. A few bacteria, called chemoautotrophs, get energy from nitrogen, sulfur, or iron compounds rather than carbon compounds or the sun (Ingham 2004; NCRS 2004). Consequently, the organic material that is rich in chemical energy is the entry point of energy to the soil ecosystem.

9.2.2 Microbial Interactions in Rhizomicrobiome

The rhizosphere microbiome (rhizomicrobiome) as part of a complex food web and energy flow in the soil ecosystem is characterized by a sophisticated interaction and interrelationship among the organisms. The main source of energy as a major driving force originates as rhizodeposits released by the plant (e.g. exudates, border cells, mucilage) and these play an important role in regulating microbial diversity and activity in the rhizosphere ecosystem (Mendes et al. 2013). Possibly, the plants may modulate the rhizosphere microbiome to their benefit by selectively stimulating microorganisms with traits that are beneficial to plant growth and health (Cook et al. 1995). Others have argued that exudates are passively "released" as overflow/waste products of the plant (Hartmann et al. 2009; Jones et al. 2009; Dennis et al. 2010). In return the soil microbe may provide essential growth substances (phytohormone) and service to facilitate or improve the nutrient availability. In addition, intensive interaction occurs among the soil microbe and other organisms in soils. Briefly, the rhizomicrobiome may directly and indirectly influence the microbial species richness belowground and plant diversity and productivity above-ground (De Deyn et al. 2004; Powell and Kalironomos 2007; Van Der Heijden et al. 2008; Lau and Lennon 2011; Wagg et al. 2011).

The possible community interaction in the rhizomicrobiome can be grouped into beneficial or positive effect (mutualism, commensalism and protocooperation), and harmful or negative (amensalism, antagonism, competition, parasitism and predation) (Atlas and Bartha 1993; Paul and Clark 1996; Mendes et al. 2013) (Table 9.4). Briefly, mutualism (symbiosis association) is reciprocal benefit to both partners, in commensalism one organism benefits and the other is neither harmed nor helped (neutral), and amensalism is the negative impact of one organism on another based on release of a specific compound. Mendes et al. (2013) proposed a diagram of possible interaction of microbial community in the rhizomicrobiome (Figure 9.4) and diversity of bacterial taxa (Table 9.5). Briefly, the rhizosphere microorganisms based on their impact on plant health are categorized as the good (beneficial impact), bad (plant pathogenic) and ugly (human pathogenic). The beneficial rhizosphere microorganisms promote plant growth and protect plants from pathogen attack by a range of mechanisms (Lugtenberg and Kamilova 2009; Raaijmakers et al. 2009). These organisms are commonly used as biofertilizers (N-fixer, phosphate solubilizing microbe, potassium solubilizing bacteria [KSB], plant growth promoting rhizobacteria [PGPR], mycorrhizal fungi, etc.) and biocontrol agents (*Trichoderma* sp. *and Gliocladium* sp., etc.) (Simarmata 2013; Srivastava and Singh 2017).

The bad organisms cause major yield reductions in the production of food, feed, fiber, and fuel crops and include soil-borne plant pathogens, such as nematodes and fungi, including the true fungi and the fungal-like oomycetes. In temperate climates, plant pathogenic fungi, oomycetes, and nematodes are agronomically more important than plant

Table 9.4 The possible microbials between A organism and B organism interaction in the rhizomicrobiome (+ = beneficial, − = detrimental or negative and 0 = no effect).

Category	Interaction	Org A	Org B	Schematic
Positive Interaction	Mutualism (Symbioss)	+	+	Obigatory 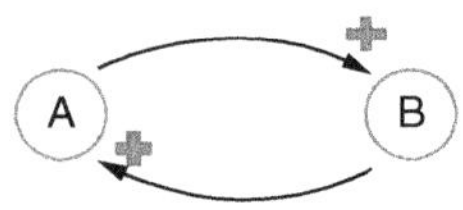
	Proto-Cooperation	+	+	Not obigatory 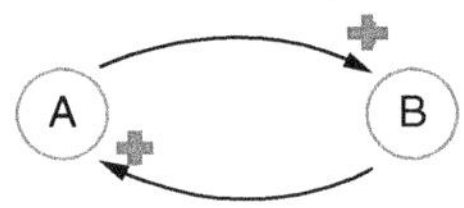
	Commensalism	0	+	
Negative Interaction	Amensalism	0	−	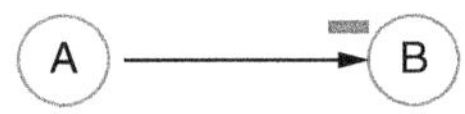
	Parasitism	+	−	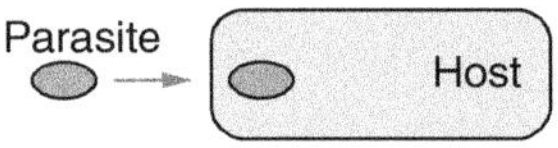
	Predation	+	−	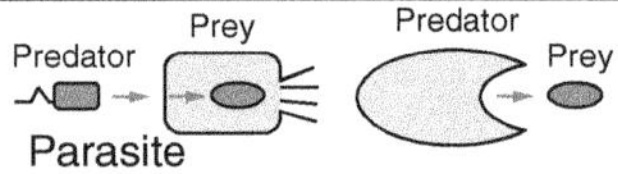
	Cannibalism	+	−	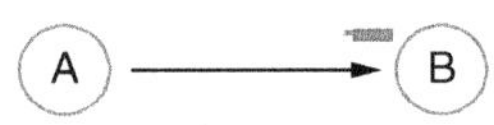
	Competitions	−	−	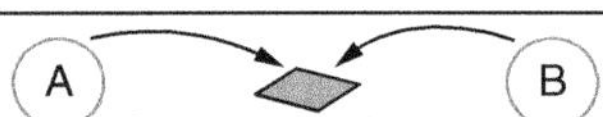

pathogenic bacteria, although some bacterial genera (i.e. *Pectobacterium and Ralstonia*) can cause substantial economic damage in some crops. In addition, the bacterial pathogens that infect plants via roots (*Agrobacterium tumefaciens*, *Ralstonia solanacearum*, *Dickeya dadanthi* and *Dickeya solani*, and *Pectobacterium carotovorum* and *Pectobacterium atrosepticum*) are among the top 10 most notorious (Mansfield et al. 2012). Also, viruses can infect plants via the roots but require vectors such as nematodes or zoosporic fungi to enter the root tissue (Campbell 1996; Macfarlane 2003). The ugly organisms cause disease outbreaks linked to human pathogen contamination of fresh plant produce and are a growing concern worldwide (Berg et al. 2005; Van Baarlen et al. 2007; Tyler and Triplett 2008; Whipps et al. 2008; Holden et al. 2009; Teplitski et al. 2009; Critzer and Doyle 2010).

Recent studies revealed that microbial diversity of soil and rhizosphere microbiomes is highly underestimated. Next-generation sequencing technologies have demonstrated that

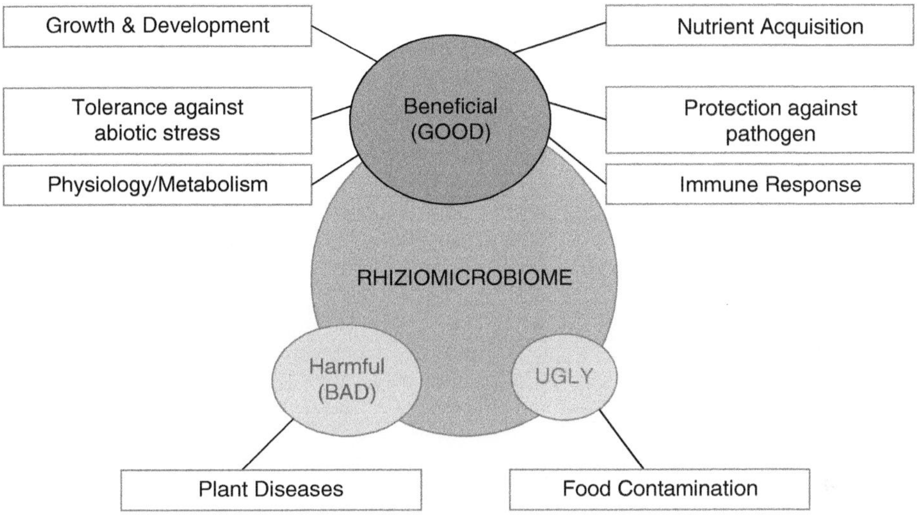

Figure 9.4 Schematic overview of microbial community interaction (the good is beneficial microbe, the bad is plant pathogenic and the ugly is human pathogenic microorganisms) in the rhizomicrobiome (Mendes et al. 2013).

Table 9.5 Number of bacterial and archaeal taxa identified in the rhizosphere microbiome (Mendes et al. 2013).

Host	Approach	Main findings related to rhizosphere microbiome composition
Erica andevalensis in a naturally metal-enriched and extremely acidic environment	16S rRNA gene clone library	Bacteria: of 101 sequenced clones, the majority was affiliated with the Actinobacteria (38 clones; 12 OTUs) followed by the Acidobacteria (21 clones; 10 OTUs), and Proteobacteria (18 clones; eight OTUs). Archaea: considering 27 clones, the community was composed by Crenarchaeota (21 clones; four OTUs) and Euryarchaeota (6 clones; two OTUs (Mirete et al. 2007).
Maize crop	nifH Cluster I clone library	*Azospirillum*, *Bradyrhizobium*, and *Ideonella* were the most abundant genera found in the rhizosphere, comprising c. 5, 21, and 11% of the clones, respectively. The portion of unidentified bacteria was of 27% (Roesch et al. 2007).
Oat microcosms	16S rRNA gene microarray	A total of 1917 taxa were detected, and the community was dominated by Proteobateria and Firmicutes. Less expected rhizosphere-competent phyla were also detected, including Actinobacteria, Verrucomicrobia, and Nitrospira (DeAngelis et al. 2009).
Deschampsia antarctica and *Colobanthus quitensis* in the Arctic	16S rRNA gene pyrosequencing	Firmicutes was the most abundant group found, and Acidobacteria was rarely detected. The predominant genera found were *Bifidobacterium* (phylum Actinobacteria), *Arcobacter* (phylum Proteobacteria), and *Faecalibacterium* (phylum Firmicutes) (Teixeira et al. 2010).

(Continued)

Table 9.5 (Continued)

Host	Approach	Main findings related to rhizosphere microbiome composition
Oak in a forest soil	16S rRNA gene pyrosequencing	In one of the rhizosphere samples, 5619 OTUs were identified in the bacterial community. The predominant phyla were Proteobacteria (38%), Acidobacteria (24%), and Actinobacteria (11%). A high proportion of unclassified bacteria (20%) were observed (Uroz et al. 2010).
Sugar beet in agricultural soil	16S rRNA gene microarray	A total of 33 346 bacterial and archaeal OTUs were detected, and the community was dominated by Proteobacteria (39%), Firmicutes (20%), and Actinobacteria (9%). The Gamma- and Betaproteobacteria and Firmicutes were identified as the most dynamic taxa associated with disease suppression (Mendes et al. 2011).
Potato in field soil	16S rRNA gene microarray	A total of 2432 OTUs were detected in at least one of the samples. The highest number of OTUs belonged to the Proteobacteria (46%), followed by Firmicutes (18%), Actinobacteria (11%), Bacteroidetes (7%), and Acidobacteria (3%). The bacterial families Streptomycetaceae, Micromonosporaceae, and Pseudomonadaceae showed the strongest response at the potato cultivar level (Weinert et al. 2011).
Rhizophora mangle and *Laguncularia racemose* in mangrove	Archaeal 16S rRNA gene pyrosequencing	About 300 archaeal OTUs were identified. Four classes were found: Halobacteria, Methanobacteria, Methanomicrobia, and Thermoprotei (Pires et al. 2012).
Potato in field soil	Pyrosequencing	A total of 55,121 OTUs were found. Actinobacteria and Alphaproteobacteria were the most abundant groups, followed by Gammaproteobacteria, Betaproteobacteria, Acidobacteria, Gemmatimonadetes, Firmicutes, Verrucomicrobia, Deltaproteobacteria, Cyanobacteria, Bacteriodetes, and the TM7 group (Ichinohe et al. 2011).
Rhizophora mangle in mangrove	16S rRNA gene pyrosequencing	Proteobacteria was the most abundant phylum in all samples covering 36–40% of the total sequencing reads (Gomes et al. 2010).
Mammillaria carnea (cactus) in semi-arid environment	16S rRNA gene pyrosequencing	Dominant bacterial groups were Acidobacteria, Actinobacteria, Proteobacteria, and Bacteroidetes (Aguirre-Garrido et al. 2012).
Arabidopsis thaliana in Cologne and Golm soils	16S rRNA gene pyrosequencing	About 1000 OTUs were estimated in the rhizosphere and 1000 OTUs in root compartments. The rhizosphere was dominated by Acidobacteria, Proteobacteria, Planctomycetes, and Actinobacteria. Proteobacteria, Actinobacteria, and Bacteroidetes were found as dominant phyla in root bacterial communities and significantly enriched compared with soil and rhizosphere (Bulgarelli et al. 2013).

Table 9.5 (Continued)

Host	Approach	Main findings related to rhizosphere microbiome composition
Arabidopsis thaliana in Mason farm and Clayton soils	16S rRNA gene pyrosequencing	18 783 bacterial OTUs were firstly detected, and 778 measurable OTUs were used for analysis. The rhizosphere microbiome was dominated by Proteobacteria, Bacteroidetes, Actinobacteria, and Acidobacteria. The endophytic compartment was dominated by Actinobacteria, Proteobacteria, and Firmicutes and was depleted of Acidobacteria, Gemmatimonadetes, and Verrucomicrobia (Lundberg et al. 2012).

OTUs, operational taxonomic units.

only a minority (up to 5%) of bacteria have been cultured by current methodologies and that a significant proportion of the bacterial phyla detected by these technologies has no cultured representative yet. The diversity of bacterial taxa in the rhizosphere depending on the techniques used range from <100 to more than 55 000 operational taxonomic units (OTUs) (Table 9.5) (Mendes et al. 2013).

9.2.3 The Role and Management of Rhizomicrobiome for Plants Health

As described before, the function of the rhizomicrobiome is like a gut for the plant (Turnbaugh et al. 2007). Consequently, the diversity of the microbial community and their activity are highly correlated with soil and plant health. The domination of beneficial organisms (the good) in the rhizomicrobiome will provide an excellent media and nutrient availability to support plant growth and development optimally. In contrast, the domination of harmful organisms will suppress or give negative impact on plant health. Therefore, the main strategy and effort are to maintain the soil food web and energy flow running efficiently in the rhizosphere microbiome. It is well-known that the beneficial microbial community promotes plant growth and protects plants from pathogen attack by a range of mechanisms (Lugtenberg and Kamilova 2009; Raaijmakers et al. 2009). In addition, members of the rhizosphere microbiome can significantly influence the nutrient status of plants. Well-known examples are the nitrogen-fixing rhizobia to increase the N availability and the mycorrhizal fungi that facilitate phosphorus uptake (Hawkins et al. 2000; Richardson et al. 2009; Miransari 2011). Various other nitrogen-fixing bacterial genera living in the rhizosphere have been identified (Zehr et al. 2003; Gaby and Buckley 2011). Based on glasshouse experiments and 16S rRNA gene sequencing, they indicated that *Bradyrhizobium*, *Rhizobium*, *Burkholderia*, and *Achromobacter* species were able to nodulate cowpea and were efficient in biological nitrogen fixation (Guimarães et al. 2012). Moreover, rhizosphere microorganisms can also facilitate the uptake of specific trace elements such as iron. Due to the scarcity of available iron in many microbial habitats as well as the toxicity of free iron at elevated concentrations, bacteria employ a variety of mechanisms to regulate intracellular iron concentrations by secretion of siderophores (Lindsay and Schwab 1982; Andrews et al. 2003; Buckling et al. 2007; Hider and Kong 2010).

It is also well known that microbial community of rhizobacteria produce metabolites that inhibit the growth or activity of competing microorganisms. Rhizosphere fungi are prolific producers of antibiotic metabolites (Hoffmeister and Keller 2007; Brakhage and Schroeckh 2011). Especially, *Trichoderma* species have received considerable attention for the production of antimicrobial compounds (Elad et al. 1983; Vyas and Mathus 2002; Harman et al. 2004; Mathivanan et al. 2005; Druzhinina et al. 2011). Most fungal and bacterial biocontrol strains produce more than one antibiotic compound with overlapping or different degrees of antimicrobial activity. Interestingly, many antibiotic compounds have different effects on other microorganisms at subinhibitory concentrations, an observation which led to an exciting new direction in research on the natural functions of antibiotics. Other natural functions attributed to antibiotics include a role in defense against predatory protozoa, motility, biofilm formation, and nutrition (Raaijmakers and Mazzola 2012).

Recently, it has become known that the rhizosphere microbiome plays an important role in promoting the induced systemic resistance and other physiological processes to support plant growth under abiotic stress, and contributes to the ability of some plant species to survive under extreme conditions (Jorquera et al. 2012). Therefore, the rhizosphere microbiome has an extremely important role in increasing the availability of nutrients and to facilitate nutrient uptake, the capability against harmful diseases (or as biocontrol agents) and contribute to adaptation against extreme conditions (salt tolerance, drought and flooding, etc.). Briefly, the role of the rhizomicrobiome to enhance soil health, plant growth and health are by: (i) releasing plant major and minor nutrients from insoluble inorganic forms; (ii) decomposing organic residues; (iii) contributing to soil humus formation; (iv) improving plant health through symbiotic or mutualistic association; (v) maintaining soil stability through biogeochemical cycling; and (vi) biological control of plant pathogens, insects, and weeds (Weller et al. 2002; Bhardwaj et al. 2014; Gerbore et al. 2014).

The key points for managing the rhizomicrobiome are as follows:

9.2.3.1 Management of Soil Organic and Energy Flow

The microbial community of the rhizomicrobiome belong to the organotrophs, and they use organic compounds as the main source of energy. Consequently, the sustainable supply of organic substances or organic compounds into the rhizosphere microbiome are the key factors for the healthy rhizomicrobiome (Mendes et al. 2013; Jegan et al. 2016). Mostly agricultural soil has limited availability and accessibility for degradable organic compounds, and becomes a common limiting factor for soil bacteria growth (Aldén et al. 2001; Demoling et al. 2007; Rousk and Bååth 2007). Therefore, the key success to improve and maintain the health of rhizomicrobiome in sustainable ways is primarily depends on integrated soil organic matter (SOM) management. Organic matter is defined as any material that is derived from living organisms (plants, animals, and soil organisms) which consists of both living parts (principally roots, soil organisms) and dead material (principally dead plant parts), including well decomposed humus. The pools of SOM can be divided into four categories: (i) living organisms and roots that makes up less than 5 per cent of the total; (ii) residue from dead plants, animals, and soil organisms that have not yet begun to decompose is less than 10 per cent; (iii) SOM being decomposed rapidly (undergoing rapid decomposition) is about 20–45 percent; and (iv) stabilized organic matter (humus) which is resistant for further decomposition is about 50–80 per cent (Moravec et al. 2009). Organic matter significantly affects the biological, chemical, and physical properties of the soil and its overall health. Properties influenced by organic matter include: soil structure; moisture holding capacity; diversity and activity of soil organisms, both those that are beneficial and harmful to crop production; and nutrient availability. Organic matter also acts as the store house for the energy and nutrients used by plants and other organisms (Ingham 2001; Sullivan 2001). Application of organic fertilizer (cow dung and compost) or other organic ameliorants improve soil health, physical, chemical and biological properties, and various crop productivity significantly (Simarmata et al. 2016; Sudjana et al. 2017). In addition, the application of compost or other organic fertilizers also increases the induced systemic resistance and reduces yield losses significantly (Purwanto et al. 2016; Simarmata et al. 2016).

Briefly, the main functions of organic compounds in soils include: (i) a revolving nutrient bank account and entry point of energy flow (supply) into soils ecosystem. SOM is derived mainly from plant residues and it contains all essential plant nutrients (macro and micronutrients) and chemical energy resulting from photosynthesis. Therefore, accumulated organic matter is a storehouse for nutrients and chemical energy. Either nutrients or energy are released by microbial processes (mineralization or decomposition and oxidation of organic substances). The released nutrients are highly available to plants (plant-available form) and the stable organic fraction (humus) adsorbs and holds nutrients in that form. Furthermore, if the rate of addition of organic matter into soil (plant residue, fertilizers and etc.) is less than the rate of decomposition, SOM will decline, and if the rate of addition is greater than the rate of decomposition, SOM will increase; (ii) an agent to activate and regulate the biological system in soils; and (iii) an agent to improve soil structure, maintain tilth and minimize erosion (Ingham 2001; NCRS 2004). Consequently, organic matter plays an important and vital role in rhizosphere ecosystems and it may be categorized as the heart of the soil ecosystem (Simarmata 2013). Soil as a living system absolutely depends on supply and availability of organic matter which acts as fuel for soil as a bioreactor. From the viewpoint of the life system, organic matter is the entry point of energy into the soil ecosystem for fueling the food web or energy flow in soils (Ingham 2001; NCRS 2004). Thus, low supply or low content of SOM will disturb the energy flow (food web) and leads to unhealthy soils (dead or sick soils). SOM is utilized by soil organisms, mainly as a source of energy. Therefore, SOM is subject to decomposition and highly affected by soil organism activity. Under favorable conditions, such suitable soil moisture content and nutrient availability, especially nitrogen the decomposition, occurs rapidly and it may cause sharp decline of SOM in soils. Consequently, one of the most important challenges is how to maintain the soil organic in safe levels by managing supply and decomposition of SOM in soils. Figure 9.5 shows declining soil health and plant yield, as well as the minimum acceptable area of soil health (threshold). Agricultural practices under continuous management cause the decrease of soil health gradually. In addition, under improper management the soil health will be decreased sharply, and in a short time the fertile and healthy soils turn and become sick or degraded soils (Bunning and Jiménez 2003).

Furthermore, Bunning and Jiménez (2003) state that the threshold value is the lower limit of system performance, at which the land management system will become unsustainable and a high investment will be needed in the recovery of the degraded land (restoration of soil properties and function). The red line is the optimal situation, or target value. In reality, the inputs and outputs vary over time (over weeks, seasons, and years) but should generally balance each other, so that the system oscillates between both limits, maintaining

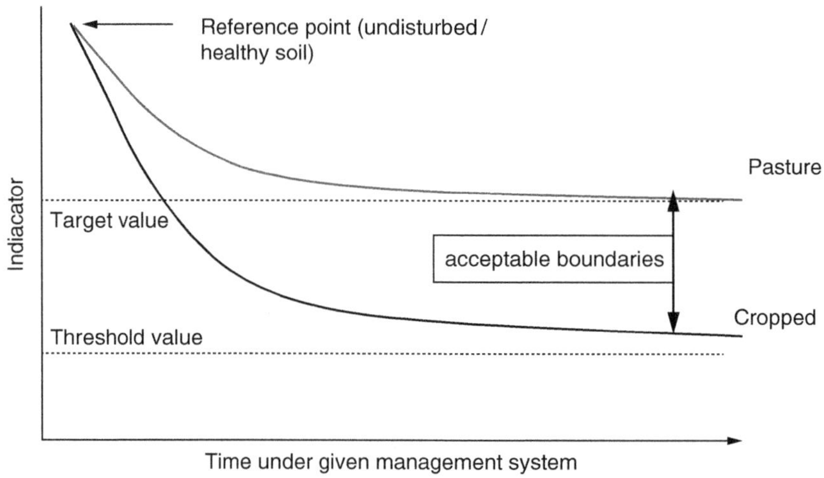

Figure 9.5 A declining yield trend over a period of time under a consistent and continuous land management system, i.e. crops or pastures, indicating a gradual loss of soil health (Bunning and Jiménez 2003).

a relatively constant value. In the case of SOM, the lower limit content of organic carbon for mineral soils is about 2 per cent (Kang et al. 2014; Simarmata et al. 2016). Consequently, the adding of organic material through organic fertilizers, plant residue or green manure is necessary for restoring and improving the soil health. Integrated soil organic management can be adopted to maintain the organic carbon at more than 2 per cent. This includes re-use of crop residue, using organic mulch or cover crops, efficient fertilizer use, crop rotation or multiple crops, conservation tillage and combined with other good agricultural practices. Practically, about 5–10 ton ha^{-1} organic fertilizers (green manure, compost, manure and plant residue) is needed to fuel energy supply into soil ecosystems. This high dosage of organic demand can be secured only by good organic management, such as the recycling process of plant residue surrounding cultivated land.

The practical guidance for soil health and quality management are as follows (NCRS 2004):

1) **Enhance organic matter**. Whether your soil is naturally high or low in organic matter, adding new organic matter every year is perhaps the most important way to improve and maintain soil quality. Regular additions of organic matter improve soil structure, enhance water and nutrient holding capacity, protect soil from erosion and compaction, and support a healthy community of soil organisms. Practices that increase organic matter include: leaving crop residues in the field, choosing crop rotations that include high residue plants, using optimal nutrient and water management practices to grow healthy plants with large amounts of roots and residue, growing cover crops, applying manure or compost, using low or no tillage systems, and mulching.
2) **Avoid excessive tillage**. Reducing tillage minimizes the loss of organic matter and protects the soil surface with plant residue. Tillage is used to loosen surface soil, prepare the seedbed, and control weeds and pests. But tillage can also break up soil structure, speed the decomposition and loss of organic matter, increase the threat of erosion, destroy the habitat of helpful organisms, and cause compaction. New equipment allows crop production with minimal disturbance of the soil.

3) **Manage pests and nutrients efficiently**. An important function of soil is to buffer and detoxify chemicals, but soil's capacity for detoxification is limited. Pesticides and chemical fertilizers have valuable benefits, but they also can harm non-target organisms and pollute water and air if they are mismanaged. Nutrients from organic sources also can pollute when misapplied or over-applied. Efficient pest and nutrient management means testing and monitoring soil and pests; applying only the necessary chemicals, at the right time and place to get the job done; and taking advantage of non-chemical approaches to pest and nutrient management such as crop rotations, cover crops, and manure management.
4) **Prevent soil compaction**. Compaction reduces the amount of air, water, and space available to roots and soil organisms. Compaction is caused by repeated traffic, heavy traffic, or traveling on wet soil. Deep compaction by heavy equipment is difficult or impossible to remedy, so prevention is essential.
5) **Keep the ground covered**. Bare soil is susceptible to wind and water erosion, and to drying and crusting. Ground cover protects soil, provides habitats for larger soil organisms, such as insects and earthworms, and can improve water availability. Ground can be covered by leaving crop residue on the surface or by planting cover crops. In addition to ground cover, living cover crops provide additional organic matter, and continuous cover and food for soil organisms. Ground cover must be managed to prevent problems with delayed soil warming in spring, diseases, and excessive build-up of phosphorus at the surface.
6) **Diversify cropping systems**. Diversity is beneficial for several reasons. Each plant contributes a unique root structure and type of residue to the soil. A diversity of soil organisms can help control pest populations, and a diversity of cultural practices can reduce weed and disease pressures. Diversity across the landscape can be increased by using buffer strips, small fields, or contour strip cropping. Diversity over time can be increased by using long crop rotations. Changing vegetation across the landscape or over time not only increases plant diversity, but also the types of insects, microorganisms, and wildlife that live on your farm.

9.2.3.2 Crop Management and Genetic Engineering

Plants play an important role in capturing solar energy or converting the solar radiation to chemical energy in organic compounds through photosynthesis. The organic substances will deliver to other parts of the plant and some release various organic substances (organic acid, protein, amino acid, vitamin, etc. into rhizosphere), known as root exudate (Neumann and Romheld 2000). The composition and the quantity released are highly influenced by the plant type and age. The higher roots exudate, the higher the microbial community in the rhizomicrobiome (Mendes et al. 2013). The sustainability of the beneficial microbial community (the good) depend on the supply of root exudate and other rhizodeposits (Table 9.6). The component and function of root exudates is presented in Table 9.7.

Consequently, crop management is needed to improve the plants' traits through plant breeding or genetic modifications that produce more root exudate. This strategy, is also referred to as "rhizosphere engineering" (Oger et al. 2004; Ryan et al. 2009; Bakker et al. 2012). Unfortunately, the breeding programs are still limited for rhizosphere-related traits and root exudation characteristics (Smith and Goodman 1999; Smith et al. 1999; Rengel 2002;

Table 9.6 Root exudates detected in higher plants (Neumann and Romheld 2000).

Class of compounds	Single components
Sugars	Arabinose, glucose, fructose, galactose, maltose, raffinose, rhamnose, ribose, sucrose, xylose
Amino acids and amides	All 20 proteinogenic amino acids, aminobutyric acid, homoserine, cysrathionine, mugineic acid phytosiderophores (mugineic acid, deoxymugineic acid, hydroxymugineic acid, epi-hydroxymugineic acid, avenic acid, distichonic acid A)
Aliphatic acids	Formic, acetic, butyric, popionic, malic, citric, isocitric, oxalic, fumaric, malonic, succinic, maleic, tartaric, oxaloacetic, pyruvic, oxoglutaric, maleic, glycolic, shikimic, cis-aconitic, trans-aconitic, valeric, gluconic
Aromatic acids	p-Hydroxybenzoic, caffeic, p-coumaric, ferulic, gallic, gentisic, protocatechuic, salicylic, sinapic, syringic
Miscellaneous phenolics	Flavonols, flavones, flavanones, anthocyanins, isoflavonoids
Fatty acids	Linoleic, linolenic, oleic, palmitic, stearic
Sterols	Campestrol, cholesterol, sitosterol, stigmasterol
Enzymes	Amylase, invertase, cellobiase, desoxyruibonuclease, ribonuclease, acid phosphatase, phytase, pyrophosphatase apyrase, peroxidase, protease
Miscellaneous	Vitamins, plant growth regulators (auxins, cytokinins, gibberellins), alkyl sulfides, ethanol, H^+, K^+, Nitrate, Phosphate, HCO_3^-

Wissuwa et al. 2009). Bakker et al. (2012) stated that currently there is no known breeding program that evaluates plant lines for their broad interaction with the soil microbiome. Smith et al. (1999) provided one of the initial studies to investigate the genetic basis in plants for interactions with rhizobacteria that are antagonistic to plant pathogens. They detected significant phenotypic variation among recombinant inbred lines of tomato and made an attempt to identify loci associated with resistance to *Pythium torulosum*, disease suppression by *Bacillus cereus*, and with growth of *B. cereus* on the seed. Some of the genetic engineering has been performed to modify the rhizosphere pH (Gévaudant et al. 2007; Yang et al. 2007), to modulate organic anion efflux (Tesfaye et al. 2001; Anoop et al. 2003; Delhaize et al. 2004; Sasaki et al. 2004; Li et al. 2005), or to facilitate citrate efflux (Furukawa et al. 2007; Wang et al. 2007).

The experimental result of manipulated transgenic plants secreted specific signal molecules to communicate with microorganisms in the rhizosphere (Oger et al. 1997; Oger et al. 2004). The release of chemical compounds may interfere with bacterial quorum sensing (Gao et al. 2003; Bauer and Mathesius 2004). GM potato plants able to interfere with bacterial quorum sensing via expression of a lactonase gene, showed a higher level of resistance to the bacterial plant pathogen *P. carotovorum* (Dong et al. 2001). Transgenic potato plants with increased production of 5-O-glucosyltransferase (Lorenc-Kukula et al. 2005) and pectate lyase (Wegener 2001) also showed increased resistance to *P. carotovorum*. Another bottleneck is the potentially high variability of root exudation across environments, soil types, and plant physiological conditions such as growth stage (Sandnes et al. 2005; Dessureault-Rompré et al. 2006; Dessureault-Rompré et al. 2007; Phillips et al. 2008; Shi et al. 2011).

Table 9.7 Component and function of roots exudate (Jones et al. 2004).

Component	Function	Component	Function
Phenolics	• Nutrient source • Chemoattractant signals to microbes • Microbial growth promoters • *Nod* gene inducers/ inhibitors in rhizobia • Resistance inducers against phytoalexins • Chelators of poorly soluble mineral • Nutrients (e.g. Fe) • Detoxifiers of Al • Phytoalexins against soil pathogens	Root border cells	• Produce signals that control mitosis • Produce signals controlling gene expression • Stimulate microbial growth • Release chemoattractants • Synthesize defense molecules for the rhizosphere • Act as decoys that keep root cap infection free • Release mucilage and proteins
Organic acids	• Nutrient source • Chemoattractant signals to microbes • Chelators of poorly soluble mineral nutrients • Acidifiers of soil • Detoxifiers of Al • *Nod* gene inducers	Enzymes	• Catalysts for P release from organic molecules • Biocatalysts for organic matter transformation in soil
Amino acids and phytosiderophores	• Nutrient source • Chelators of poorly soluble mineral nutrients • Chemoattractant signals to microbes	Purines	• Nutrient source and nutrient uptake
Vitamins	• Promoters of plant and microbial growth • Nutrient source	Sugars	• Nutrient source • Promoters of microbial growth

9.2.3.3 Engineering of the Rhizomicrobiome with Biofertilizers

The composition and structure of the microbial community in the rhizomicrobiome can be influenced directly by introducing beneficial microorganisms or inoculation into soils as seed treatments or soil dressing and indirectly by stimulating or promoting indigenous beneficial rhizosphere microorganisms through soil or crop management (organic manure application and crop rotation, etc.). Manipulating the rhizosphere microbiome by introducing beneficial microorganisms that protect the host plant against pathogen infections is in many ways comparable with the use of probiotics in humans. The effective beneficial microorganisms should proliferate and survive in the rhizomicrobiome and attain cell densities above a specific threshold density at a time and place that is critical for pathogen infection. When introduced into new environments, however, many microbial strains do not survive or cannot establish densities in the rhizosphere that are necessary to control soil-borne pathogens (Raaijmakers et al. 2009). One approach to meet these criteria is to develop so-called designer probiotics, which are genetically engineered microbial strains (Paton et al. 2006; Picard and Bosco 2008; Picard et al. 2008; Berlec 2012).

The domination of beneficial microbial communities in the rhizomicrobiome during crop growth play an important part in: (i) biochemical transformation and nutrient availability; (ii) root growth and development; and (iii) nutrients. Consequently, plant growth and productivity are highly correlated with management or engineering to keep and maintain the rhizomicrobiome in balance and in healthy condition. In addition, engineering the composition of the microbial community in the rhizomicrobiome is done by inoculation of microbial fertilizers which are well-known as biofertilizers (Singh and Purohit 2011; Board and NIIR 2012; Simarmata 2013; Bhardwaj et al. 2014). Biofertilizers as low cost, renewable sources of plant nutrients and environmentally friendly fertilizers contain a large population of specific or groups of beneficial microorganisms for enhancing the productivity of soil. They do this either by fixing atmospheric nitrogen, by solubilizing soil phosphorus, by stimulating plant growth through synthesis of growth promoting substances or latent cells that activate the biological process. These are rendered to form a fertilizer compound, to make the unavailable form of elements to be available, or to facilitate nutrient availability for plants (Gupta and Roget 2004; Singh and Purohit 2011; Simarmata 2013; Jehangir et al. 2017).

Briefly, biofertilizers (Table 9.8) include: (i) nitrogen fixing microorganisms; (ii) phosphate solubilizing microorganisms; (iii) cellulolytic microorganisms (decomposer); (iv) PGPR; and (v) mycorrhizal fungi.

The research results have revealed that the application of biofertilizers under proper application may increase the productivity of various crops by 25–50 per cent and reduce the inorganic fertilizers by 25 per cent (Simarmata 2013). Rhizobia are well-known as symbiotically N_2-fixing bacteria of legume plants and able to contribute up to 90% of the N requirements of the host (Singh and Purohit 2011).The free-living N-fixing bacteria species, known as non-symbiotically rhizobacteria with non-legume crops (maize, wheat, rice, canola, tomatoes, papers, etc.) are able to enhance N uptake of plants and they contribute up to 25 per cent of plant N requirement. In addition, they are also produce a relative high amount of phytohormone (IAA, gibberellins [GAs]) which are important for the cell division and roots growth (Purwanto et al. 2016; Manasa et al. 2017). These bacteria are well known as PGPR (Table 9.9). Moreover, Cyanobacteria (*Anabaena*, *Aulosira*, and *Nostoc*), as free-living or in symbiosis with Azolla (floating fresh water fern) are able to fix the N up to 100–300 kg N ha^{-1} and produce the relative high amount of Azolla biomass up to 10–15 ton ha^{-1} which it can be used as organic fertilizers (Board and NIIR 2012). In addition, biofertilizers can improve soil health and provide protection against drought and some soil-borne diseases (Subba Rao 1988; Singh and Purohit 2011). Moreover, some beneficial microbes act as biocontrol agents in controlling pathogens (Jetiyanon and Kloepper 2002). Mendes et al. (2011) detected more than 33,000 bacterial and archaeal species in the rhizosphere of sugar beet plants grown in a soil suppressive to *Rhizoctonia* damping-off disease. In addition, soils suppressive to black root rot of tobacco showed that multiple bacterial genera, including *Azospirillum*, *Gluconacetobacter*, *Burkholderia*, *Comamonas*, *Sphingomonadaceae*, and the fluorescent

Table 9.8 The contribution of beneficial microbes used as a biofertilizers (Subba Rao 1988: Singh and Purohit 2011, Simarmata 2013).

Microorganism	Contribution (ha year^{-1})	Suitable crops
Bacterial nitrogen biofertilizers		
• *Rhizobium* sp.	50–300 kg N	Pulse or legume crops
• *Bradyrhizobium* sp.	50–100 kg N	Soybean
• *Azotobacter* sp.	1–20 kg N	Maize, rice, and sugarcane
• *Azosprillum* sp.	10–20 kg N	Maize, rice, sugarcane, and sorghum
• *Azorhizobium caulidans*	50–100 kg	Sersbania
Algal biofertilizers		
• Blue green algae (BGA)	25 kg N	Rice and maize
• *Azolla–Anabaena* association	300–900 kg N	Paddy rice
Phosphate biofertilizers		
• PSB (*Bacillus* sp. *and Pseudomonas* sp.)	Solubilize about 50% soil P	All crops
• PSF (*Aspergillus* sp. *and Penicillium* sp.)	Solubilize about 50% soil P	All crops
Mycorrhizal biofertilizers		
• Endomycorrhizae (AMF: *Glomus* sp., *Gigaspora* sp., *Acaulospora* sp., *Scutellospora* sp. and *Sclerocystis* sp.)	Solubilize of 60% P in soils	All crops
• Ectomycorrizae: *Laccaria* sp., *Pisolithus* sp., *Boletus* sp., *Amanita* sp.	Solubilize of 60% P in soils	Tree crops
Actinomycetes biofertilizers		
• Frankia	150 kg N	Tree crops (e.g. Casuarina)
PGPR biofertilizers		
Pseudomonas, Bacillus, Azotobacter and Azospirillum	Growth substances or phytohormone producers in rhizosphere; indole-3-acetic acid (IAA), giberrelic acid	All crops

PSB = phosphorus solubilizing bacteria, FSF = phosphorus solubilizing fungi, AMF = arbuscular mycorrhizal fungi and PGPR = plant growth promoting rhizobacteria.

Pseudomonas, were more prevalent on roots of tobacco plants grown in disease suppressive than in conducive (non-suppressive) soils (Kyselková et al. 2009).

Arbuscular mycorrhizal fungi (AMF) are able to enhance nutrient and water availability. The availability of P in acid soils ecosystem can be increased up to 75 per cent. In addition, AMF are also known as root bio protectors and are highly important in maintaining soil and

Table 9.9 Important plant growth promoting rhizobacteria (PGPR) (Parmar and Dufresne 2011).

PGPR	Agricultural crop	References
Pseudomonas fluorescens PCL1606	Avocado	Cazorla et al. (2006)
P. fluorescens CHA0	*Arabidopsis* sp.	Iavicoli et al. (2003)
Bacillus subtilis FB17	*Arabidopsis thaliana*	Rudrappa et al. (2008)
Collimonas fungivorans	Tomato	Kamilova et al. (2006)
Bacillus subtilis, B. amyloliquefaciens IN 937 *and Enterobacter cloaca*	*Arabidopsis* sp.	Ryu et al. (2003)
P. putida KD	Tomato and cucumber	Rezzonico et al. (2005)
P. fluorescens WCS 365	Tomato	Kamilova et al. (2006)
Comamonas acidovorans	Kiwi	Erturk et al. (2010)
Bacillus cereus UW 85	Grain legumes	Vessey and Buss (2002)
Bradyrhizobium and PGPR	Mungbean	Shaharoona et al. (2006)
Pseudomonas BA-8, *Bacillus* OSU-142, *Bacillus* M-3	Strawberry	Pirlak and Kose (2009)
Agrobacterium amazonense	Rice	Rodrigues et al. (2008)
Bacillus cepacia strain OSU-7	Stored potatoes	Recep et al. (2009)
Pseudomonas brassicacaerum, P. Marginali, P. oryzihabitans, P. putida and *Alcaligenes xylosoxidans*	Indian mustard and rape	Belimov et al. (2007)
Rhizobium sp.	Rice	Puwanto et al. (2016)
Acinetobacter sp. *and Pseudomonas* sp.	Rice	Setiawati et al. (2015)
Pseudomonas mallei, *P. cepaceae, Aspergillus niger* and *Penicillium* sp.	Maize	Fitriatin et al. (2014)

plant health (Brundrett 2009). Recent studies revealed that a wide range of bacterial genera (e.g. *Pseudomonas*, *Burkholderia*, *Acidothiobacillus*, *Bacillus*, and *Paenibacillus*) are able to release potassium from minerals such as mica, illite, muscovite, biotite, and orthoclases (Liu et al. 2014 and with K availability up to 15 per cent (Bhattacharjee and Dey 2014; Supanjani et al. 2006). KSB has been isolated from the rhizosphere of maize using Aleksandrov agar selective media with K-feldspar as a potassium source, namely: J36-1, J25-2, and J15-1 KSB isolates. Isolate KSB J25-2 significantly increased the potassium available, while isolate J36-1 gave the highest maize potassium absorption value compared to other treatments on Inceptisols Jatinangor (Herdiyantoro 2013; Herdiyantoro et al. 2014, 2015).

The current status of biofertilizer application is much lower than its potential. In general, the utilization of inorganic fertilizers reaches a million tons per year, whereas the biofertilizers are used only in small amounts (hundreds of tons per year) in Asian countries (India, Thailand, Indonesia, Philippine) (Singh and Purohit 2011; Simarmata 2013; Bhardwaj et al. 2014).

The main problems are correlated with the availability and accessibility of inoculants at a reasonable price, lack of knowledge and information, quality and life-time of inoculants and other obstacles. Basically, the need of nutrients for plants, such N, P, K, and others can be supplied by using natural resources and optimizing the recycling process. Bacterial nitrogen biofertilizers can be used to fix nitrogen in the air and become available to plants, and may substitute the utilization of chemical fertilizers. The phosphorus solubilizing microorganisms (PSM) and KSB combined with decomposer can be used to provide these nutrients up to 90 per cent of total its requirement. Thus, biofertilizers and other biological process have great prospects for supplying essential plant nutrients in sustainable ways.

9.3 Challenges and Future Research

The analogy concept of rhizosphere microbiome (rhizomicrobiome) as a gut for the plant genome requires new insights to approach and redefine the core rhizosphere microbiome as a home to the microbial community and important aspects of soil and plant health. The comprehensive understanding of major biochemical and chemical transformation of essential elements, biological process of organic compounds, biological interaction and interrelations and chemical reaction in the rhizomicrobiome is urgently required to maintain the energy flow and food web in sustainable ways. The rhizomicrobiome as a gut for plants is highly sensitive and influenced by external factors (climate, edaphic, and plants). In contrast, the human or animal guts are well-protect in the body, and are less influenced by external factors (Turnbaugh et al. 2007; Mendes et al. 2013). The coupling traditional approaches and modern sequencing techniques to assess organismal or community ecology and physiology make it possible to understand the microbial activity and biodiversity in soil ecosystems. Identification of the exudates, signals, and key players in the rhizosphere microbiome will provide chemical and microbial markers to elucidate whether and how plants recruit and stimulate beneficial (micro)organisms. Much research has revealed that the rhizomicrobiome has a key role and contributes to crop growth and soil health (Chaparro et al. 2014; Mendes et al. 2013; Quiza et al. 2015; Rout 2014). They are involved in:

1) Biogeochemical process and transformation, including cycle of nutrients (carbon, nitrogen, phosphate, and other elements) which are essential for nutrient availability and plant growth. Mineralization of organic compounds (organic matter) is the key role to change the nutrient status from non-usable form to usable form (available nutrients) to the plants, while nutrient immobilization is an important process to bind the element in organic compounds.
2) Improving soil physical, chemical, and biological properties which are important for soil fertility, soil health, soil quality and water retention. The improving of soil structure and soil aeration through biological processes, including the production of humic substance are highly important in soil ecosystems.
3) Improving the resistance of plants against the disease by out-competing pathogenic microbes and stimulating complex biochemical plant defenses, including the induced systemic resistance which is an important mechanism to reduce the yield losses and to increase the crop productivity. The microbial community is well-known to act as

bioprotectors in the rhizomicrobiome. Rhizobacteria are reported to influence growth, development, health, and nutrient status of plants under salinity and drought stress through several mechanisms like production or degradation of ethylene, EPS, osmoprotectant, GAs, jasmonic acid (JA), and salicylic acid (SA), and to regulate multiple physiological processes in the plant, including root initiation, elongation, and root hair formation (Bulgarelli et al. 2013; Quiza et al. 2015; Smith et al. 2015). Generally bacteria belonging to different genera including *Rhizobium*, *Bacillus*, *Pseudomonas*, *Pantoea*, *Paenibacillus*, *Burkholderia*, *Achromobacter*, *Serratia*, *Azospirillum*, *Azotobacter*, *Microbacterium*, *Methylobacterium*, *Variovorax*, *Enterobacter*, *Brevibacterium*, *Brachybacterium*, *Acinetobacter*, *and Streptomyces*, etc. have been reported to provide tolerance to host plants under different abiotic stress environments (Grover et al. 2011).

4) Improving the resilience of plants to environmental or abiotic stresses, such as fluctuations in temperature and moisture. The production of slime and some exudates, and to some extent the presence of fungal hyphae as mycorrhizal biofertilizers, are able to increase the root capability or zones up to 10 times to exploit water and nutrients for plant growth and health. Abiotic stress includes salinity, drought, flooding, high and low temperatures, ultraviolet light, air pollution (ozone), and heavy metals, all of which affect the yield, quality, quantity, physiology, and biochemistry of plants. The yield losses associated with abiotic stress is predicted as 50–82 per cent, depending on the crop type and stress period (Christensen et al. 2007).
5) Enhancing root growth and nutrient uptake are included in the essential role of the rhizomicrobiome to enhance plant growth and health. The production of some phytohormones (IAA, GAs, and other growth substances) are highly important to promote the cell division in root growth and development. In addition, the slime produced by microbes is important to support the roots' growth and health in the soil ecosystem. Rhizobacteria stimulate plant growth under harsh conditions like pest and pathogen infection (Somers et al. 2004). Classified PGPR based on their functional activities include: (i) biofertilizers (increasing the availability of nutrients to plant); (ii) phytostimulators (plant growth promotion by phytohormones); (iii) rhizoremediators (degrading organic pollutants); and (iv) biopesticides and thus exhibit multiple functions (Lugtenberg and Kamilova 2009). Certain PGPR isolates produce quorum sensing signaling molecules and induce a plant immune response (Hartmann and Schikora 2012; Viswanath et al. 2015).
6) Endophytes as plant growth promoters under stress. Recent research reveals that apart from the rhizobiome, other plant microbiomes such as endophytes and the phyllosphere bacteria are involved in the induction of plant growth, disease protection and productivity enhancement (Ryan et al. 2008; Delmotte et al. 2009; Dias et al. 2012; Knief et al. 2012; Gopal et al. 2013; Berg et al. 2014).The endophytes colonizing the internal tissues of their host plant constitute a range of different relationships, including symbiotic, mutualistic, commensalistic, and trophobiotic. Endophytic bacteria have been recognized as symbionts with a unique and intimate interaction with the plant (Ryan et al. 2008; Reinhold-Hurek and Hurek 2011) with several beneficial roles in plant growth, soil health and phytoremediation, though they were considered as contaminants for a long time.

Briefly, the rhizomicrobiome plays a huge role in soil heath and quality, plant health and productivity. Therefore, intensive research of rhizomicrobiome is urgently required for better understanding of the traits, biodiversity of the microbial/organism community, structure, and complexity of the microbial community, the energy flow, dynamic food web, and dynamics of the biogeochemical and biological system of rhizomicrobiome. Moreover, a comprehensive understanding and evaluation of the health of rhizomicrobiomes are needed as a basic knowledge and action plan or design to maintain or to improve the health of the rhizomicrobiome in sustainable ways. The research includes:

1) The compressive understanding of genetic and functional capacity of plant diversity or aspects, including important aspects of metabolism and physiology on the microbial community, root exudates and secretion and other aspects related to health and traits of the rhizomicrobiome. Specific plant attributes and environmental factors that increase the plant's ability to select their rhizomicrobiome communities through the exudate currency shared among the plant and its colonizers (Doornbos et al. 2012). One potential approach is to initiate plant breeding programs that are directed towards unraveling the molecular basis of interactions between plant lines and beneficial members of the rhizosphere microbiome (Mendes et al. 2013). This approach of going "back to the roots" will most likely also lead to the identification of new rhizosphere microorganisms, genes, and traits that can be exploited for other applications. To reduce the impact of plant diseases, we propose to design a "core microbiome" that is effective against soil-borne pathogens in different agro-ecosystems.
2) Identification of the compounds present in the root exudates that influence the soil microbial community structure and function would help to build novel strategies for improving plant performance and for increasing crop yield and sustainability. The majority of studies analyze how plant root exudates attract and regulate these microbial interactions, but knowledge is lacking on how specific microbes modulate these interactions, especially at the community level and how root-associated microbial communities influence plant root exudation. Further research is needed to identify the microbial factors influencing the host root exudation process. This will help to develop strategies to engineer microbes to manipulate plant root exudation and in turn the microbial communities in the rhizosphere.
3) The better understanding and knowledge about the dynamic of interactions and how shifts in biodiversity impact functions in the context of ecosystem services (plant productivity, biogeochemical pools, and fluxes) will be a critical factor for elucidating plant microbiome growth and gene expression patterns.
4) Formulation and production consortia of beneficial microbial as biofertilizers and biocontrol agents may be able to increase the nutrient supply and plant growth, and to control the soil-borne diseases. Developing an appropriate technology for mass production of biofertilizers and to make access easy to users at a reasonable price will need to be worked out.

9.4 Conclusion

The rhizomicrobiome as the gut of the plant is a highly complex microbial community that influences the plant growth and health require a comprehensive and holistic study to investigate dynamic of diversity of microbial and its relation to soil health and the plant growth. The health of rhizomicrobiome as a part of soil food web or energy in soil ecosystem are highly depend on roots exudate, SOM and plant management. The rhizomicrobiome enhance the soil health, plant growth and health by releasing the major nutrients (macro- and micro-elemental) from insoluble inorganic forms or non-usable form to usable form, decomposing organic residues, contributing to soil humus formation, improving plant health through symbiotic or mutualistic association (maintaining soil stability through biogeochemical cycling and biological control of plant pathogens, insects, and weeds). A comprehensive and holistic study for better understanding of dynamic of microbial community (soil microbiome diversity and function), and its traits under different plants or soils ecosystem are urgently require to develop the new technologies and products such as biofertilizers and biopesticides. These can be exploited from natural alliances among beneficial microbes and plants to enhance the soil health, plant productivity and stress tolerance in sustainable ways. In addition, the plant breeding with root systems that selectively encourage beneficial crop-microbe associations to create the health rhizomicrobiome. Moreover, the research to optimize soil management, crop management (crop rotation and intercropping) are need to enhance beneficial soil microbiome diversity and function in rhizomicrobiome and to develop the methods to evaluate the health of the rhizomicrobiome which can be used as key factor of soil and crop health management in sustainable ways. The assessment of rhizomicrobiome can be done by adopting the soil health evaluation using the physical, chemical, and biological indicator. Based on the assessment, the health of rhizomicbiome can be categorized as very healthy, healthy, fairly healthy, sick, and very sick (unhealthy).

Acknowledgments

We thank and acknowledge the Universitas Padjadjaran-Bandung-Indonesia for financial support through an Academic Leadership Grant (ALG). Finally, we apologize to those authors whose work could not be discussed because of space limitations.

Glossary

Biocontrol The use of microorganisms to suppress diseases and improve overall plant health.

Biofilms Assemblage of many microbial cells bound together by an adhesive substance (often polysaccharides) and attached to a surface.

Chelate A molecule with multiple binding sites that attaches to a metal ion.

Chemoattractant Compounds released by plants or microorganisms which attract microbes to the plant or direct their movement often toward one another.

Cortex The tissue of a root confined externally by the epidermis (the "skin" of the root) and internally by the endodermis (the tissues encircling the vascular tissues).

Ectorhizosphere The area surrounding (extracellularly) the root where microbes fed by root derived compounds may colonize.

Ectomycorrhizae	A plant-fungus relationship where the fungal hyphae extend into the plant root and occupy the area between the cortical cells, forming a Hartig net.
Endodermis	A single layer of cells enclosing the vascular tissue of the root.
Endomycorrhizae	A plant–fungus relationship where the fungal hyphae extend into the plant root and occupy the area between and within (intracellularly) the cortical cells. Fed by root derived compounds, and may colonize including root cortex, epidermis, and root hairs.
Exudates	Compounds released into the soil/rhizosphere by plant roots.
Induced systemic resistance (ISR)	The protection of the whole plant induced by an external of biotic and abiotc nature.
Mucilage	A high molecular weight mixture of various polysaccharides primarily secreted from the root cap and border cells.
Mycorrhizae	A term used to describe a symbiotic association between root colonizing fungi and plants.
Phytohormones	Plant-derived hormones which regulate physiological processes.
Phytosiderophores	Low molecular weight compounds (e.g., mugienic acid) released from plant roots which chelate Fe, thereby, making it more available for the plant.
Quorum sensing	A phenomenon whereby the gene regulation of an individual microbial cell is influenced in a cell-population density dependent manner through the release of signaling chemicals which increase as the population grows.
Rhizodeposits	Root-derived carbon sources imparted to soils from growing plants.
Rhizomicrobiome	The community of organisms at the rhizosphere.
Rhizoplane	The root surface including associated soil particles.
Rhizosphere	The zone of chemical, biological, and physical influence generated by root growth and activity. The concept usually pertains to the soil–root interface, but is sometimes extrapolated to other media–root interfaces.
Secretions	A subset of exudates which are actively released from the root.
Siderophore	Compounds released from bacteria, fungi, and grasses with an extremely high affinity for iron.

References

Aguirre-Garrido, J.F., Montiel-Lugo, D., Hernández-Rodríguez, C. et al. (2012). Bacterial community structure in the rhizosphere of three cactus species from semi-arid highlands in central Mexico. *Antonie van Leeuwenhoek*: 891–904. https://doi.org/10.1007/s10482-012-9705-3.

Aldén, L., Demoling, F., and Bååth, E. (2001). Rapid method of determining factors limiting bacterial growth in soil. *Applied and Environmental Microbiology* 67 (4): 1830–1838.

Andrews, S.C., Robinson, A.K., and Rodríguez-Quiñones, F. (2003). Bacterial iron homeostasis. *FEMS Microbiology Reviews* 27 (2–3): 215–237.

Anoop, V.M., Basu, U., McCammon, M.T. et al. (2003). Modulation of citrate metabolism alters aluminum tolerance in yeast and transgenic canola overexpressing a mitochondrial citrate synthase. *Plant Physiology* 132 (4): 2205–2217.

Aranda, E., Scervino, J.M., Godoy, P. et al. (2013). Role of arbuscular mycorrhizal fungus *Rhizophagus custos* in the dissipation of PAHs under root-organ culture conditions. *Environmental Pollution* 181: 182–189.

Arumugam, M., Raes, J., Pelletier, E. et al. (2011). Enterotypes of the human gut microbiome. *Nature* 473 (7346): 174.

Atlas, R.M. and Bartha, R. (1993). *Microbial Ecology Fundamentals and Applications*. California: The Benjamins/Cummings Publishing Co., Ins.

Bais, H.P., Weir, T.L., Perry, L.G. et al. (2006). The role of root exudates in rhizosphere interactions with plants and other organisms. *Annual Review of Plant Biology* 57: 233–266.

Bakker, M.G., Manter, D.K., Sheflin, A.M. et al. (2012). Harnessing the rhizosphere microbiome through plant breeding and agricultural management. *Plant and Soil* 360 (1–2): 1–13.

Bashir, O., Khan, K., Hakeem, K.R. et al. (2016). Soil microbe diversity and root exudates as important aspects of rhizosphere ecosystem. In: *Plant, Soil and Microbes* (eds. K.R. Hakeem et al.), 337–357. Cham: Springer.

Bauer, W.D. and Mathesius, U. (2004). Plant responses to bacterial quorum sensing signals. *Current Opinion in Plant Biology* 7: 429–433.

Berlec, A. (2012). Novel techniques and findings in the study of plant microbiota: search for plant probiotics. *Plant Science* 193: 96–102.

Berg, G., Eberl, L., and Hartmann, A. (2005). The rhizosphere as a reservoir for opportunistic human pathogenic bacteria. *Environmental Microbiology* 7 (11): 1673–1685.

Berg, G., Grube, M., Schloter, M., and Smalla, K. (2014). Unraveling the plant microbiome: looking back and future perspectives. *Frontiers in Microbiology* 5: 148.

Bernedsen, R.L., Pieterse, C.M.J., and Bakker, A.H.M. (2012). The rhizosphere microbiome and plant health. *Trends in Plant Science* 17: 478–486.

Bhardwaj, D., Ansari, M.W., Sahoo, R.K., and Tuteja, N. (2014). Biofertilizers function as key player in sustainable agriculture by improving soil fertility, plant tolerance and crop productivity. *Microbial Cell Factories* 13 (1): 66.

Bhattacharjee, R. and Dey, U. (2014). Biofertilizer, a way towards organic agriculture: a review. *African Journal of Microbiology Research* 8 (24): 2332–2343.

Bhattarai, A., Bhattarai, B., and Pandey, S. (2015). Variation of soil microbial population in different soil horizons. *Journal of Microbiology & Experimentation* 2 (2): 00044.

Belimov, A.A., Dodd, I.C., Safronova, V.I. et al. (2007). *Pseudomonas brassicacearum* strain Am3 containing 1-aminocyclopropane-1-carboxylate deaminase can show both pathogenic and growth-promoting properties in its interaction with tomato. *Journal of Experimental Botany* 58 (6): 1485–1495.

Board, N. NIIR (2012). *The Complete Technology Book on Biofertilizers and Organic Farming*. Niir Project Consultansy Services. Kamla Nagar, Delhi-110007 (India). ISBN: 978-93-81039-07-6.

Bollen, W.B., 1959. Microorganisms and Soil Fertility. Oregon State College. Oregon State Monographs. Studies in Bacteriology, (1), p.22.

Brakhage, A.A. and Schroeckh, V. (2011). Fungal secondary metabolites – strategies to activate silent gene clusters. *Fungal Genetics and Biology* 48 (1): 15–22.

Bron, P.A., Van Baarlen, P., and Kleerebezem, M. (2012). Emerging molecular insights into the interaction between probiotics and the host intestinal mucosa. *Nature Reviews Microbiology* 10 (1): 66.

Brundrett, M.C. (2009). Mycorrhizal associations and other means of nutrition of vascular plants: understanding the global diversity of host plants by resolving conflicting information and developing reliable means of diagnosis. *Plant and Soil* 320 (1–2): 37–77.

Buckling, A., Harrison, F., Vos, M. et al. (2007). Siderophore-mediated cooperation and virulence in *Pseudomonas aeruginosa*. *FEMS Microbiology Ecology* 62 (2): 135–141.

Buee, M., De Boer, W., Martin, F. et al. (2009). The rhizosphere zoo: an overview of plant-associated communities of microorganisms, including phages, bacteria, archaea, and fungi, and of some of their structuring factors. *Plant and Soil* 321: 189–212.

Bulgarelli, D., Schlaeppi, K., Spaepen, S. et al. (2013). Structure and functions of the bacterial microbiota of plants. *Annual Review of Plant Biology* 64: 807–838.

Bunning, S. and Jiménez, J.J., 2003. Indicators and assessment of soil biodiversity/soil ecosystem functioning for farmers and governments. In: Proceedings "OECD Expert Meeting on Soil erosion and Soil Biodiversity Indicators", Rome, March 2003 (pp. 1–16).

Campbell, R.N. (1996). Fungal transmission of plant viruses. *Annual Review of Phytopathology* 34 (1): 87–108.

Cazorla, F.M., Duckett, S.B., Bergström, E.T. et al. (2006). Biocontrol of avocado dematophora root rot by antagonistic *Pseudomonas fluorescens* PCL1606 correlates with the production of 2-hexyl 5-propyl resorcinol. *Molecular Plant-Microbe Interactions* 19 (4): 418–428.

Chapin, F.S. III, Matson, P.A., and Vitousek, P. (2011). *Principles of Terrestrial Ecosystem Ecology*. Springer Science & Business Media.

Chaparro, J.M., Sheflin, A.M., Manter, D.K., and Vivanco, J.M. (2012). Manipulating the soil microbiome to increase soil health and plant fertility. *Biology and Fertility of Soils* 48 (5): 489–499.

Chaparro, J.M., Badri, D.V., and Vivanco, J.M. (2014). Rhizosphere microbiome assemblage is affected by plant development. *The ISME Journal* 8 (4): 790.

Chinen, T. and Rudensky, A.Y. (2012). The effects of commensal microbiota on immune cell subsets and inflammatory responses. *Immunological Reviews* 245 (1): 45–55.

Christensen, J.H., Hewitson, B., Busuioc, A. et al. (2007). Regional climate projections. In: *Climate Change, 2007: The Physical Science Basis. Contribution of Working Group I to the Fourth Assessment Report of the Intergovernmental Panel on Climate Change*, 847–940. Cambridge: University Press.

Cook, R.J., Thomashow, L.S., Weller, D.M. et al. (1995). Molecular mechanisms of defense by rhizobacteria against root disease. *Proceedings of the National Academy of Sciences of the United States of America* 92 (10): 4197–4201.

Critzer, F.J. and Doyle, M.P. (2010). Microbial ecology of foodborne pathogens associated with produce. *Current Opinion in Biotechnology* 21 (2): 125–130.

Clark, F.E. (1949). Soil micro-organisms and plant roots communication: acyl-homoserine lactone quorum sensing. *Annual Review of Genetics* 35: 439–468.

Callaway, R.M. and Rout, M.E. (2011). Soil biota and plant invasions: biogeographic effects on plant-microbe interactions. In: *Fifty Years of Invasion Ecology: The Legacy of Charles Elton* (ed. D.H. Richardson), 131–142. West Sussex: Wiley.

De Deyn, G.B., Raaijmakers, C.E., Van Ruijven, J. et al. (2004). Plant species identity and diversity effects on different trophic levels of nematodes in the soil food web. *Oikos* 106 (3): 576–586.

DeAngelis, K.M., Brodie, E.L., DeSantis, T.Z. et al. (2009). Selective progressive response of soil microbial community to wild oat roots. *The ISME Journal* 3 (2): 168.

Demoling, F., Figueroa, D., and Bååth, E. (2007). Comparison of factors limiting bacterial growth in different soils. *Soil Biology and Biochemistry* 39 (10): 2485–2495.

Dennis, P.G., Miller, A.J., and Hirsch, P.R. (2010). Are root exudates more important than other sources of rhizodeposits in structuring rhizosphere bacterial communities? *FEMS Microbiology Ecology* 72 (3): 313–327.

Derrien, M., van Passel, M.W., van de Bovenkamp, J.H. et al. (2010). Mucin-bacterial interactions in the human oral cavity and digestive tract. *Gut Microbes* 1 (4): 254–268.

Dessureault-Rompré, J., Nowack, B., Schulin, R., and Luster, J. (2006). Modified micro suction cup/rhizobox approach for the in-situ detection of organic acids in rhizosphere soil solution. *Plant and Soil* 286 (1–2): 99–107.

Dessureault-Rompré, J., Nowack, B., Schulin, R., and Luster, J. (2007). Spatial and temporal variation in organic acid anion exudation and nutrient anion uptake in the rhizosphere of *Lupinus albus* L. *Plant and Soil* 301 (1–2): 123–134.

Delhaize, E., Ryan, P.R., Hebb, D.M. et al. (2004). Engineering high-level aluminum tolerance in barley with the ALMT1 gene. *Proceedings of the National Academy of Sciences* 101 (42): 15249–15254.

Dias, A.C.F., Taketani, R.G., Andreote, F.D. et al. (2012). Interspecific variation of the bacterial community structure in the phyllosphere of the three major plant components of mangrove forests. *Brazilian Journal of Microbiology* 43 (2): 653–660.

Delmotte, N., Knief, C., Chaffron, S. et al. (2009). Community proteogenomics reveals insights into the physiology of phyllosphere bacteria. *Proceedings of the National Academy of Sciences of the United States of America* 106 (38): 16428–16433.

Dong, Y.H., Wang, L.H., Xu, J.L. et al. (2001). Quenching quorum-sensing-dependent bacterial infection by an N-acyl homoserine lactonase. *Nature* 411 (6839): 813.

Doornbos, R.F., van Loon, L.C., and Bakker, P.A. (2012). Impact of root exudates and plant defense signaling on bacterial communities in the rhizosphere. A review. *Agronomy for Sustainable Development* 32 (1): 227–243.

Druzhinina, I.S., Seidl-Seiboth, V., Herrera-Estrella, A. et al. (2011). *Trichoderma*: the genomics of opportunistic success. *Nature Reviews Microbiology* 9 (10): 749.

Elad, Y., Barak, R., Chet, I., and Henis, Y. (1983). Ultrastructural studies of the interaction between *Trichoderma* spp. and plant pathogenic fungi. *Journal of Phytopathology* 107 (2): 168–175.

Egamberdieva, D., Kamilova, F., Validov, S. et al. (2008). High incidence of plant growth-stimulating bacteria associated with the rhizosphere of wheat grown on salinated soil in Uzbekistan. *Environmental Microbiology* 10 (1): 1–9.

Erturk, Y., Ercisli, S., Haznedar, A., and Cakmakci, R. (2010). Effects of plant growth promoting rhizobacteria (PGPR) on rooting and root growth of kiwifruit (*Actinidia deliciosa*) stem cuttings. *Biological Research* 43 (1): 91–98.

Fageria, N.K. and Stone, L.F. (2006). Physical, chemical, and biological changes in the rhizosphere and nutrient availability. *Journal of Plant Nutrition* 29 (7): 1327–1356.

Fagundes, C.T., Amaral, F.A., Teixeira, A.L. et al. (2012). Adapting to environmental stresses: the role of the microbiota in controlling innate immunity and behavioral responses. *Immunological Reviews* 245 (1): 250–264.

Fitter, A.H., Gilligan, C.A., Hollingworth, K. et al. (2005). Biodiversity and ecosystem function in soil. *Functional Ecology* 19 (3): 369–377.

Fitriatin, B.N., Yuniarti, A., Turmuktini, T., and Ruswandi, F.K. (2014). The effect of phosphate solubilizing microbe producing growth regulators on soil phosphate, growth and yield of maize and fertilizer efficiency on Ultisol. *Eurasian Journal of Soil Science* 3 (2): 101–107.

Furukawa, J., Yamaji, N., Wang, H. et al. (2007). An aluminum-activated citrate transporter in barley. *Plant and Cell Physiology* 48 (8): 1081–1091.

Gopal, M., Gupta, A., and Thomas, G.V. (2013). Bespoke microbiome therapy to manage plant diseases. *Frontiers in Microbiology* 4: 355.

Gaby, J.C. and Buckley, D.H. (2011). A global census of nitrogenase diversity. *Environmental Microbiology* 13 (7): 1790–1799.

Gao, M., Teplitski, M., Robinson, J.B., and Bauer, W.D. (2003). Production of substances by *Medicago truncatula* that affect bacterial quorum sensing. *Molecular Plant-Microbe Interactions* 16 (9): 827–834.

Gerbore, J., Benhamou, N., Vallance, J. et al. (2014). Biological control of plant pathogens: advantages and limitations seen through the case study of *Pythium oligandrum*. *Environmental Science and Pollution Research* 21 (7): 4847–4860.

Gévaudant, F., Duby, G., von Stedingk, E. et al. (2007). Expression of a constitutively activated plasma membrane H^+-ATPase alters plant development and increases salt tolerance. *Plant Physiology* 144 (4): 1763–1776.

Gomes, C.M., Flocco, C.M., Costa, R. et al. (2010). Mangrove microniches determine the structural and functional diversity of enriched petroleum hydrocarbon-degrading consortia. *FEMS Microbiology Ecology* 74: 276–290.

Grover, M., Ali, S.Z., Sandhya, V. et al. (2011). Role of microorganisms in adaptation of agriculture crops to abiotic stresses. *World Journal of Microbiology and Biotechnology* 27 (5): 1231–1240.

Guimarães, A.A., Jaramillo, P.M.D., Nóbrega, R.S.A. et al. (2012). Genetic and symbiotic diversity of nitrogen-fixing bacteria isolated from agricultural soils in the western Amazon by using cowpea as the trap plant. *Applied and Environmental Microbiology* 78 (18): 6726–6733.

Gupta, V.V.S.R. and Roget, D.K., 2004. Understanding soil biota and biological functions: Management of soil biota for improved benefits to crop production and environmental health. In: *Principles and Methods for Sustainable Disease Management in Rainfed Agricultural Systems* (pp.149–184). doi: 10.10071/978-1-4020-9132-2_6.

Harman, G.E., Howell, C.R., Viterbo, A. et al. (2004). *Trichoderma* species – opportunistic, avirulent plant symbionts. *Nature Reviews Microbiology* 2 (1): 43.

Hartmann, A. and Schikora, A. (2012). Quorum sensing of bacteria and trans-kingdom interactions of N-acyl homoserine lactones with eukaryotes. *Journal of Chemical Ecology* 38 (6): 704–713.

Hartmann, A., Rothballer, M., and Schmid, M. (2008). Lorenz Hiltner, a pioneer in rhizosphere microbial ecology and soil bacteriology research. *Plant and Soil* 312 (1–2): 7–14.

Hartmann, A., Schmid, M., Van Tuinen, D., and Berg, G. (2009). Plant-driven selection of microbes. *Plant and Soil* 321 (1–2): 235–257.

Hawkins, H.J., Johansen, A., and George, E. (2000). Uptake and transport of organic and inorganic nitrogen by arbuscular mycorrhizal fungi. *Plant and Soil* 226 (2): 275–285.

Herdiyantoro, D. 2013. Isolation of potassium solubilizing bacteria from rhizosphere plant in Inceptisols Jatinangor. Proceedings of the Indonesian-Malaysian Culture Symposium XIII Volume 13 November 2013. p. 76.

Herdiyantoro, D., Setiawati, M. R., and Hudaya, R. 2014. Selection of potassium solubilizing bacteria isolates and utilization in the potassium supply for plant growth. Proceedings of the National Seminar on Indonesian Soil Science Association. ISBN: 978-602-1270-17-2; Theme: Sustainable Land Management to Support National Food Security; Banda Aceh, 16-17 September 2014. pp.152-159.

Herdiyantoro, D., Setiawati, M. R., and Hudaya, R. 2015. The effect of potassium solubilizing bacteria (KSB) dossage to available K, K absorption, total population of KSB, growth and yield of maize on Inceptisols Jatinangor. Proceedings of HITI 2015. National Seminar. pp. 279-286.

Helgason, T., Downie, A, Prosser, J., Sen, R, Hirsch, P., East, A and P. Poole (2015). Food Security from the Soil Microbiome. Microbiology Society, Charles Darwin. London.

Hiltner, L. (1904). U¨ ber neuere erfahrungen und probleme auf dem gebiet der bodenbakteriologie und unter besonderer berucksichtigung der grundungung und brache. *Arbeiten der Deutschen Landwirtschaftlichen Gesellschaft* 98: 59–78.

Hider, R.C. and Kong, X. (2010). Chemistry and biology of siderophores. *Natural Product Reports* 27 (5): 637–657.

Hoffmeister, D. and Keller, N.P. (2007). Natural products of filamentous fungi: enzymes, genes, and their regulation. *Natural Product Reports* 24 (2): 393–416.

Holden, N., Pritchard, L., and Toth, I. (2009). Colonization outwith the colon: plants as an alternative environmental reservoir for human pathogenic enterobacteria. *FEMS Microbiology Reviews* 33 (4): 689–703.

Hoorman, J.J., 2010. Understanding soil microbes and nutrient recycling. Agriculture and Natural Resources, pp.1–5. Ohio State University [on-line resource].

Hinsinger, P., Bengough, A.G., Vetterlein, D., and Young, I.M. (2009). Rhizosphere: biophysics, biogeochemistry and ecological relevance. *Plant and Soil* 321 (1–2): 117–152.

Huang, X.F., Chaparro, J.M., Reardon, K.F. et al. (2014). Rhizosphere interactions: root exudates, microbes, and microbial communities. *Botany* 92 (4): 267–275.

Iavicoli, A., Boutet, E., Buchala, A., and Metraux, J.P. (2003). Induced systemic resistance in *Arabidopsis thaliana* in response to root inoculation with *Pseudomonas fluorescens* CHA0. *Molecular Plant Microbe Interactions* 16: 851–858.

Ichinohe, T., Pang, I.K., Kumamoto, Y. et al. (2011). Microbiota regulates immune defense against respiratory tract influenza A virus infection. *Proceedings of the National Academy of Sciences of the United States of America* 108 (13): 5354–5359.

Ingham ER. (2001). The food web and soil health. Soil Biology Primer [online]. www.statlab.iastate.edu/survey/SQI/soil_biology_primer.htm/ Accessed 06/02/20.

Ingham, E.R. 2004. The Soil Biology Primer. Published by the Soil and Water Conservation Society (SWCS). Available at: http://soils.usda.gov/sqi/concepts/soil_biology/index.html/ accessed 06/02/20.

Jegan, S., Baskaran, V., Ganga, V. et al. (2016). Rhizomicrobiome – a biological software to augment soil fertility and plant induced systemic tolerance under abiotic stress. In: *Microbes for Plant Stress Management* (ed. D.J.B. Jamaluddin), 25–53. New Dehli, India: New India Publishing Agency.

Jehangir, I.A., Mir, M.A., Bhat, M.A., and Ahangar, M.A. (2017). Biofertilizers an approach to sustainability in agriculture: a review. *International Journal of Pure & Applied Bioscience* 5 (5): 327–334.

Jetiyanon, K. and Kloepper, J.W. (2002). Mixtures of plant growth-promoting rhizobacteria for induction of systemic resistance against multiple plant diseases. *Biological Control* 24 (3): 285–291.

Jones, D.L., Hodge, A., and Kuzyakov, Y. (2004). Plant and mycorrhizal regulation of rhizodeposition. *New Phytologist* 163 (3): 459–480.

Jones, D.L., Nguyen, C., and Finlay, R.D. (2009). Carbon flow in the rhizosphere: carbon trading at the soil–root interface. *Plant and Soil* 321 (1–2): 5–33.

Jorquera, M.A., Shaharoona, B., Nadeem, S.M. et al. (2012). Plant growth-promoting rhizobacteria associated with ancient clones of creosote bush (*Larrea tridentata*). *Microbial Ecology* 64 (4): 1008–1017.

Kamilova, F., Kravchenko, L.V., Shaposhnikov, A.I. et al. (2006). Effects of the tomato pathogen *Fusarium oxysporum* f. sp. *radicis-lycopersici* and of the biocontrol bacterium *Pseudomonas fluorescens* WCS365 on the composition of organic acids and sugars in tomato root exudate. *Molecular Plant-Microbe Interactions* 19 (10): 1121–1126.

Kang, S.M., Khan, A.L., Waqas, M. et al. (2014). Plant growth-promoting rhizobacteria reduce adverse effects of salinity and osmotic stress by regulating phytohormones and antioxidants in *Cucumis sativus*. *Journal of Plant Interactions* 9: 673–682.

Knief, C., Delmotte, N., Chaffron, S. et al. (2012). Metaproteogenomic analysis of microbial communities in the phyllosphere and rhizosphere of rice. *The ISME Journal* 6 (7): 1378.

Kyselková, M., Kopecký, J., Frapolli, M. et al. (2009). Comparison of rhizobacterial community composition in soil suppressive or conducive to tobacco black root rot disease. *The ISME Journal* 3 (10): 1127.

Lambers, H., Raven, J.A., Shaver, G.R., and Smith, S.E. (2008). Plant nutrient-acquisition strategies change with soil age. *Trends in Ecology & Evolution* 23 (2): 95–103.

Lathrop, S.K., Bloom, S.M., Rao, S.M. et al. (2011). Peripheral education of the immune system by colonic commensal microbiota. *Nature* 478 (7368): 250.

Lau, J.A. and Lennon, J.T. (2011). Evolutionary ecology of plant–microbe interactions: soil microbial structure alters selection on plant traits. *New Phytologist* 192 (1): 215–224.

Ley, R.E., Peterson, D.A., and Gordon, J.I. (2006). Ecological and evolutionary forces shaping microbial diversity in the human intestine. *Cell* 124 (4): 837–848.

Li, J., Xu, R., Xiao, S., and Ji, G. (2005). Effect of low-molecular-weight organic anions on exchangeable aluminum capacity of variable charge soils. *Journal of Colloid and Interface Science* 284 (2): 393–399.

Liu, N.N., Tian, Q.Y., and Zhang, W.H. (2014). Comparison of adaptive strategies to phosphorus-deficient soil between dominant species *Artemisia frigida* and *Stipa krylovii* in typical steppe of Nei Mongol. *Chinese Journal of Plant Ecology* 38: 905–915.

Lundberg, D., Lebeis, S., Paredes, S. et al. (2012). Defining the core *Arabidopsis thaliana* root microbiome. *Nature* 488: 86–90.

Lugtenberg, B. and Kamilova, F. (2009). Plant-growth-promoting rhizobacteria. *Annual Review of Microbiology* 63: 541–556.

Lynch, J.M. (1989). *The Rhizosphere*. Chichester: Wiley.

Lynch, J.M. (1990). *The Rhizosphere*. Chichester: Wiley.

Macfarlane, S.A. (2003). Molecular determinants of the transmission of plant viruses by nematodes. *Molecular Plant Pathology* 4 (3): 211–215.

Manasa, K., Subhash Reddy, R., Triveni, S. et al. (2017). Characterization of *Rhizobium* isolates and their potential PGPR characteristics of different rhizosphere soils of Telangana region. *International Journal of Current Microbiology and Applied Sciences* 6 (5): 2808–2813.

Manoharachary, C. and Mukerji, K.G. (2006). Rhizosphere biology – an overview. In: *Microbial Activity in the Rhizosphere* (eds. K.G. Mukerji et al.), 1–15. Berlin, Heidelberg: Springer.

Mansfield, J., Genin, S., Magori, S. et al. (2012). Top 10 plant pathogenic bacteria in molecular plant pathology. *Molecular Plant Pathology* 13 (6): 614–629.

Mathivanan, N., Prabavathy, V.R., and Vijayanandraj, V.R. (2005). Application of talc formulations of *Pseudomonas fluorescens* Migula and *Trichoderma viride* Pers. ex SF Gray decrease the sheath blight disease and enhance the plant growth and yield in rice. *Journal of Phytopathology* 153 (11–12): 697–701.

Mendes, R., Kruijt, M., De Bruijn, I. et al. (2011). Deciphering the rhizosphere microbiome for disease-suppressive bacteria. *Science* 332 (6033): 1097–1100.

Mendes, R., Garbeva, P., and Raaijmakers, J.M. (2013). The rhizosphere microbiome: significance of plant beneficial, plant pathogenic, and human pathogenic microorganisms. *FEMS Microbiology Reviews* 37 (5): 634–663.

Miransari, M. (2011). Arbuscular mycorrhizal fungi and nitrogen uptake. *Archives of Microbiology* 193 (2): 77–81.

Mirete, S., De Figueras, C.G., and González-Pastor, J.E. (2007). Novel nickel resistance genes from the rhizosphere metagenome of plants adapted to acid mine drainage. *Applied and Environmental Microbiology* 73 (19): 6001–6011.

Moravec, M., D. Whiting, A. Card and C. Wilson. (2009). The Living Soil. Colorado State University. http://cmg.colostate.edu gardennotes/212.pdf/ accessed 07/02/20.

NCRS (2004). Soil Biology and Land Management. Soil Quality – Soil Biology Technical Note No. 4. Available on: http://soils.usda.gov/sqi/ accessed 07/02/20.

Neumann, G. and Romheld, V. (2000). The release of root exudates as affected by the plant's physiological status. In: *The Rhizosphere* (eds. S. Willig et al.), 57–110. CRC Press.

Oger, P., Petit, A., and Dessaux, Y. (1997). Genetically engineered plants producing opines alter their biological environment. *Nature Biotechnology* 15 (4): 369.

Oger, P.M., Mansouri, H., Nesme, X., and Dessaux, Y. (2004). Engineering root exudation of lotus toward the production of two novel carbon compounds leads to the selection of distinct microbial populations in the rhizosphere. *Microbial Ecology* 47 (1): 96–103.

Parmar, N. and Dufresne, J. (2011). Beneficial interactions of plant growth promoting rhizosphere microorganisms. In: *Bioaugmentation, Biostimulation and Biocontrol* (eds. A. Singh et al.), 27–42. Berlin, Heidelberg: Springer.

Paton, A.W., Morona, R., and Paton, J.C. (2006). Designer probiotics for prevention of enteric infections. *Nature Reviews Microbiology* 4 (3): 193–200.

Paul, E.A. and Clark, F.E. (1996). *Soil Microbiology and Biochemistry*. San Diego: Academic Press, Inc.

Phillips, R.P., Erlitz, Y., Bier, R., and Bernhardt, E.S. (2008). New approach for capturing soluble root exudates in forest soils. *Functional Ecology* 22 (6): 990–999.

Picard, C. and Bosco, M. (2008). Genotypic and phenotypic diversity in populations of plant-probiotic *Pseudomonas* spp. colonizing roots. *Naturwissenschaften* 95 (1): 1–16.

Picard, C., Baruffa, E., and Bosco, M. (2008). Enrichment and diversity of plant-probiotic microorganisms in the rhizosphere of hybrid maize during four growth cycles. *Soil Biology and Biochemistry* 40 (1): 106–115.

Pii, Y., Mimmo, T., Tomasi, N. et al. (2015). Microbial interactions in the rhizosphere: beneficial influences of plant growth-promoting rhizobacteria on nutrient acquisition process. A review. *Biology and Fertility of Soils* 51 (4): 403–415.

Pires, A.C., Cleary, D.F., Almeida, A. et al. (2012). Denaturing gradient gel electrophoresis and barcoded pyrosequencing reveal unprecedented archaeal diversity in mangrove sediment and rhizosphere samples. *Applied and Environmental Microbiology* 78 (16): 5520–5528.

Powell, J. and Kalironomos, J. (2007). The ecology of plant–microbial mutualisms. In: *Soil Microbiology, Ecology and Biochemistry* (ed. E. Paul), 257–281. Academic Press.

Pinton, R., Varanini, Z., and Nannipieri, P. (2000). The rhizosphere as a site of biochemical interactions among soil components, plants, and microorganisms. In: *The Rhizosphere*, 17–34. CRC Press.

Purwanto, P., Yuwariah, Y., Sumadi, S., and Simarmata, T. (2016). Nitrogenase activity and IAA production of indigenous diazotroph and its effect on rice seedling growth. *AGRIVITA, Journal of Agricultural Science* 39 (1): 31–37.

Quiza, L., St-Arnaud, M., and Yergeau, E. (2015). Harnessing phytomicrobiome signaling for rhizosphere microbiome engineering. *Frontiers in Plant Science* 6: 507.

Raaijmakers, J.M. and Mazzola, M. (2012). Diversity and natural functions of antibiotics produced by beneficial and plant pathogenic bacteria. *Annual Review of Phytopathology* 50: 403–424.

Raaijmakers, J.M., Paulitz, T.C., Steinberg, C. et al. (2009). The rhizosphere: a playground and battlefield for soil-borne pathogens and beneficial microorganisms. *Plant and Soil* 321 (1–2): 341–361.

Recep, K., Fikrettin, S., Erkol, D., and Cafer, E. (2009). Biological control of the potato dry rot caused by *Fusarium* species using PGPR strains. *Biological Control* 50 (2): 194–198.

Reinhold-Hurek, B. and Hurek, T. (2011). Living inside plants: bacterial endophytes. *Current Opinion in Plant Biology* 14 (4): 435–443.

Rengel, Z. (2002). Breeding for better symbiosis. In: *Food Security in Nutrient-Stressed Environments: Exploiting Plants' Genetic Capabilities* (ed. J.J. Adu-Gyamfi), 245–260. Dordrecht: Springer.

Rezzonico, F., Binder, C., Défago, G., and Moënne-Loccoz, Y. (2005). The type III secretion system of biocontrol *Pseudomonas fluorescens* KD targets the phytopathogenic Chromista *Pythium ultimum* and promotes cucumber protection. *Molecular Plant-Microbe Interactions* 18 (9): 991–1001.

Richardson, A.E., Barea, J.M., McNeill, A.M., and Prigent-Combaret, C. (2009). Acquisition of phosphorus and nitrogen in the rhizosphere and plant growth promotion by microorganisms. *Plant and Soil* 321 (1–2): 305–339.

Rodrigues, E.P., Rodrigues, L.S., de Oliveira, A.L.M. et al. (2008). *Azospirillum amazonense* inoculation: effects on growth, yield and N_2 fixation of rice (*Oryza sativa* L.). *Plant and Soil* 302 (1–2): 249–261.

Roesch, L.F., Fulthorpe, R.R., Riva, A. et al. (2007). Pyrosequencing enumerates and contrasts soil microbial diversity. *The ISME Journal* 1 (4): 283.

Rousk, J. and Bååth, E. (2007). Fungal and bacterial growth in soil with plant materials of different C/N ratios. *FEMS Microbiology Ecology* 62 (3): 258–267.

Rout, M.E. (2014). The plant microbiome. In: *Advances in Botanical Research*, vol. 6 (ed. J.A. Callow). Elsevier Ltd. J.A. Callow.

Rudrappa, T., Czymmek, K.J., Paré, P.W., and Bais, H.P. (2008). Root-secreted malic acid recruits beneficial soil bacteria. *Plant Physiology* 148 (3): 1547–1556.

Ryan, R.P., Germaine, K., Franks, A. et al. (2008). Bacterial endophytes: recent developments and applications. *FEMS Microbiology Letters* 278 (1): 1–9.

Ryan, P.R., Dessaux, Y., Thomashow, L.S., and Weller, D.M. (2009). Rhizosphere engineering and management for sustainable agriculture. *Plant and Soil* 321 (1–2): 363–383.

Ryu, C.M., Farag, M.A., Hu, C.H. et al. (2003). Bacterial volatiles promote growth in *Arabidopsis*. *Proceedings of the National Academy of Sciences of the United States of America* 100 (8): 4927–4932.

Sandnes, A., Eldhuset, T.D., and Wollebæk, G. (2005). Organic acids in root exudates and soil solution of Norway spruce and silver birch. *Soil Biology and Biochemistry* 37 (2): 259–269.

Sasaki, T., Yamamoto, Y., Ezaki, B. et al. (2004). A wheat gene encoding an aluminum-activated malate transporter. *The Plant Journal* 37 (5): 645–653.

Setiawati, M.R., Suryatmana, P., Herdiyantoro, D., and Primadi, U. (2015). Isolation and bioassay screening of biofertilizer diazotroph bacteria from paddy field. *Academic Journal of Science* 18 (2): 159–164.

Singh T. and Purohit S.S. (2011). Biofertilizers Technology. Agrobios: New Delhi. India.

Shaharoona, B., Arshad, M., and Zahir, Z.A. (2006). Effect of plant growth promoting rhizobacteria containing ACC-deaminase on maize (*Zea mays* L.) growth under axenic conditions and on nodulation in mung bean (*Vigna radiata* L.). *Letters in Applied Microbiology* 42 (2): 155–159.

Shi, S., Condron, L., Larsen, S. et al. (2011). *In situ* sampling of low molecular weight organic anions from rhizosphere of radiata pine (*Pinus radiata*) grown in a rhizotron system. *Environmental and Experimental Botany* 70 (2–3): 131–142.

Simarmata, T., 2013. Tropical bioresources to support biofertilizer industry and sustainable agriculture in Indonesia. In: International Seminar on Tropical Bioresources for Sustainable Bioindustry; from Basic Research to Industry. 30e, 31st October 2013 in ITB. Bandung. Indonesia.

Simarmata, T., Turmuktini, T., Fitriatin, B.N., and Setiawati, M.R. (2016). Application of bioameliorant and biofertilizers to increase the soil health and rice productivity. *HAYATI Journal of Biosciences* 23 (4): 181–184.

Smith, K.P. and Goodman, R.M. (1999). Host variation for interactions with beneficial plant-associated microbes. *Annual Review of Phytopathology* 37 (1): 473–491.

Smith, K.P., Handelsman, J., and Goodman, R.M. (1999). Genetic basis in plants for interactions with disease-suppressive bacteria. *Proceedings of the National Academy of Sciences of the United States of America* 96 (9): 4786–4790.

Smith, D.L., Praslickova, D., and Ilangumaran, G. (2015). Inter-organismal signaling and management of the phytomicrobiome. *Frontiers in Plant Science* 6: 722.

Somers, E., Vanderleyden, J., and Srinivasan, M. (2004). Rhizosphere bacterial signalling: a love parade beneath our feet. *Critical Reviews in Microbiology* 30 (4): 205–240.

Srivastava, R. and Singh, A. (2017). Plant Growth Promoting Rhizobacteria (PGPR) for sustainable agriculture. *International Journal of Agricultural Science and Research* 7 (4): 505–510.

Subba Rao, N.S. (1988). *Biofertilizers in Agriculture*, 2e, 1–208. Oxford and IBH Publishing Co. Pvt, Ltd.

Sudjana, B., Jingga, A., and Simarmata, T. (2017). Enriched rice husk biochar ameliorant to increase crop productivity on typic hapludults. *Global Advanced Research Journal of Agricultural Science* 6 (5): 108–113.

Sullivan, P. (2001). Sustainable Soil Management. www.attra.org/attra-pub/pdf/soilmgmt.pdf/ accessed 06/02/20.

Sullivan, P. (2004). Sustainable Soil Management. National Sustainable Agriculture Information Service. Availble at: www.attra.ncat.org/ accessed 06/02/20.

Teixeira, L.C., Peixoto, R.S., Cury, J.C. et al. (2010). Bacterial diversity in rhizosphere soil from Antarctic vascular plants of Admiralty Bay, maritime Antarctica. *The ISME Journal* 4 (8): 989–1001.

Teplitski, M., Barak, J.D., and Schneider, K.R. (2009). Human enteric pathogens in produce: un-answered ecological questions with direct implications for food safety. *Current Opinion in Biotechnology* 20 (2): 166–171.

Tesfaye, M., Temple, S.J., Allan, D.L. et al. (2001). Overexpression of malate dehydrogenase in transgenic alfalfa enhances organic acid synthesis and confers tolerance to aluminum. *Plant Physiology* 127 (4): 1836–1844.

Turnbaugh, P.J., Ley, R.E., Hamady, M. et al. (2007). The human microbiome project. *Nature* 449 (7164): 804.

Tyler, H.L. and Triplett, E.W. (2008). Plants as a habitat for beneficial and/or human pathogenic bacteria. *Annual Review of Phytopathology* 46: 53–73.

Uroz, S., Buée, M., Murat, C. et al. (2010). Pyrosequencing reveals a contrasted bacterial diversity between oak rhizosphere and surrounding soil. *Environmental Microbiology Reports* 2 (2): 281–288.

Van Baarlen, P., Van Belkum, A., Summerbell, R.C. et al. (2007). Molecular mechanisms of pathogenicity: how do pathogenic microorganisms develop cross-kingdom host jumps? *FEMS Microbiology Reviews* 31 (3): 239–277.

Van Der Heijden, M.G., Bardgett, R.D., and Van Straalen, N.M. (2008). The unseen majority: soil microbes as drivers of plant diversity and productivity in terrestrial ecosystems. *Ecology Letters* 11 (3): 296–310.

Van der Ent, S., Van Hulten, M., Pozo, M.J. et al. (2009). Priming of plant innate immunity by rhizobacteria and β-aminobutyric acid: differences and similarities in regulation. *New Phytologist* 183 (2): 419–431.

Vessey, J.K. and Buss, T.J. (2002). *Bacillus cereus* UW85 inoculation effects on growth, nodulation, and N accumulation in grain legumes: controlled-environment studies. *Canadian Journal of Plant Science* 82 (2): 282–290.

Vyas, R.K. and Mathus, K. (2002). Trichoderma spp. in cumin rhizosphere and their potential in suppression of wilt. *Indian Phytopathology* 55: 455–457.

Wagg, C., Jansa, J., Schmid, B., and van der Heijden, M.G. (2011). Belowground biodiversity effects of plant symbionts support aboveground productivity. *Ecology Letters* 14 (10): 1001–1009.

Wang, J., Raman, H., Zhou, M. et al. (2007). High-resolution mapping of the Alp locus and identification of a candidate gene HvMATE controlling aluminium tolerance in barley (*Hordeum vulgare* L.). *Theoretical and Applied Genetics* 115 (2): 265–276.

Weinert, N., Piceno, Y., Ding, G.C. et al. (2011). PhyloChip hybridization uncovered an enormous bacterial diversity in the rhizosphere of different potato cultivars: many common and few cultivar-dependent taxa. *FEMS Microbiology Ecology* 75 (3): 497–506.

Wegener, C.B. (2001). Transgenic potatoes expressing an *Erwinia* pectate lyase gene – results of a 4-year field experiment. *Potato Research* 44 (4): 401–410.

Whipps, J., Hand, P., Pink, D., and Bending, G.D. (2008). Phyllosphere microbiology with special reference to diversity and plant genotype. *Journal of Applied Microbiology* 105 (6): 1744–1755.

Wissuwa, M., Mazzola, M., and Picard, C. (2009). Novel approaches in plant breeding for rhizosphere-related traits. *Plant and Soil* 321 (1–2): 409–430.

Weller, D.M., Raaijmakers, J.M., Gardener, B.B.M., and Thomashow, L.S. (2002). Microbial populations responsible for specific soil suppressiveness to plant pathogens. *Annual Review of Phytopathology* 40 (1): 309–348.

Yang, H., Knapp, J., Koirala, P. et al. (2007). Enhanced phosphorus nutrition in monocots and dicots over-expressing a phosphorus-responsive type I H^+-pyrophosphatase. *Plant Biotechnology Journal* 5 (6): 735–745.

Zamioudis, C. and Pieterse, C.M. (2012). Modulation of host immunity by beneficial microbes. *Molecular Plant-Microbe Interactions* 25 (2): 139–150.

Zehr, J.P., Jenkins, B.D., Short, S.M., and Steward, G.F. (2003). Nitrogenase gene diversity and microbial community structure: a cross-system comparison. *Environmental Microbiology* 5 (7): 539–554.

10

Mycorrhizal Microbiome

An Ideal Association in Sustainable Agriculture

Baby Summuna[1], Sachin Gupta[2], and Moni Gupta[3]

[1] *Division of Plant Pathology, Sher-e-Kashmir University of Agricultural Sciences and Technology of Kashmir, Jammu and Kashmir, India*
[2] *Division of PlantPathology, Sher-e-Kashmir University of Agricultural Sciences and Technology of Jammu, Jammu and Kashmir, India*
[3] *Division of Biochemistry, Sher-e-Kashmir University of Agricultural Sciences and Technology of Jammu, Jammu and Kashmir, India*

10.1 Introduction

In order to meet the future needs of a burgeoning human population and to achieve food security in the context of climate change, food production needs to increase with minimum negative environmental impact. Sustainable intensification of agriculture (Andres and Bhullar 2016), also called ecological intensification, is likely to include key aspects of conservation agriculture (Giller et al. 2015). Appreciation of the integral importance of soil life in agricultural sustainability including plant-symbiotic associations is steadily growing (Bender et al. 2016). Plants are colonized by millions of microbes which form complex communities on plant surfaces and in plant tissues. The plant and its microbiota can interact in various ways ranging from parasitism to mutualism and outcome of their interaction can be pivotal for plant performance. Plant fitness can be influenced by the associated microbes by modulating plant growth, root architecture, nutrient acquisition or drought and disease resistance. Thus, the plant microbiota can be seen as an extension of the plant genome in the sense that it can increase the plant's adaptation capacity. A prominent player among these symbioses is mycorrhiza, the widespread symbiotic association of fungi with plant roots. This is thought to have facilitated the adaptation of plants to a terrestrial life. Soil microorganisms are fundamental for soil health and provide ecosystem services that are essential for plant production. Eighty percent of the vascular plant species is estimated to receive phosphorus and other nutritional elements from these fungi in exchange for photosynthates. The continuous growth of societal awareness about agricultural sustainability makes it increasingly important to unravel the biotic and abiotic factors in different soils supporting the beneficial components of their microbiome.

There is tremendous microbial diversity associated with plant roots: the greatest reservoir of biological diversity known in the world so far likely represents the soil microbial community (Roesch et al. 2007). The collective genome of this rhizosphere microbial

The Plant Microbiome in Sustainable Agriculture, First Edition. Edited by Alok Kumar Srivastava, Prem Lal Kashyap and Madhumita Srivastava.

community is composed of a distinct community of interacting microbes that compete and cooperate with each other and the host plant, which is typically much larger than that of the host plant (Denison et al. 2003; Berendsen et al. 2012). Hosts can be benefited by this complex network of species by boosting immunity and stimulating growth via providing nutrients, increased stress resistance and pathogen exclusion. The fundamental component of nearly all terrestrial ecosystems is mycorrhizal fungi, creating the critical link between both above- and below-ground systems by establishing an intimate association with the roots of more than 90 percent of land plants. This relationship enables the fungal partners to obtain the carbon sources they need to survive and reproduce which is photosynthetically fixed by plants. Mycorrhizal fungi impact the function and biodiversity of entire ecosystems, connect unrelated individual plant species, and function as a sink for atmospheric carbon dioxide. Mycorrhizal fungi are now widely used in organic agriculture, plant nurseries, and restoration of mine tailings to improve the growth of economically important species because of the many benefits they provide to plants. Although the factors that drive fungi to establish and maintain mycorrhizal associations are critically important for understanding their function in governing biodiversity, nutrient cycling, and plant nutrition, they are still very poorly understood. While the economic and ecological value of plant–mycorrhizal interactions has been identified and understood as important from a human and ecosystem perspective, interactions between fungi and endobacteria are now at the frontier of understanding plant–mycorrhizal interactions.

10.2 Mycorrhizal Networks

A scientist in Germany named A.B. Frank coined the term "mycorrhiza," more than 100 years ago. Its literal meaning is fungus-root which describes the mutualistic association existing between a group of soil fungi and higher plants. The association is based on the plant component providing carbohydrates and other essential organic compounds to the fungi. In return, the fungal component, which colonizes both the root and the adjacent soil, helps the plant to take up nutrients by extending the reach of its root system. The cortex of plant roots, the surface of the root or the epidermal cells of the root are colonized by mycorrhizal fungi. The fungi acquire nutrients through their hyphae which grow out from the roots into the soil where they forage for nutrients that are limiting to plant growth, especially nitrates and phosphates, but organically bound nutrients are also acquired by some mycorrhizal types [ectomycorrhizae (ECM) and ericoid mycorrhizal (EM) fungi]. In exchange for carbohydrates, these nutrients as well as other benefits are then delivered to their host plants. Consequently, a strong influence is exerted by the mycorrhizal symbiosis on plant growth and fitness.

Although mycorrhizal associations were discovered over 100 years ago, their role in plant productivity did not receive attention until the past 30 years. Based on their structure and function, four major mycorrhizal types have been described, *viz.*, arbuscular mycorrhiza (AM), ectomycorrhiza (ECM), orchid mycorrhiza, and ericoid mycorrhiza (EM).

Multiple mycorrhizal fungi colonize most plant roots and are not host specific, colonizing various host plants at the same time. As a result, plants are usually interconnected by mycelial networks of mycorrhizae in so called "wood wide webs," e.g., in some temperate forests, trees like oak, pine, birch, are interconnected by ECM fungal networks, while understory shrubs, grasses, and herbs are interconnected by AM fungi. In some communities, other networks are formed between ericoid and orchid mycorrhizal plants. The possible hubs between these networks, e.g. ECM fungi may form ericoid mycorrhizae and some trees/tree seedlings form symbiotic relation with both ECM and AM fungi. The estimates suggest that mycorrhizal associations are formed by 40,000–50,000 fungal species which represents 0.5–10 per cent of the total number of 0.5–10 million fungal species estimated to be present on earth (Taylor et al. 2014). Plants also associate with a wide range of beneficial fungal endophytes (e.g. dark septate endophytes and some Sebacinales). These fungal endophytes do not form a specialized plant–fungal interface for resource exchange although they improve plant growth and provide resistance to stress and pathogens, and currently these are not considered as mycorrhizal associates.

10.3 Biodiversity of Mycorrhizal Associations

In the plant kingdom, mycorrhizal associations are extremely abundant. About 74 per cent of all plant species form AMs with fungi of the Glomeromycota clade, 2 per cent of plants form ECM associations, 9 per cent of plants form orchid mycorrhizae, and 1 per cent of plants form ericoid mycorrhizae (Brundrett 2009). Dual symbiotic associations, e.g., with AM and ECM fungi, are formed by some plant species, such as poplars and eucalyptus. Mycorrhizal plants dominate almost all ecosystems with the exception of early successional communities, intensively managed arable fields and extremely P impoverished soils that are dominated by plants with cluster roots. AM fungi are a monophyletic group (phylum Glomeromycota) of asexual and obligate symbiotic fungi having a unique combination of traits regarding morphology, genomic structure and ecology. The co-existence of multiple, potentially and genetically divergent nuclei within their coenocytic mycelia and spores, makes it difficult to delineate an individual even with molecular methods. Through intraradical and extraradical mycelia, they occupy a dual niche, the plant root and soil. Arbuscular mycorrhizal fungi (AM fungi) are present in most soils and form symbiotic associations with the majority of crop plants. The multiple beneficial impacts of mycorrhizal symbiosis on nutrient cycling and plant stress tolerance represent under-exploited potential to increase global food security. Hence, a well-developed extra-radical mycelium (ERM) that proliferates in the bulk soil beyond the rhizosphere is important for plant P uptake, while the AM fungi depend on their host plant for carbon nutrition. The large surface area of the ERM provides nutrient-rich niches for colonization and growth of other soil microorganisms, particularly in bacteria. Hence, the abundance of Firmicutes, Streptomycetes, and members of the family Oxalobacteraceae on hyphae, or in soil affected by hyphae is increased by several members of the Glomeraceae. The bacteria associated with the hyphae may in return affect the AM fungi as exemplified by mycorrhizal helper bacteria that promote AM hyphal growth and root colonization.

It is now firmly established for many plant species, whether they form mycorrhizal associations or not. Studies showed that lower land plants, in particular species of hornworts and liverworts, associate with AM, ECM, or EM fungi (Pressel et al. 2014). The number of fungal partners involved in the symbiosis is less clear and varies depending on mycorrhizal type. Based on morphological characteristics of the spores, 244 species of Glomeromycota have been described so far (Schüssler 2010). Richness and composition of AM fungal communities depend on host plant, climate, and soil conditions. AM fungal richness is reduced due to land use intensification. The number of ECM fungal species is thought to be higher than AM fungi and estimates suggest that there are 20,000 ECM fungi. The basis of this estimate depends on a range of traits, including morphological, molecular, and isotope studies that show the morphological diversity of ECM. There might be a higher number of ECM fungi based on high sequence diversity of EM roots in several field studies. It has been estimated that ECM fungi form associations with about 6000 plant species. Many ECM fungi have a broad host range while others are more specific and colonize certain hosts or host genera.

In contrast to AM fungi, most ECM fungi are, at least in part saprotrophic as they can be grown on artificial agar media without host plants. However, the loss of lignocelluloses degrading enzymes in most ECM fungi makes many of them dependent on their host plant photo assimilates under field conditions. It is not unusual that temperate and boreal forests are dominated by a few tree species (e.g. pines or oaks) while diverse ECM fungal communities occur below ground, and several hundreds of fungal species can coexist in a single forest.

Fungi forming mycorrhizae with orchids and ericoid plants have been given less attention. Fungi forming mycorrhizae with orchids typically live as saprotrophs in the soil or form endophytic/ECM associations with neighboring trees. Orchid seeds are extremely small (0.3–14 μg) and in natural ecosystems seedlings of most orchids are completely dependent on colonization by fungi. Because of lack of chlorophyll in orchid seedlings (protocorms), these rely on nutrients and carbon that they obtain from these fungi. In recent years, molecular identification techniques have shown that many orchids have host specific fungal associates (usually between one and 10 taxa per orchid), although higher numbers have also been reported (Kartzinel et al. 2013). New fungal taxa colonizing orchids are continuously being described (e.g. Atractiellales) and the total number of fungi forming orchid mycorrhizae may be as much as 25,000 or even more.

The fungi forming ericoid mycorrhizae with species of the Ericaceae, such as *Erica*, *Calluna*, *Rhododendron*, or *Vaccinium* (e.g. cranberry), are mostly common under acid and infertile heathland conditions. Well known ericoid fungi belong to the Helotiales (Ascomycetes) and are soil saprotrophs. Recent evidence suggests that some ericoid mycorrhizal fungi may act as plant endophytes and some Basidiomycetes are also thought to form ericoid mycorrhizal associations. Both ericoid and ECM associations are formed by some fungi with different host plants. Interestingly, it has been found that a sebacinoid fungus forming ericoid associations with *Calluna* also colonized neighboring gametophytes of a lycopod, pointing to even more complex associations than previously thought. Walker et al. (2011) indicated in his study that there are many more fungi colonizing plants of the Ericaceae, with 129 fungal Helotiales operational taxonomic units (OTUs) detected in three ericoid plant species at one location. Moreover, Sebacinales (Basidiomycetes) were also discovered to be widespread ericoid mycorrhizal fungi and there may be many more ericoid fungal partners, e.g. in Trechisporales.

10.4 Mycorrhizae As a Member of the Rhizosphere Microbiome

In agroecosystems, the role of mycorrhiza, in particular AM, formed by fungi in the phylum Glomeromycota has been well-studied. The interaction of AM fungi with their hosts usually occurs through hormonal crosstalk and actively shape the soil as ecosystem engineers.

AM fungi are obligate symbionts with tight regulation of carbon-for-nutrients exchange between the host and the fungus. AM fungi have lost the enzymatic ability to degrade carbon compounds during their evolution (Tisserant et al. 2013), which prevents them from becoming necrotrophic pathogens. The AM symbiosis has been much discussed in the context of agriculture because of the following important reasons: (i) AM is the dominant mycorrhiza formed by most crops; (ii) AM have a potentially positive, multifunctional role in plant nutrition, pathogen protection, stress tolerance and soil structure provision (Leifheit et al. 2014); (iii) agricultural practices (tillage, fertilization and non-host crops) tend to negatively affect AM fungal abundance and diversity, thus potentially affecting functioning; and (iv) easy management of AM fungi. Mycorrhizal fungi influence plant productivity and plant diversity and connect plants below ground via a hyphal network allowing the movement of resources among coexisting plants. Therefore, mycorrhizal symbiosis is of key interest to biologists and ecologists. Additionally, the symbiosis plays a key role in the cycling of carbon (C), nitrogen (N), and phosphorus (P) in ecosystems. Knowing that the vast majority of the Earth's plant species form at least one of the four mycorrhizal associations and given the amount of resources that can move above or below ground because of the mycorrhizal symbiosis, it is likely that the role of symbiosis in global nutrient cycling is significant.

10.5 Mycorrhizae in the Plant–Soil System

The coupling of plants with rhizosphere microbial processes is optimized by AM fungi in all ecosystems. A symbiotic relationship is formed by these fungi with plant roots, and the fungal symbiont becomes a major interface or connection between the soil and plant. The germination of large AM spores, or other forms of inoculum propagules, such as root fragments, that occur in most soils as a result of their production and release into the soil from previous host roots starts the AM symbiosis. Specialized haustoria-like structures called arbuscules are formed within the cells as the spores germinate and penetrate the root cortical cell. These haustoria serve as the sites where metabolite exchanges take place between fungus and host cytoplasm. Characteristic vesicles usually form in the cortical cells which function as nutrient storage organs or as propagules in root fragments. The AM-fungal hyphae extend from the root out into the soil where they interface with soil particles. Soil (extraradical) hyphae function as absorptive structures for mineral elements and water as they can extend out several centimeters from the root, and effectively bridge over the zone of nutrient depletion around roots and absorb immobile elements from the bulk soil. Other microbes are also attracted by soil hyphae, and together they form water-stable aggregates necessary for good soil tilth. The internal colonization of roots by AM fungi plus the nutrient and water uptake from soil by the soil hyphae of the fungal symbiont cause pervasive physiological changes in the host plant. Such changes make the AM plant grow and respond to environmental stresses differently from a non-AM plant. Not all AM fungi induce the same kind and extent of change, however. Nevertheless, the hypothesis that AM plants grow better than non-AM plants in agricultural soils that may have been degraded by current cultural practices is worthy of validation. Many of the reactions and interactions of the microflora and fauna that occur in the soil around roots are mediated by AM fungi that function to deliver mineral nutrients to the host plant in return for a sustained C supply. They also impart other benefits to plants including stimulation of growth regulating substances and alteration of other chemical constituents in the plant, increased rate of photosynthesis, osmotic adjustment under drought stress, enhancement of N_2 fixation by symbiotic or associative N_2-fixing bacteria, increased resistance to pests and tolerance to environmental stresses and improved soil aggregation and thus improved soil physical properties and stability. Because of the improved nutritional status of AM plants, membrane permeability is altered and this change can greatly alter the quality and quantity of root exudates, resulting in a new microbial equilibrium in the soil surrounding that root, now appropriately called the mycorrhizosphere. The chemical and physical effects of the fungal symbiont extending out into the soil creates a whole new dimension, both spatially and biologically. Specific organisms establish special relationships with mycorrhizal fungi that have a net beneficial effect on plant growth. Such discoveries open possibilities for understanding the role of mycorrhizae in N cycling and developing management strategies for improved N economies of agricultural crop plants.

10.6 Plant Productivity, Ecosystem Functioning, and Multifunctionality

A key role is being played by mycorrhizal fungi in ecosystems and also influence various important ecosystem functions. They are well-known to enhance plant productivity, although negative effects on plant biomass have also been repeatedly reported in natural and agricultural ecosystems. Growth responses are also plant species dependent: some plant species, especially those with relatively thick roots rely more on mycorrhizal fungi than do plants with fine roots, such as grasses. Growth responses to AM fungi fall along a continuum from mutualism to parasitism, and even within the life cycle of a plant, the benefit obtained from the symbiosis can vary. Seedlings benefit more from the symbiosis than do adult plants.

A wide range of other ecosystem functions are provided by mycorrhizal fungi and have a large impact on seedling establishment, litter decomposition, soil formation, and soil aggregation. Furthermore, resistance to drought, heavy metals, disease and stress is also provided by mycorrhizal fungi. It has also been proposed that AM fungi extend the niche of plants, and many plants would not be able to coexist with other plants without AM fungi.

It has been shown by a wide range of studies that mycorrhizal fungi modify competitive interactions between plants. Consequently, plant community structure and diversity are altered depending on the presence and/or composition of mycorrhizal fungal communities. A study by Yang et al. (2014) showed that the temporal stability of a plant community is influenced by AM fungi by differentially influencing plant species and reducing temporal variability in productivity. Some studies also showed that the introduction of mycorrhizal fungi into new habitats supports plant invasion. Moreover, the suppression of mycorrhizal networks by some invasive plants can modify plant community structure and impair seedling establishment of mycorrhizal plants.

The overall impact of mycorrhizal fungi on ecosystems is difficult to summarize, because so many variables are influenced simultaneously. This problem can be solved by summarizing the effects of a range of ecosystem functions and calculate an overall response index. In biodiversity research, multiple ecosystem functions are summarized into a so-called ecosystem multifunctionality index. It was found that, mycorrhizal fungi were one of the drivers of the positive effects of soil biodiversity on ecosystem multifunctionality, especially through their positive effects upon plant diversity.

It has been found that many endobacteria are harbored by arbuscular mycorrhizal fungi (AM fungi) in their hyphae and spores. In Glomeromycota, two types of endobacteria are known *viz.*, rod-shaped Gram-negative *Candidatus Glomeribacter gigasporarum* limited in distribution to members of the Gigasporaceae family, and coccoid Mollicutes-related endobacteria, widely distributed across different lineages of AM fungi. A recent breakthrough in understanding that allows fungi to establish mycorrhizal associations with plants has come from the discovery of bacteria living within fungal cells (endobacteria), which are now proposed as drivers of plant–fungus symbiosis. There is increasing evidence that bacteria living within fungal cells are widespread among mycorrhizal fungi and that their dynamics may be crucial to development and function of mycorrhizal associations. It was recently revealed that some endobacterial genes are critical to the establishment of mycorrhizal associations and that endobacteria alter the metabolic profile and growth of host fungi, such that without endobacteria fungi respond inappropriately to plant exudates.

Despite these advances, the overall picture of how endobacteria and fungi interact in the mycorrhizal association is still limited and fragmentary. Some types of mycorrhizal fungi, such as orchid mycorrhizal fungi, are only facultatively associated with plants, allowing the potential to generate a much wider array of fungi free of endobacteria. Such a system would facilitate a wide range of experiments elucidating the roles endobacteria play in affecting how fungi function in ecosystems and in interactions with their plant hosts.

10.7 Mineral Nutrients and Mycorrhiza

The role of mycorrhizal fungi in acquisition of mineral nutrients by host plants is examined for ectomycorrhizae (ECM), the ericoid mycorrhizae (EM), and the vesicular-arbuscular mycorrhizae (VAM). The mineral nutrition of the host plant may directly be affected by mycorrhizal infection enhancing plant growth through nutrient acquisition by the fungus, or indirectly by modifying transpiration rates and the composition of rhizosphere microflora.

A capacity for the external hyphae to take up and deliver nutrients to the plant has been demonstrated for the following nutrients and mycorrhizae; P (VAM, EM, ECM), NH_4^+ (VAM, EM, ECM), NO_3^- (ECM), K (VAM, ECM), Ca (VAM, EM), SO_4^{2-} (VAM), Cu (VAM), Zn (VAM) and Fe (EM). In experimental chambers, the external hyphae of AM fungi can deliver up to 80 per cent of plant P, 25 per cent of plant N, 10 per cent of plant K, 25 per cent of plant Zn and 60 per cent of plant Cu. Knowledge of the role of mycorrhiza in the uptake of nutrients other than P and N is limited because definitive studies are few, especially for the ECM. Although further quantification is required, it is feasible that the external hyphae may provide a significant delivery system for N, K, Cu, and Zn in addition to P in many soils. Proposals that ECM and AM fungi contribute substantially to the Mg, B, and Fe nutrition of the host plant have not been substantiated.

Ectoenzymes produced by ECM and EM fungi provide host plants with the potential to access organic N and P forms that are normally unavailable to AM fungi or to non-mycorrhizal roots. The relative contribution of these nutrient sources requires quantification in the field.

It has been shown that AM fungi forming 80–90 per cent associations with all terrestrial plants can contribute to alleviate problems related to intensive cultivation practices. A broad spectrum of beneficial functions have been shown by AM fungi for their host plants as well as for their environments like improved nutrient and water uptake, enhanced tolerance against biotic and abiotic stresses and improved soil structure to counteract soil erosion (Leifheit et al. 2014). At the same time, it has been observed that AM fungi and other fungal symbionts are severely affected by intensive agricultural practices like the use of large amounts of mineral fertilizers and pesticides and consequently can negatively impact plant nutrition and growth. Loss of fungal diversity disrupts major ecosystem services such as ecosystem variability and productivity and might decrease plant biodiversity (Wagg et al. 2014). On the other hand, there are several agricultural practices that can stimulate the presence of beneficial soil biota, such as crop rotation, diversification, or reduction of tillage intensity. However, no data is available for comparing the impact of different agricultural management practices such as conventional versus organic and permaculture systems on fungal communities in tropical rain forest ecosystems. Besides the wide range of beneficial properties, AMF can be used as inoculant and act as a biological fertilizer and disease control agent.

10.8 Factors Affecting AM Fungal Community Assembly

10.8.1 Regional Pool

According to paleocontinents, there are species pools of AM fungi with distinct composition, although endemic species are rare (Davison et al. 2015). Regionally, observed AM fungal communities are spatially heterogenous, but temporally stable, suggesting a fairly constant soil species pool from which mycorrhizae form during the season.

10.8.2 Propagules and Vectors of Dispersal in AM Fungi

The spores of AM fungi are often large, and many species are distributed by zoochory, earthworms, and collembolans or on the hooves of bison as opposed to wind, where their spores are detected rarely. Thus, AM fungal species are mostly limited to short-distance dispersal. However, over long time spans, these limited dispersal capabilities allow for a surprisingly efficient spread of taxa (Davison et al. 2015).

10.8.3 Spatial Community Structure, Dispersal Limitation and Other Stochastic Processes

Fungal communities of AM fungi are spatially structured, and patchily distributed even in relatively homogenous local environments, suggesting that there are other processes beyond environmental filtering that contribute to the structure of AM fungal communities, for example, dispersal limitation. The relative importance of dispersal to environmental filtering is scale-dependent and varies (soil type, soil pH, C/N ratio, phosphorus, soil temperature, plant biomes and dispersal). Dispersal and other neutral processes thus exhibit an effect size spectrum that can be completely masked by extreme environmental heterogeneity or anthropogenic disturbance, resulting in communities more dissimilar than expected under the assumptions of neutral theory. On the other hand, stochastic effects are also limited under very homogeneous environmental conditions because of niche effects.

10.8.3.1 Environmental Filter

10.8.3.1.1 Niche Partitioning Along Environmental Gradients Various soil environmental variables, such as pH, soil type, soil chemistry and nutrient availability influence the assembly process and the coexistence of AM fungi. As nutrient transport is a function of AM fungi, the effect of nutrient availability is well-studied. The filtering role of the environment, when some species from the regional pool are not present under certain soil conditions, was shown in fertilizer addition experiments: AM fungal phylotype diversity decreased with increasing N and P availability and some AM fungi were only found in specific soil nutrient conditions (Liu et al. 2015).

10.8.3.1.2 Seasonality AM fungi show temporal niche partitioning over the course of the year. Previously rare types might replace the dominant species. As possible explanations for this shift, both changing environment, for example, changes in temperature and sunshine hours (which influence the soil carbon pool), and the seasonal cycle of the plant community and phenology were suggested.

10.8.3.1.3 Disturbance Increasing agricultural land-use intensity selectively removes rare AM fungal species from the local community. The number of AM fungal species, their abundance and root colonization are lowered by heavy anthropogenic disturbance, such as plowing, tillage, and fungicide treatment, while favoring generalist. However, disturbance does not always shift communities in a predictable way, probably because of the dominance of stochastic effects.

10.8.3.2 Host Filter

One of the particular features of AM fungal community assembly is the importance of the host filter compared with free-living or facultative symbiotic organisms. AM fungal diversity in roots is restricted by plants and they also differentially influence sporulation. Given the obligatory symbiotic AM fungal lifestyle, the existence of host effects could be obvious. The non-evident detail in the host–AM fungal relationship is the apparent lack of species-level specificity. As an explanation, it was proposed that being able to colonize and to be colonized by a wider range of partners has an evolutionary benefit, and that environmental conditions affect the ability of plants to differentially reward their symbionts. In the field, different plant species, and even plants of the same species at different growth stages, associate with different fungal communities from the same soil and some AM fungi do not colonize certain plants. AM fungi differ regarding how beneficial they are for hosts, and plants are able to reward better fungal partners with photosynthates. The solution might be that host specificity does not happen at the species level, but on an ecological level, where generalist AM fungi interact with generalist plants while specialists tend to occur in the roots of specialist plants. Furthermore, pairings of hosts and symbionts with similar life history strategies (competitive, stress tolerant and ruderal) are likely more beneficial. The functional traits defining these strategies are often conserved at a higher taxonomic level. Not only the host itself but also neighboring plants and plant species richness influence fungal communities. In addition, AM fungal preference regarding hosts also exists.

10.8.4 Non-Host Biotic Interactions and Feedbacks

10.8.4.1 Horizontal Interactions Between Members of Local AM Fungal Communities

As opposed to root colonization, the ability of AM fungal species to colonize soil did not influence coexistence (Maherali and Klironomos 2012). Phylogenetic overdispersion promotes coexistence: communities of more distantly related and functionally different species showed higher realized species richness. Conserved differences in other functional traits, such as timing of spore production and hyphal growth rate, metabolism of photosynthates, P and N uptake, might alleviate competition as well.

Despite the potential importance for commercial use of fungal inocula, the effect of arrival order in AM fungi is not well-understood. Priority effects were shown; however, it was observed that the resident AM fungi did not suffer from reduced growth despite being invaded, which makes competition for space an unlikely explanation, and suggesting downregulation by the host instead (Werner and Kiers 2015).

10.8.4.2 Interactions With Other Non-Host Organisms

Negative interactions with consumers (fungal grazers), pathogens, and parasites could reduce competition between AM fungi. However, collembola feeding on AM fungi had no effect on the community composition, and parasitism has not yet been conclusively shown to exist in AM fungi. Either these interactions are really not important for AM fungal communities or we are limited by data.

AM fungi harbor bacteria associated with their spores. These bacteria promote hyphal growth and stimulate nutrient biodynamics. They might facilitate not only the fungus, but the whole mycorrhizal system by contributing to the suppression of soil-borne plant pathogens and by adding nitrogen fixation to the benefits of the plant.

10.9 Carbon and Nutrient Cycling and Ecosystem Multifunctionality

10.9.1 Effects on C Cycling

Mycorrhizal fungi play a key role in the biogeochemical cycles in terrestrial ecosystems. It has been shown that plants allocate between 10 and 20 per cent of their photosynthates to AM fungi. Approximately 20 per cent, and sometimes up to 50 per cent, of assimilates can be allocated to ECM and EM fungi. This, together with the fact that almost all terrestrial ecosystems, and some agricultural ecosystems, are dominated by ECM, EM or AM forming trees, shrubs and herbs, indicates that mycorrhizal fungi probably play a key role in the global C cycle. Heavily managed agricultural ecosystems are not included because the abundance of AM fungi is often reduced as a result of heavy fertilization, soil disturbance, and cultivation of non-mycorrhizal crops (e.g. oil rapeseed and sugar beet). Still, the majority of crops (e.g. maize, cereals, soybean, potato and rice grown under non-paddy conditions) are colonized by AM fungi in the field and thus allocate C to the fungal compartment below ground. The recent observations that mycorrhizal fungi are important regulators of C dynamics because of impaired degradation of fungal residues (Clemmensen et al. 2013) and that C storage is increased in ECM dominated versus AM dominated ecosystems (Averill et al. 2014) further support the role of mycorrhizal fungi in the terrestrial C cycle.

10.9.2 Effects on N and P Cycles

Significant amounts of N and P are provided by mycorrhizal fungi to their host plants in natural ecosystems, especially those with reduced soil nutrient availability. Experiments with single plants and plant communities have shown that AM fungi contribute up to 90 per cent of plant P. The contribution of AM fungi to plant N nutrition is less pronounced, often negligible and depends on factors such as soil water content, soil pH, and soil type. Moreover, AM fungi can immobilize significant amounts of N in mycelia.

Significant amounts of (organic) N and P can be acquired by ectomycorrhizal fungi and ericoid fungi, sometimes representing up to 80 per cent of plant N and plant P. A number of important tools and techniques have been developed to provide evidence for the role of mycorrhizal fungi in natural ecosystems, including hyphal bags or, comparison of wild type and mycorrhiza deficient plant mutants, tree trenching, or isotopes. Furthermore, stable isotope probing can be used to follow the fate of N and/or C through mycorrhizal networks.

While many studies have focused on nutrient uptake, relatively few have considered that mycorrhizal fungi could also contribute to the reduction of nutrient losses (e.g. efficient nutrient uptake reduces the risk of nutrient loss as a result of leaching or denitrification). Mycorrhizal fungi can significantly reduce N (up to $70\,kg\,N\,ha^{-1}\,year^{-1}$) and P (up to $150\,g\,P\,ha^{-1}\,year^{-1}$) leaching losses (Bender et al. 2015). AM fungi have been shown to reduce nutrient leaching losses even in the absence of a plant growth response. Interestingly, AM fungi can reduce leaching losses of both organic and inorganic nutrients (Bender et al. 2015). This might indicate that AM fungi acquire organic nutrients in line with a few earlier observations, and this is important because significant amounts of organic nutrients can be lost through leaching. Nutrient leaching losses also depend on the dominant mycorrhizal type (AM or ECM) of the vegetation. Mycorrhizal fungi can enhance nutrient use efficiency and ecosystem sustainability by preventing nutrient losses. Such ecosystem services are especially important in nutrient poor ecosystems where plant productivity is nutrient limited. AM fungi were shown to reduce emissions of N_2O, an important greenhouse gas with a 300 times higher global warming potential than CO_2. Such effects may be indirect, as AM fungi influence bacterial communities inhabiting the rhizosphere and mycorrhizosphere, and AM fungi may induce shifts in denitrifying communities.

10.10 Evolutionary Stability and Maintenance of Mutualism in Mycorrhizal Symbiosis

The symbiotic relation between plants and mycorrhizal fungi is widespread and very old. Thus, this symbiosis can be considered to be evolutionarily stable. However, the mechanisms contributing to evolutionary stability and to plant–fungal coexistence are only partly understood. Key questions are whether plants and fungi have the ability to regulate resource exchange (C and nutrients) and whether they can detect and sanction non-beneficial partners or specifically select for beneficial ones. Also, did plants or fungi evolve any specific mechanisms to control the exchange of C for nutrients? Furthermore, how do plants interact with multiple fungi at a molecular level? For instance, is the establishment of a particular fungus in a root segment influenced by the presence of other established fungi? The evolutionary biology of the mycorrhizal symbiosis is a research area with many unanswered questions.

Mycorrhizal fungi vary in effectiveness and some fungi deliver many more nutrients to plants compared with other plants. Vice versa, plant species differ in the amounts of C they can deliver to mycorrhizal fungi. Thus, selection forces should exist that favor the detection of beneficial partners and the efficiency of the symbiosis. Recent work revealed that AM fungi allocated more mineral nutrients to the most C rewarding plants, while, reciprocally, plants allocated more C to the fungus providing the highest mineral nutrition i.e. P or N. Thus, these studies indicate that plants and fungi both have the ability to detect and selectively reward beneficial symbionts. This analogy with a "biological market" has been used to explain the evolutionary stability of the symbiosis between plants and AM fungi. It is still unclear whether preferential resource allocation to beneficial symbionts is also occurring in complex mycorrhizal networks with many different symbionts or in other mycorrhizal associations (e.g. ECM, ericoid or orchid mycorrhizas) and this is an issue that deserves attention. The fact that plants usually have reduced root colonization levels when soil fertility is high (e.g. when fungal symbionts are not beneficial for the plant) or when light intensities and C availability are low further demonstrates that plants, at least in part, have the ability to regulate this symbiosis.

Several observations question the view of a biological market where partners control the exchange of C against mineral nutrients as a sole mechanism for understanding plant–fungal coexistence in mycorrhizal associations. First, mycoheterotrophic and mixotrophic plants bypass the previous market logic. The fungi reverse the C flow, and provide no sugar for the mineral nutrients they obtain. Moreover, mycoheterotrophic plants feeding on ECM networks extract N more efficiently from their fungus than autotrophic plants. It is possible that mycoheterotrophic plants provide other benefits to the fungi (e.g. vitamins or protection), but this has not been convincingly proven. Second, a meta-analysis by Hoeksema et al. (2010) revealed that in 10 per cent of the studies, mycorrhizal fungi reduced plant growth. This suggests that plants do not necessarily benefit from the symbiosis and are unable to exclude fungal symbionts under deleterious conditions. It also indicates that mycorrhizal fungi are regularly "cheating" and that the plant is not in full control of the symbiosis. Third, many plant species are not C limited, but grow in nutrient limited soils, and plant photosynthates are a luxury commodity under these conditions. Hence, for such plants there is no strong selection pressure to reward beneficial fungi or develop defense mechanisms against less effective fungi (e.g. because C delivery to less beneficial symbionts does not directly reduce plant fitness). Moreover, the biological market theory offers a view where each partner acts independently, but there is evidence for manipulations of the host gene expression by the colonizing symbionts through effector proteins. These effector proteins counteract the plant immune system and further work should test whether such effector proteins facilitate the establishment of other (non-beneficial) symbionts. Plants can control their N-fixing bacterial symbionts by using secreted effector polypeptides, and the possibility that host plants release effector-like proteins to control the penetration of mycorrhizal fungi deserves further study. Clearly, resource exchange regulation and biological market theory are not the only factors that explain plant–fungal coexistence and evolutionary stability in the mycorrhizal symbiosis.

10.11 Mycorrhizal Competition with Disease Organisms

Mycorrhizae can be beneficial to crops grown in soils with low nutrient supply. Mycorrhizal associations can lead to increase nutrient concentrations in plant tissue, particularly nutrients (e.g. P and Zn) which are dependent on diffusion toward roots for uptake. Therefore, management of mycorrhizae is becoming a potential method for crop nutrient management, although there remain a number of practical problems in this application.

In addition to nutritional benefits of mycorrhizal associations with plants, mycorrhizae may also compete with fungal pathogens for infection of plant roots. For example, ectomycorrhizal associations with pine seedlings (*Pinus sylvestris*) reduced disease severity of the pathogens, *Fusarium moniliforme* and *Rhizoctonia solani*. Inoculation of rice with VAM fungi is a promising management technique for management of Khaira disease. VAM also reduces *Fusarium* wilt severity in cumin.

10.12 Role of Mycorrhiza in Plant Disease Management

Soil-borne diseases are one of the important limiting factors for the commercial cultivation of various crops. The management of these diseases with chemical pesticides has exhibited numerous ill effects to the biotic system and environment. The efficacy of pesticides in the soil system is restricted and an effective delivery system is cumbersome. To provide effective and durable protection to the root system, the role of mycorrhiza against pathogenic population is well-established. The benefits of these associations have been found to reduce the vulnerability of the root system by decreasing abiotic stresses besides improving the ecological fitness of plant species in the soil environment. Balanced plant nutrition with macro/micronutrients attributes to host endurance against different biotic and abiotic stresses. Plant species are benefited from mycorrhizal association because of greater efficiency in nutrient and water uptake. Abiotic stresses predispose the host to pathogenic infections. Some of the diseases are more in nutrient-poor or moisture-deficient soils. Mycorrhizal plants are often more competitive and better equipped to tolerate environmental stresses than the non-mycorrhizal plants. Several studies have proved that seedlings with symbiotic associations exhibit more resistance to the host system against pathogenic fungi, bacteria and nematodes with various mechanisms. Symbiotic fungi utilize surplus carbohydrates from the root exudates and transform the monosaccharide to less soluble sugars which discourage the germination of propagules and attractiveness of the roots to the pathogens. In the case of ECM, the presence of fungal mantle of varied thickness creates a physical barrier to prevent the entrance or spread of a pathogen and nematode. Mycorrhizosphere is the root region where interactions between plants and microorganisms are more intense leading to the production of different types of antibiotics and metabolites exhibiting inhibition to pathogenic organisms. Symbiotic associations have been found to enhance the concentration of these inhibitors many times greater than non-symbiotic roots and protecting the plant species against infection. For artificial inoculation, mass multiplication procedure of ectomycorrhizal fungi and AM fungi differ because of their growth characteristics. For large scale production of inoculum, ECM fungi can be cultured on synthetic and natural media since these fungi rarely sporulate on these culture media, therefore a grain inocula (mycelium impregnated grain spawn) can be prepared by using locally available grains. AM fungi require a living host to grow which are known as nurse host species. Fibrous roots of these nurse host species are utilized as inoculum along with chlamydospores produced in the mycorrhizosphere. A new AM fungus isolated from desert of Rajasthan (*Piriformospora indica*) has been successfully cultured and multiplied on synthetic media. Considering multiple benefits, mycorrhiza incorporation can be considered as one of the important component of integrated disease management strategies particularly for organic production systems.

Mycorrhizosphere refers to the zone of soil influence by mycorrhizal association. The first report about the role of the mycorrhizosphere in biocontrol of pathogens was given by Meyer and Linderman (1986). They found that extracts of rhizosphere soil from mycorrhizal plants reduced sporangia formation of *Phytophthora cinnamomi* in comparison with extracts of rhizosphere soil from non-mycorrhizal plants. These authors postulated that either the sporulation-inducing microorganisms were missing or that the number of sporulation-inhibiting microorganisms increased. The changes in root exudates affect the microbial communities around the roots leading to the formation of mycorrhizosphere.

10.13 Conclusion

The species richness and diversity of the plant microbiota are greater than was believed just a few years ago. A new field of research into the intricacies of the plant holobiont is opening up. Certain key issues still needing to be addressed include: (i) description of the core microbiome and rules of community assemblage; (ii) the functions of the microbiota; (iii) the molecular interactions occurring between the host plant and its symbionts; and (iv) the link between symbiont diversity and functions. Mycorrhizal fungi occur as communities of organisms, embedded in complex biotic interactions in the soil, responding dynamically to an array of environmental cues and management factors – this is a complex ecological and also evolutionary situation that lacks a "one size fits all" solution. An appropriate technology needs to be developed, with a view toward social sustainability, and matching this complexity to better harness the mycorrhizal symbiosis for a sustainable functioning of agroecosystems.

References

Andres, C. and Bhullar, G.S. (2016). Sustainable intensification of tropical agro-ecosystems: need and potentials. *Frontiers in Environmental Science* 4: 5.

Averill, C., Turner, B.L., and Finzi, A.C. (2014). Mycorrhiza-mediated competition between plants and decomposers drives soil carbon storage. *Nature* 505 (7484): 543.

Bender, S.F., Conen, F., and Van der Heijden, M.G. (2015). Mycorrhizal effects on nutrient cycling, nutrient leaching and N_2O production in experimental grassland. *Soil Biology and Biochemistry* 80: 283–292.

Bender, S.F., Wagg, C., and van der Heijden, M.G. (2016). An underground revolution: biodiversity and soil ecological engineering for agricultural sustainability. *Trends in Ecology & Evolution* 31 (6): 440–452.

Berendsen, R.L., Pieterse, C.M., and Bakker, P.A. (2012). The rhizosphere microbiome and plant health. *Trends in Plant Science* 17 (8): 478–486.

Brundrett, M.C. (2009). Mycorrhizal associations and other means of nutrition of vascular plants: understanding the global diversity of host plants by resolving conflicting information and developing reliable means of diagnosis. *Plant and Soil* 320 (1–2): 37–77.

Clemmensen, K.E., Bahr, A., Ovaskainen, O. et al. (2013). Roots and associated fungi drive long-term carbon sequestration in boreal forest. *Science* 339 (6127): 1615–1618.

Davison, J., Moora, M., Öpik, M. et al. (2015). Global assessment of arbuscular mycorrhizal fungus diversity reveals very low endemism. *Science* 349 (6251): 970–973.

Denison, R.F., Bledsoe, C., Kahn, M. et al. (2003). Cooperation in the rhizosphere and the "free rider" problem. *Ecology* 84 (4): 838–845.

Giller, K.E., Andersson, J.A., Corbeels, M. et al. (2015). Beyond conservation agriculture. *Frontiers in Plant Science* 6: 870.

Hoeksema, J.D., Chaudhary, V.B., Gehring, C.A. et al. (2010). A meta-analysis of context-dependency in plant response to inoculation with mycorrhizal fungi. *Ecology Letters* 13 (3): 394–407.

Kartzinel, T.R., Trapnell, D.W., and Shefferson, R.P. (2013). Highly diverse and spatially heterogeneous mycorrhizal symbiosis in a rare epiphyte is unrelated to broad biogeographic or environmental features. *Molecular Ecology* 22 (23): 5949–5961.

Leifheit, E.F., Veresoglou, S.D., Lehmann, A. et al. (2014). Multiple factors influence the role of arbuscular mycorrhizal fungi in soil aggregation – a meta-analysis. *Plant and Soil* 374 (1–2): 523–537.

Liu, Y., Johnson, N.C., Mao, L. et al. (2015). Phylogenetic structure of arbuscular mycorrhizal community shifts in response to increasing soil fertility. *Soil Biology and Biochemistry* 89: 196–205.

Maherali, H. and Klironomos, J.N. (2012). Phylogenetic and trait-based assembly of arbuscular mycorrhizal fungal communities. *PLoS One* 7 (5): e36695.

Meyer, J.R. and Linderman, R.G. (1986). Selective influence on populations of rhizosphere or rhizoplane bacteria and actinomycetes by mycorrhizas formed by *Glomus fasciculatum*. *Soil Biology and Biochemistry* 18 (2): 191–196.

Pressel, S., Bidartondo, M.I., Ligrone, R., and Duckett, J.G. (2014). Fungal symbioses in bryophytes: new insights in the twenty first century. *Phytotaxa* 9 (1): 238–253.

Roesch, L.F., Fulthorpe, R.R., Riva, A. et al. (2007). Pyrosequencing enumerates and contrasts soil microbial diversity. *The ISME Journal* 1 (4): 283.

Taylor, D.L., Hollingsworth, T.N., McFarland, J.W. et al. (2014). A first comprehensive census of fungi in soil reveals both hyperdiversity and fine-scale niche partitioning. *Ecological Monographs* 84 (1): 3–20.

Tisserant, E., Malbreil, M., Kuo, A. et al. (2013). Genome of an arbuscular mycorrhizal fungus provides insight into the oldest plant symbiosis. *Proceedings of the National Academy of Sciences of the United States of America* 110 (50): 20117–20122.

Wagg, C., Bender, S.F., Widmer, F., and van der Heijden, M.G. (2014). Soil biodiversity and soil community composition determine ecosystem multifunctionality. *Proceedings of the National Academy of Sciences of the United States of America* 111 (14): 5266–5270.

Walker, J.F., Aldrich-Wolfe, L., Riffel, A. et al. (2011). Diverse Helotiales associated with the roots of three species of Arctic Ericaceae provide no evidence for host specificity. *New Phytologist* 191 (2): 515–527.

Werner, G.D. and Kiers, E.T. (2015). Order of arrival structures arbuscular mycorrhizal colonization of plants. *New Phytologist* 205 (4): 1515–1524.

Yang, G., Liu, N., Lu, W. et al. (2014). The interaction between arbuscular mycorrhizal fungi and soil phosphorus availability influences plant community productivity and ecosystem stability. *Journal of Ecology* 102 (4): 1072–1082.

11

Microbiome-Driven Nutrient Fortification in Plants

The Role of Microbiota in Chemical Transformation and Nutrient Mobilization

Irina Sidorova and Elena Voronina

Lomonosov Moscow State University, Moscow, Russia

11.1 Introduction

Symbiotic and associative interactions with beneficial microorganisms (MO) are the keystone of plant performance. The plant microbiome (treated here as sum of bacteria and fungi) plays crucial roles in plant life such as nutrient fortification, defense against pathogens and pests and mediating of multiple abiotic stresses (Doty 2017). It comprises the above-ground (e.g. phylloplane MO and stem and leave endophytes) and underground parts (root-associated soil-borne MO, root endophytes and mycorrhizal fungi). This chapter is focused on the microbiome's role in plant nutrition facilitation, and only the underground part will be considered further. Root-associated MO are of particular importance in various compounds' (both nutritional and toxic for plants) chemical transformation, are able to change the soil structure by both physical and chemical mechanisms, and to provide defense against deleterious organisms and stresses (Turner et al. 2013). Thus, the underground microbiome contributes largely to the plant performance and requires much attention from the position of sustainable agriculture. The current opinion on the plant microbiome's structure and functions are outlined in Doty (2017), Kashyap et al. (2018) and other recent reviews.

Soil is a well-acknowledged "hot spot" for numerous activities for a plethora of various living organisms interacting in many ways, both synergistic and antagonistic (Buscot and Varma 2005). The zone close to the plant roots exhibiting the prominent soil-engineering action has the highest density of biotic interactions compared with bulk soil, involving the plants themselves and a range of MO, vertebrates and invertebrates interconnected by trophic and topical links (Cardon and Whitbeck 2007). The remarkable complexity of these relations and interferences and their potential for application in plant growth promotion heightened the researchers' interest and so gave rise to masses of concepts and interpretations of terms (Summerbell 2005). To avoid likely ambiguity here we present the definition

The Plant Microbiome in Sustainable Agriculture, First Edition. Edited by Alok Kumar Srivastava, Prem Lal Kashyap and Madhumita Srivastava.

for the most widespread (and the most varied) terms related to the chapter's objective as we address them.

- **Endophyte** – a fungus or bacterium, inhabiting plant inner tissues both without pathogenesis symptoms and defined contact zone formation. An endophyte's impact on the host plant is considered as beneficial or neutral, and the relations between two partners are not always recognized.
- **Hyphosphere** – a zone of influence of extraradical mycorrhizal or free-living substrate hyphal fungal structures, including hyphal cords, rhizomorphs, and sclerotia.
- **Mycobiont, or mycorrhizal fungus** – a fungus engaged in any type of mycorrhizal interactions with the host plant. The interface between symbionts is arranged in a contact zone of typical structure, where transport processes take place. The mycobionts' impact on the host plant is considered as beneficial.
- **Mycorrhizosphere** – a zone similar to the rhizosphere for mycorrhizal roots (excluding the vast extraradical mycelial network), where both plant and fungal symbionts interfere with soil microbiota.
- **Rhizosphere** – a non-mycorrhizal root zone of influence, presumed to be governed only by the plant with its root exudates and bioactive compounds, contrary to mycorrhizosphere, where mycobionts' metabolites are not less significant than those of plants.

11.2 The Microbiome Contribution to Plant Nutrient Fortification

Soil microbiota can influence plants in numerous ways, both beneficial and deleterious. The most well-recognized and evident positive MO effect is the facilitation of plant nutrition which can lead to higher plant fitness and its lesser susceptibility to pathogen and pests attacks, drought, salinity, intoxication, and other stresses too. Microbiome-driven plant nutrient fortification is based on two main principles, overviewed below. The first implies both physical and biochemical rearrangement of soil structure by fungal hyphae, the possibility for MO to exploit soil microzones unavailable for more ponderous plant roots *per se*, and the ability of mycobionts' mycelia attached to roots to help the host root systems out of the depletion zones, inevitably pursuing the plants while they grow. The second mode of action considers nutrient mobilization from recalcitrant substrates and plant nutritional fortification via MO by various chemical activities and metabolic pathways not inherent for the plant themselves.

11.3 Microbiome-Driven Rearrangement of Soil Structure and Enhancing Soil Exploitation for Plant Nutrient Supply

Soil fertility is crucial for plant survival and health. One of the approaches to its maintenance and improvement targets the soil physical properties, especially the aggregate formation and stabilization. Soil aggregation, dependent both on root and MO activities, influence plant growth, soil resistance to erosion, soil organic compound turnover, and is critical for a wide range of soil biota groups (Chotte 2005). The role of mycorrhizal fungi in this process was revealed in the early 1900s and is now widely acknowledged. The mechanisms underlying it are not completely elucidated yet, but can be divided roughly into biophysical, biochemical, and biological factors, often acting together (Lehmann et al. 2017).

Arbuscular mycorrhizal fungi (AMF), the most studied group in this regard, possess the ability to influence soil structure in different ways. The excretion of a stable hydrophobic glycoprotein glomalin and, probably, some other related proteins and polysaccharides is the most studied AMF effect on soil aggregation (Mickan 2017). Glomalin was shown to bind soil particles facilitating the soil aggregation and corresponding water retention capacity, but its previous report of high contribution to soil organic matter is currently considered as controversial, due to the difficulties in defining its relative importance in the different soil types (Lehmann et al. 2017). Glomalin-related soil proteins are shown to be involved in soil aggregation too, but its origin from AMF remains arguable (Mickan 2017). The less recognized impact is drastic pH change in the hyphosphere leading to local acidification (Bago et al. 1998). Biophysical effects on soil structure imply soil particle entanglement and mechanical pressure exerted by hyphae along with mycelial-caused local water regime alterations (Lehmann et al. 2017). Physical binding (entanglement) of soil particles by ramifying AMF hyphae can stabilize soil aggregates, especially under disturbance, but this effect is strongly dependent on mycelial network persistence and extension, a species-specific branching pattern and other variable parameters (Rillig and Mummey 2006). Biological AMF effect on soil aggregation involves hyphosphere bacteria utilizing mycelial exudates. Some of them, such as *Paenibacillus* spp. are known to enhance microaggregate formation (Mickan 2017).

The potential effect of ectomycorrhizal fungi (EMF) on soil aggregation has derived far less attention. The possible reason is the presumed lesser significance of this process for forest biomes, inhibited by EMF and their host trees, compared to AMF-dominated grasslands and fields. However, there are evidences for EMF-dependent soil structure improvement (Lehmann et al. 2017). The biophysical effect is the main case for EMF mycelial structures possessing longevity significantly higher than AMF. The most prominent biological effect suggests the well-known ability of EMF to change the root system architecture during symbiosis formation (Smith and Read 2008), thus altering root contribution to soil aggregation. The beneficial effect was shown for a range of EMF and turned out to be species-specific. For the detailed discussion of mycorrhiza-dependent soil aggregation refer Lehmann et al. (2017).

The mycelial mode of life predetermines the prosperity of fungi in the soil. Fungal hyphae, much narrower than plant absorbing roots, can invade soil microzones unavailable for plants alone to obtain nutrient supply, and in the case of mycobionts, to transport them to their hosts, thus improving plant nutrition. Given the large amount of mycorrhizal external mycelia, especially in EMF (2000–8000 m/m of colonized root) (Smith and Read 2008), its role in enhancing the contact between roots and soil should not be underestimated. Due to slow diffusion coupled with rapid absorption typical for some key nutrients (such as P in the most common for soil in the form of orthophosphate), depletion zones are generated by roots during their growth. Both long lateral roots with abundant hairs and mycobionts' hyphae help plants to overcome the problem, but extending the symbiotic mycelial network proved to be the most efficient (and less "costly" for the plant) strategy for bypassing the nutrient deficient zone (Smith and Read 2008; Mickan 2017). Finally, the architecture of mycorrhizal mycelial networks themselves, the presence of mycelial aggregates such as cords or rhizomorphs determine the space and pattern of substrate occupying by fungus (so called, foraging strategy) and therefore mediate the host's nutrient supply (Agerer 2001).

11.4 Microbiome-Driven Chemical Transformation and Nutrient Mobilization

Fungi and bacteria are able to utilize a wide range of substrates, both natural and anthropogenic, to extract nutrients immobilized in the recalcitrant minerals, complex organic compounds and synthetic polymers too. As it was brilliantly pointed out by Jenkinson (1990), MO are "the eye of the needle through which virtually all nutrients must pass." Diverse chemical activities of microbes are the reason for the key role of MO litter mineralization processes, bioweathering and C and N cycling facilitation. The plant-associated MO inhabiting the mycorrhizosphere and rhizosphere are the most promising group for agricultural application. Their numbers increase in the root zone, and there microbial-driven chemical transformation and nutrient mobilization are more prominent than in bulk soil. Plant microbiomes can significantly contribute to plant nutrition via chemical transformation of a range of both organic and inorganic soil-borne compounds inaccessible for plants alone. It should be noted, that microbes have an integrated effect on plant nutrition for a single MO can possess, e.g. both nitrogen-fixing and phosphate-solubilizing activities. Often it is unclear, which nutritional (or non-nutritional) MO effect provides the plant growth promotion observed under experimental conditions. The main effects of plant-associated MO on plant nutrition are discussed below.

11.4.1 Phosphorus Nutrition

Phosphorus (P) is an essential macronutrient for plants as well as other living organisms. It presents in soil both in inorganic (P_i) and organic (P_o) forms. P_o pool is studied insufficiently compared to P_i; however it contributes on average 30–65 per cent of total P in mineral soils (Jones and Oburger 2011). Plants are often subject to P limitation even if its overall content in soil is high, for only 0.1 per cent of P is available (Sindhu et al. 2014b). Prevailing natural P forms are solid, low-labile and thus inaccessible for plant roots absorbing nutrients primary from soil solution. Available P supply is determined both by pH-dependent dissolution of P-bearing minerals and sorption–desorption solution equilibrium, and organic P-containing compounds cleavage during MO-driven mineralization and enzymatic hydrolysis (Jones and Oburger 2011). Another challenge for plants concerning P nutrition is the patchy distribution pattern of the element and depletion zone formation near the roots (Smith and Read 2008).

P-solubilizing activity was detected in bacteria for the first time in 1903 and since then was recognized in a range of MO, both bacteria and fungi (Fomina et al. 2006; Richardson and Simpson 2011). At present, we have an ample body of evidences for the significant role of MO-driven P solubilization, but the data generally concern either the effect observed *in vitro* (the solubilization zone detection), or the moieties/compounds potentially contributing to the process derived from a certain MO (reviewed in Sindhu et al. 2014b; Kaur and Reddy 2017). Such research papers are useful indeed, but in many cases the mechanisms underlying remain obscure and it is difficult to estimate the true contribution of the target MO to P-solubilization under realistic soil conditions.

The main mechanisms proposed for microbial P_i solubilization imply pH changes (predominantly, acidification due to ion exchange processes) and chelation (bonding metal cations) (Khan et al. 2007). The moieties/compounds involved include organic acid anions (generally referred to as organic acids), siderophores, protons, hydroxyl ions and CO_2 (Jones and Oburger 2011).

The acidification as a driving force for P_i mobilization is widely acknowledged both for fungi and bacteria (Uroz et al. 2009). The main active moieties are protons, originating from N (NH_4^+) – as was shown for *Penicillium* spp., or C (glucose) – for *Pseudomonas fluorescens*, supply (reviewed in Jones and Oburger 2011). In more rare cases the acidification was caused by inorganic acids, such as nitric and sulfuric, as was shown for bacterial genera *Nitrosomonas* and *Thiobacillus* (Kaur and Reddy 2017). Organic acid production, often accompanying P_i solubilization, is generally of little importance in the pH shifting under natural conditions, but this effect was clearly observed at least for 2-ketogluconic acid produced by nodule bacterium *Rhizobium leguminosarum* (Khan et al. 2007). The role of pH changes for MO P-solubilization was revealed mostly *in vitro*. Under field conditions the significant acidification can be expected only in "hotspots" of MO and/or root activity, such as (mycor-)rhizo- or hyphosphere, due to the bulk soil limitation in labile N and C and often shows high pH buffer capacity (Uroz et al. 2009; Jones and Oburger 2011).

The microbiome P-solubilizing effect cannot be explained by pH changes only (Whitelaw 1999). The organic acid anion production was detected in plenty of research, considering P_i mobilization for many MO species. These compounds are generally described as "acids," but as Jones and Oburger (2011) pointed out, "anions" is the more suitable term for its dissociated state both within cells and under soil conditions. They attracted a lot of attention from the researchers and topical reviews can be found in Khan et al. (2007), Sindhu et al. (2014b) and Kaur and Reddy (2017). The predominant MO-derived organic acids contributing to P solubilization are gluconic, 2-ketogluconic, oxalic, citric, succinic, and tartaric. Two former compounds are often reported from bacteria, oxalic and citric acids are more widespread within fungi. The P-solubilizing effect of organic acid anions is based primary on their metal complexation ability, and tri-carboxylic anions are presumed to be more efficient than di-carboxylic ones (Jones and Oburger 2011; Kaur and Reddy 2017). Similarly as for pH change effects, it should be remembered, that organic acid production in the soil may be dissimilar from the patterns observed under laboratory conditions due to lack of available C and other soil properties (Jones and Oburger 2011). Another promising group of MO-derived compounds mediating P-solubilization to be considered are siderophores. They possess high affinity for iron (Fe) and are widespread within MO due to Fe limitation common for soil habitats (Crowley 2006). Having in mind the frequency of Fe phosphates in soils and MO requirements for iron, it is necessary to enhance siderophore research with relation not only to Fe, but to P plant nutrition also (Jones and Oburger 2011).

MO-driven P_o mobilization is accomplished both by extracellular enzymatic activity for nutrient release and during substrate degradation aimed on the energy acquirement, and results in P_o mineralization (McGill and Cole 1981). The enzymes involved comprise phosphatases (predominantly, phosphomonoesterases including phytases), phosphonatases and C-P lyases (Sindhu et al. 2014b; Kaur and Reddy 2017). Phosphatases act via dephosphorylating of phosphoester or phosphoanhydride bonds. These enzymes are split into acid and alkaline phosphatases, but whereas the latter group is present in MO, it is scarce in plant roots thus heightening the nutrient-fortifying role of plant microbiome, especially in neutral and basic soils (Criquet et al. 2004). Acid phosphatases are produced by plant roots, but the similar microbial enzymes exhibit the higher affinity for P_o (Richardson et al. 2009). Phosphatase activity was detected for a range of soil MO, e.g. species of *Aspergillus, Penicillium,* and *Trichoderma* (Li et al. 2015; Kaur and Reddy 2017), but its correlation with final P_i concentrations in soil solution is yet unclear, and the data obtained are controversial (Jones and Oburger 2011). Phytases are a kind of phosphatase catalyzing the hydrolysis of phytate (Richardson and Simpson 2011). As this compound represents the major P_o matter suggested for the soils but is not readily available to plants, MO phytases have derived much attention and are often discussed apart from other phosphatase types. Other microbial enzymes contributing to P_o mobilization are phosphonatases and C-P lyases cleaving organophosphonates by breaking C-P bonds (Kaur and Reddy 2017).

Phosphate solubilizing MO are mostly ubiquitous soil inhabitants, regularly isolated from different soil types and microhabitats within. Up to 50 per cent of bacteria and up to 0.5 per cent of fungi from microbiota of a given soil possess this activity (Wakelin et al. 2004). The most frequently mentioned bacterial genera are *Alcaligenes, Arthrobacter, Azotobacter, Bradyrhizobium, Bacillus, Burkholderia, Chromobacterium, Erwinia, Flavobacterium, Micrococcus, Pseudomonas, Rhizobium, Serratia, Streptomyces,* and *Thiobacillus*; fungi include *Aspergillus, Fusarium, Mucor, Penicillium, Rhizopus, Trichoderma* (Sindhu et al. 2014b). P-solubilizing bacteria outnumber the matching fungi by 2–150 times, while it was shown that fungi act more efficiently than bacteria in acidic soils and generally exhibit higher organic acid production (Khan et al. 2007). Bacteria *(Pseudomonas striata, Serratia marcescens,* and *Bacillus polymyxa*) and fungal species *(Aspergillus awamori, Aspergillus niger,* and *Penicillium digitatum*) known for prominent organic acid production, are considered as the most efficient within soil-borne MO (Ben Farhat et al. 2009; Kaur and Reddy 2017). Bacterial P-solubilizing ability is known to be strain-specific and can be based on more than one mechanism of action. The main fungal modes of P mobilization are acidification, organic acids and extracellular phosphatase production (Jones and Oburger 2011).

The plant microbiome is known for more evident P-solubilization microbial activity compared to bulk soil, for some MO tend to increase their numbers in the root zone, or rhizosphere strains exhibit higher activity than their matches isolated from other soil microhabitats (Vazquez et al. 2000; Barreto et al. 2008). Plant roots provide a unique niche for MO rhizosphere, both in chemical and physical properties, as distinct from the bulk soil. For the most of the plant species involved in mycorrhizal symbiosis, this particular zone should rather be referred to as the mycorrhizosphere, and mycobionts' soil-engineering impact should be considered as well as that of plants (Voronina and Sidorova 2017).

The most apparent part of a mycorrhizal plant microbiome is the mycobiont itself. The mycorrhiza-driven plant nutrient fortification was recognized in the nineteenth century, and since then mycobionts' role in P and N plant supply has been widely discussed (Smith and Read 2008). AMF are the most important mycobiont group for agriculture due to their symbiosis with most crops, but their ability for P chemical transformations is rather limited. AMF are presumed to uptake predominantly dissolved P_i and only recently were shown to produce acid phosphatases, this fact requiring more detailed research (Saito and Ezawa 2017). So the main AMF contribution to plant P nutrition is determined by more efficient absorbing via an extended hyphal network beyond the depletion zone (Smith and Read 2008). EMF, on the contrary, possess significant ability to litter decomposing and are able to extract P and N from organic matter and to transport the nutrients to host plants (Read et al. 2004). Hydrolyzing capacity varies within the EMF strains and species, but there are evidences of their phosphatase activity being comparable, or even surpassing those in saprobic fungi (Tunlid et al. 2017).

The most controversial point is EMF P_i-solubilizing capacity. The idea of "rock-eating" mycorrhizal fungi, able to dissolve recalcitrant minerals by organic acids secreting (H^+ releasing) and thus to enhance plant nutrient uptake, became popular since the 1990s (Jongmans et al. 1997). EMF are well-acknowledged agents of bioweathering, i.e. degrading minerals, but so far it remains unknown, if this process answers fungal nutritional demands and it is impossible to separate host and mycobiont effects even under experimental conditions (Rosenstock 2009; Jansa et al. 2011). One of the witnesses for "rock-eating" fungi benefits, are the tunnels in minerals claimed to be drilled by fungi; in the opinion of some authors these are of "totally abiotic origin" and ascribing these "holes and tracks" to the action of living organisms is based on "a lack of understanding of crystallography and chemical dissolution mechanisms" (Sverdrup 2009). The microcosm-based studies demonstrating H^+-mediated P release from minerals, such as apatite (Rosling et al. 2004), do not allow extrapolation of the data obtained to environmental processes, and besides, the occurrence of MO other than the mycobiont is often probable in the experimental system.

Currently mycorrhiza is considered as the multitrophic symbiotic complex containing an ample suite of associative MO along with the mycobiont and its host plant (Timonen and Marschner 2006; Duponnois et al. 2008). Many functions in plant nutrient fortification and plant protection, previously ascribed directly to mycorrhizal fungus, are indeed mediated by accompanying mycorrhizosphere MO (Voronina and Sidorova 2017). This is the case for P supply too, for P-solubilizing bacteria accumulate in the ectomycorrhizosphere, and there is evidence for their significant increase in the symbiotrophic fungi hyphosphere compared to saprobic fungi from the same habitats (Sidorova et al. 2014). Mycorrhizosphere was shown to favor MO beneficial for plants, so in the case of *Pseudotsuga menziesii* – *Laccaria bicolor* EM, the P-solubilizing strains of *P. fluorescens* increased in the root zone compared to the bulk soil (Frey-Klett et al. 2005). Thus current opinion on EMF-mediated P_i-solubilization and other mineral-dissolving processes accompanied with nutrient release suggests a significant role of bacteria as primary drivers of bioweathering, and mycobionts themselves are considered rather as providers of niches for MO, than direct agents of solubilization (Koele et al. 2009; Uroz et al. 2009). The unequivocal evidence of P-solubilizing mycobionts activity not interfered with by other MO could be provided only by ericoid mycorrhizal fungi known for their high saprobic capacity and ability to be maintained in axenic cultures without host plants (Martino et al. 2003; Gibson and Mitchell 2004). Bacteria contribute to P_o biological mycorrhizal mineralization too by producing phytases, so EMF can further take up the products of phytate cleavage and transport them to the plants (Saito and Ezawa 2017).

11.4.2 Nitrogen Nutrition

Nitrogen (N) is one of essential macronutrients for plants, more abundant in soils than P, but required by plants about 10-fold comparable to P (Smith and Read 2008). Plants are able to acquire N predominantly in the forms of ammonium (NH_4^+), nitrate (NO_3^-), or nitrite (NO_2^-). Soil nitrogen (N) pools are dominated by recalcitrant organic N compounds (N_o) unavailable to plants except for the simple forms, so MO activity is crucial for its decomposition and mineralization resulting in available N release. The continuous process of mineral N (N_i) incorporation in soil MO biomass and its subsequent release into the soluble pool, so called "mineralization–immobilization turnover" is highly important for plants both under natural conditions and in agroecosystems without application of large amounts of N fertilizers (McNeill and Unkovich 2007). In boreal forest soils N acquired by plants consists of about 80 per cent amino acids, 5–15 per cent ammonium, and 5–11 per cent nitrate (Pena 2017).

Relative inert atmospheric N_2 fixation implying both biological and non-biological pathways is considered as one of the essential "entry-points" for N into terrestrial ecosystems. Biological N_2 fixation (reducing N_2 to NH_3 by nitrogenase enzymatic complex) accomplished by diazotrophic MO at present contributes no lesser than 25 per cent of the global N input (Galloway et al. 1995). Diazotrophs comprise about 90 prokaryotic genera belonging to actinomycetes, cyanobacteria, proteobacteria, etc. (McNeill and Unkovich 2007). Some diazotrophs are free-living MO (*Acetobacter*, *Azospirillum*, *Beijerinckia*, *Herbaspirillum*, *Klebsiella* and *Pseudomonas*), but the majority of them are associated, or even engaged in symbiotic relationships with plants (James 2000). Even seemingly asymbiotic N_2-fixers, e.g. *Azospirillum*, are stimulated in the (mycor-)rhizosphere zone compared to the bulk soil (Bloemberg and Lugtenberg 2001). They can establish tripartite associations with mycorrhizal mycobionts and host plants, trading N for organic compounds excreting by roots and hyphae and acquiring a low-oxygen microhabitat suitable for nitrogenase activity (Li et al. 1992). The rhizospheric diazotrophs are vital for cereals for these plants naturally are not involved in N_2-fixing symbiosis (Santi et al. 2013). The proteobacteria *Azospirillum*, *Azotobacter*, *Burkholderia*, *Enterobacter*, and *Pseudomonas* lacking obligatory dependence on plants and more closely plant-associated *Azoarcus*, *Herbaspirillum and Glucenobacter* are considered as plant-growth-promoting rhizobacteria due to their multiple beneficial effects on plants (Richardson et al. 2009). Thus, *Azospirillum* was shown to facilitate cereal growth by production of phytohormones rather than by N_2-fixation (Santi et al. 2013). It is not always easy to delimit free-living and associated diazotrophs in the natural environment because of continual transition from free-living to rhizospheric or even endophytic forms.

Symbiotic N_2-fixation in general is more efficient than its match in free-living and associative MO due to the C and P supply from the host. Besides, the plant provides diazotrophs with protection from oxygen-caused deactivation of nitrogenase complex (McNeill and Unkovich 2007). The most environmentally significant and well-studied symbiotic diazotrophs are so called "rhizobia," proteobacteria nodulating predominantly leguminous plants (Fabaceae) with one exception of the genus *Parasponia* from Cannabaceae (Sprent

2001). Initially all N_2-fixing legume-associated bacteria were considered as nodule-making Rhizobiaceae (α-proteobacteria), but later β-proteobacteria, such as *Ralstonia*, and other MO were discovered in root nodules and their diazotrophic capability was shown (Chen et al. 2003). Now a wide range of legume-nodulating bacterial genera are recognized, i.e. *Azorhizobium, Bradyrhizobium, Mesorhizobium, Ralstonia and Rhizobium* in tree species and *Blastobacter, Burkholderia, Devosia, Methylobacterium and Sinorhizobium* in herbs and shrubs (Chen et al. 2003; Sprent and James 2007). Their numbers are steadily increasing, but there is an urgent need to delve into rhizobial ecology to obtain a realistic estimation of N amounts and various MO fix for their hosts (Graham 2008).

Diazotrophic symbiosis of non-leguminous plants derived far less attention from the researchers compared to legumes–rhizobia interactions despite of their noteworthy agricultural potential (McNeill and Unkovich 2007). The main MO groups involved are heterotrophic actinomycetes of the genus *Frankia* nodulating so-called actinorhizal plants (220 species from 8 families of orders Fagales, Rosales and Cucurbitales) and autotrophic cyanobacteria (often *Nostoc* species) establishing permanent symbiosis or associations with liverworts, hornworts, ferns (*Azolla*), gymnosperm (Cycadaceae) and angiosperm (Gunneraceae) plants (Rai et al. 2002; Santi et al. 2013).

Actinorhizal plants are mainly shrubs and trees with worldwide distribution, with both temperate (*Alnus*) and tropical (*Casuarina*) species. Some of them are recruited in mycorrhizas thus maintaining tripartite symbiotic systems adopted to dwell in poor soils (Dawson 2008). These plants are used in revegetation of disturbed areas or to prevent desertification. Actinorhizas are highly efficient in N_2-fixing; in temperate biomes their activity matches with those of rhizobia estimated as 300 kg ha^{-1} $year^{-1}$ (Santi et al. 2013).

Cyanobacterial associations are extremely variable and occur within a wide range of living organisms, both aquatic and terrestrial. A comprehensive topical overview is provided by Rai et al. (2002). The most common agricultural N biofertilizer exploiting cyanobacteria is the aquatic fern *Azolla* (Azollaceae) distributed from warm temperate to tropic regions and containing the obligatory symbiotic diazotroph *Nostoc azollae* (Santi et al. 2013).

Fungi should be anticipated as active N miners due to their additional requirement of this nutrient for chitin synthesis. Both AMF and EMF are known to utilize a range of N sources, but for the first, the ability to transfer its significant amounts to plants was debated. Since the role of AMF in plant P nutrition is well-documented, their observed effect on plant N status varied from positive to negative (Giovannetti et al. 2017). Until recently, AMF contribution to plant N budget was considered generally as some "side-effect" of P supply, the most obvious in the case of legumes involved a tripartite plant–AM–rhizobial symbiosis (Sprent and James 2007). Another suggested indirect effect results in AMF influence on mycorrhizosphere populations of N-transforming (ammonium oxidizers, ammonifying, and denitrifying) MO (Amora-Lazcano et al. 1998). However, recent advances in molecular and genomic approaches allowed recognition of a series of fungal as well as plant AMF-inducible N transporters, therefore indicating the existence of direct N supply provided by the mycobiont to its host (see discussion in Giovannetti et al. 2017). Another noteworthy recent disclosure is the ability of at least some AMF to transfer substantial amounts of N derived from organic sources to the plant causing enhancement of its N content and growth promotion. Unlike N_i, N_o fertilization did not suppress the symbiotic activity (Thirkell et al. 2016). However, N metabolism of AM roots needs further study.

On the other hand, the role of EMF extensive mycelial networks in plant N acquisition is proverbial and considered as an essential host benefit in this symbiosis (Smith and Read 2008). The functionality of N uptake greatly varies even between strains, characterized by switching between different sources and depends on the particular soil environment. EMF can utilize both NH_4^+ and NO_3^- N_i forms with some preference toward the first and has potential to mobilize N from complex organic substrates, such as chitin (one of the major N_o compounds in soil), proteins and phytate (Hodge 2017). It is known that EMF are well adapted to N-deficiency and conversely are in general suppressed by redundant industrially originated N (Smith and Read 2008). About 87 per cent of species examined under laboratory conditions demonstrated protease releasing and the ability to utilize proteins as the sole N source (Pena 2017). According to genome-based data, EMF have a limited degrading enzymes set compared to saprobes, but at least one species (*Paxillus involutus*) exhibited non-enzymatic mineralization capacity similar to the brown-rot fungi glucose-depending Fenton-reaction, implying highly-oxidizing free radical release (Rineau et al. 2013). Though the inherent EMF ability to degrade organic matter is not very high, it can significantly facilitate soil MO-driven N_o transformation for plant benefit via delivering host-derived carbon to the microbial community (Pena 2017). So EMF-mediated N supply is crucial for plants, but there is an urgent need for accurate estimation of the nutrient amounts transferred under realistic and often N-limited conditions. To obtain a clearer understanding of MO-driven plant nutrient fortification, the environmental dynamics, the diversity of fungal responses to N as well as the links between different nutrient availability should be considered.

11.4.3 Potassium Uptake

Potassium (K) is the third main plant nutrient after P and N. It is essential for intracellular enzyme activity, membrane transport processes and cell osmotic potential maintenance. K uptake and assimilation in plants is tightly interconnected with other nutrients, such as P and N, metabolic pathways (Sindhu et al. 2014a). K decrease with large-scale biomass removal with crop harvesting and leaching losses due to anthropogenic soil acidification can cause its deficiency; and there are already large agricultural areas in the tropics with very low K status (Meena et al. 2014).

In view of the above, the MO capacity to release K from minerals which are unavailable for plants, comprising more than 90 per cent of the total soil K pool, attracted attention of the researchers after long neglect. The main K-bearing minerals in the soil are the silicates of feldspar and mica groups, such as biotite, illite, muscovite, orthoclase, phlogopite and vermiculite (Römheld and Kirkby 2010). Similar to P-solubilization, their bioweathering can be accomplished by joint bacterial–fungal chemical activity via the same mechanisms, including acidification and chelation, and acidolysis. The organic acids, especially citric, oxalic, and tartaric, production was shown to facilitate K release (Uroz et al. 2009; Sindhu et al. 2014a). MO-mediated K-solubilization in rhizosphere is exercised by a large number of soil bacteria (*Acidothiobacillus ferrooxidans*, *Bacillus mucilaginosus*, *Bacillus edaphicus*, *Bacillus circulans*, *Burkholderia* spp. *and Paenibacillus* spp.) and microfungi from the genus *Aspergillus*. Some strains are already successfully applied for soil K fertility amendment (Meena et al. 2014).

Until recently the role of mycorrhizal fungi in K supply to the host plant was generally neglected in favor of N and P nutrition studies. Taking the overall nutrient content increase for the main estimator of mycobiont efficiency proved to be irrelevant due to poor reflection of real nutrient absorption and its role in plant performance (Smith and Read 2008). The application of the stable isotope technique to nutrient transfer research provided insights in mycorrhizal-driven plant support (Finlay 1992). There are some reports of AMF-mediated increase of plant K uptake, based both on direct and indirect (P-dependent) effects. However, as K accumulation is linked with plant P metabolism and is strongly influenced by the N forms, the AMF role in K nutrition needs further investigation (Smith and Read 2008).

EMF are known as efficient bioweathering agents, so their K-solubilizing activity was tested in a wide range of *in vitro* and microcosm-based studies (Rosling et al. 2004; Rosenstock 2009). *P. involutus* and *Laccaria laccata* exhibited an ability to obtain K from phlogopite, and the effect was enhanced by dual inoculation with *Agrobacterium*, presumably due to acidification increase. Intense oxalate production in *P. involutus* was the main reason suggested for its high capacity for muscovite solubilizing (Smith and Read 2008). However, the hyphal proliferation and utilization of feldspar by *Hebeloma crustuliniforme* and *Piloderma fallax* observed in microcosms was not the case under natural conditions (Rosling et al. 2004). In the EM system potassium co-transported to plants with P, so mycobionts are able to enhance their host K nutrition, but the effect relies on P availability and certain fungal properties. As for efficiency of K-derived supply to the host, the relation between weathering activity and plant nutrient demand remains ambiguous (Rosenstock 2009).

11.4.4 Sulfur Uptake

Sulfur (S) is a macronutrient required for the plant cell wall and plasmalemma formation, and some essential amino acids, glutathione and a range of secondary metabolites biosynthesis (Giovannetti et al. 2017). S deficiency in soils became a problem since the second part of twentieth century when the control of S-rich industrial air pollutant emission was reinforced, thus sharply decreasing the oxidized S supply to the soil with precipitation. The S pool is dominated by organic compounds (S_o) and comprises only about 5 per cent of that directly available for root inorganic sulfate (S_i) crucial for plants as a cysteine precursor (Kertesz and Mirleau 2004). S cycling is strongly mediated by MO activity. Sulfate-reducing bacteria transform sulfate into hydrogene sulfide, thus decreasing S availability for plant uptake. Sulfur bacteria (*Acidithiobacillus*, *Beggiatoa*, *Thiobacillus*, *Thioploca and Thiothrix*) and some fungi (*Aspergillus and Penicillium*), reversely, are capable of S_i compound oxidizing (Buscot and Varma 2005). Sulfuric acid production by *Acidithiobacillus* can cause soil acidification up to 2.0 pH value, thus having a strong potential for mineral bioweathering (Uroz et al. 2009). *A. thiooxidans* became a base for organic fertilizer aimed to improve plant P nutrition by rock phosphate acid-mediated dissolving (Buscot and Varma 2005). The common rhizosphere bacterium, *Pseudomonas putida*, was shown to mineralize a wide range of S_o compounds such as sulfonates and sulfate esters. Therefore, the strain exhibited an additional plant growth promotion effect along with acknowledged abilities to produce phytohormones and to suppress pathogens (Kertesz and Mirleau 2004).

Despite the crucial role of S for plant survival and plant production of human-utilized nutritional and health-related compounds, the research into the mycobiont role in host S deficiency alleviation is only at its start. EMF are suggested to alleviate S deficiency in their hosts, but little is known of the transfer process (Smith and Read 2008). AMF were demonstrated to translocate both S_i and S_o forms to the plants under limiting conditions (Allen and Shachar-Hill 2009). Besides the direct absorption and transfer, mycobionts are able to mediate the plant sulfate transporters expression to improve plant S nutrition (Giovannetti et al. 2017). Mycobionts profoundly affect their hosts' primary and secondary S metabolism, but its mechanisms are far from elucidated. Under the progressive S limitation, this gap should be filled (Bitterlich et al. (2017).

11.4.5 Iron Uptake

Iron (Fe) is often a limiting tracing element because of Fe (III) precipitation as hydroxides and association with P and colloids (Vansuyt et al. 2007). In response to Fe deficiency the majority of MO produces siderophores, compounds with a high affinity for Fe; about 500 siderophores are currently recognized, and their number are probably underestimated (Crowley 2006). These compounds can promote plant growth both by enhancing Fe availability for the plant and by iron removal from the rhizosphere to make it unavailable to pathogens, thus protecting the plant. Siderophores can also contribute to P plant nutrition by chelation of Fe from phosphates (Jones and Oburger 2011). Bacteria are known to produce phenol catecholates, hydroxamates, carboxylates, and pyoverdines. The last group is produced by rhizospheric pseudomonads and plays a more significant role in plant defense rather than in Fe nutrition (Crowley 2006; Vansuyt et al. 2007). Fungi produce about 150 structurally different siderophores; the main classes are ferrichromes, coprogens, fusarinines, and polycarboxylates. One fungal species can possess a range of siderophores with different properties (Winkelmann 2007).

As Crowley (2006) pointed out, the key question on the role of MO siderophores in plant nutrition, is whether siderophore quantity is sufficient to respond to the plant's Fe requirements. The substantial quantities of these compounds can be expected in the zones of high MO activity, such as mycorrhizo- or rhizosphere. In these microhabitats siderophores play a key role in Fe mobilization and consequently in rhizosphere community sustainability. The second point of siderophore efficiency is their stability and resistance against degrading by other MO-derived metabolites (Winkelmann 2007). The plant's ability to use MO siderophores as Fe sources remains controversial, despite numerous studies (Crowley 2006). The reports on evident plant Fe nutrition benefits from inoculation of bacterial strains rich in siderophores are scarce; moreover, even the pathways of Fe acquisition of such compounds are often obscure (Vansuyt et al. 2007). The activity in Fe chelating alone should not be treated as a marker of potential growth promoters; for many plants deleterious MO are known to possess this capacity.

Mycorrhizal fungi were occasionally reported to enhance Fe plant nutrition under limiting conditions (Smith and Read 2008), but the mechanisms and compounds involved are poorly known. There were no data at all on AMF siderophores until 2017, when the first chelator glomuferrin of the rhizoferrin family was revealed (Winkelmann 2017). EMF (*Cenococcum geophilum*, *H. crustuliniforme*) were shown to produce ferricrocin engaged in bioweathering (Haselwandter 2008), but the number of fungal species examined is rather small. Further research is vital, especially in the sight of the tentative role of mycobionts chelating agents in plant protection against metal intoxication (Smith and Read 2008).

11.5 Plant Microbiome Prospects and Limitations in Sustainable Agriculture

Progressively increasing at an alarming rate, the global human population is endangered by such world-wide problems as potential food shortages and multiple consequences of the environmental pollution. These risks have stimulated the shift to eco-friendly organic farming and sustainable agriculture, which can be considered as "going back to the roots," speaking both literally and figuratively. Currently the profound role of soil MO in plant performance is undeniable, but people applied manure and green fertilizers, such as *Azolla* and legumes (Santi et al. 2013; Jacoby et al. 2017) in their agricultural practices for thousands of years, being totally unaware of diazotrophic (as well as other) microbial activity and the microbes themselves.

It is well-acknowledged that "the role of the microbiome and its relationship to plant health should be considered as much as the plant itself" (Turner et al. 2013), and the research and review papers on MO-mediated plant growth promotion are countless. All of them can be roughly split into "pro" and "contra" groups; sometimes tending either to glorify MO beyond all measure, or, conversely, to underestimate its role to the brink of neglect. To achieve a reasonable middle way, it is necessary to realize two main equally significant gaps: the gap between restricted research and large-scale application and between effects observed under controlled conditions and in the field.

The most "tamed" and well-studied promising agricultural MO group are the diazotrophic rhizobia. The efficient combinations of the plant genotypes and MO strains have been already selected for different regions and soil types (Jacoby et al. 2017). The focus on N nutrition research should be shifted to non-leguminous plants and associated diazotrophs, including those existing in tripartite symbiosis with mycorrhizal fungi (Santi et al. 2013). As for rhizobia, their demands for P and other nutrient supplies for efficient N_2-fixing should be better concentrated upon, for data on their interactions with other MO are insufficient and controversial. Besides, the presence of N_i compounds in soils suppressing biological N_2-fixation should be taken into account (McNeill and Unkovich 2007). The N losses from the ecosystem can sharply reduce its productivity and indirectly affect the adjacent ecosystems. The role of AMF in preventing N losses has been revealed (Cavagnaro et al. 2015). The maintenance of this fungal group strongly affected by fungicides and fertilizers in agricultural soils is crucial for plant P and N nutrition.

Phosphatic fertilizers are the second largest group of agricultural chemicals on a global scale due to all-round low access of plants to natural forms of P. The application is often surplus, due to the rapid immobilization of up to 90 per cent of readily reactive soluble phosphate anions in soil by metals, thus requiring an enormous fertilizer supply to support the concentration of plant-available P (Sindhu et al. 2014b). Large amounts of low-efficient chemicals endanger the environment with soil pollution and water eutrophication and affect soil health. Rock phosphates were recently considered as less harmful promising P fertilizers, but with soil pH value exceeding 6.0 they are not generally available for plants and disadvantageous when compared with soluble P (Khan et al. 2007). The MO solubilization can assist with this problem in neutral and basic soils. There are numerous data on different crops (e.g. legumes, maize, and wheat) yield increase with inoculation rock phosphate-fertilized soils with P-solubilizing bacteria *Bacillus megaterium* and *P. striata* and fungi from the *Aspergillus* genus (Kaur and Reddy 2017). It is noteworthy that combined treatment including both P-solubilizing MO inoculation and rock phosphate application resulted in higher plant growth promotion compared with utilization of MO only (Singh and Reddy 2012). Such an ecofriendly and cost-effective alternative for chemical fertilizers is promising indeed in sustainable agriculture.

The MO contributing to other than N and P nutrient supply for benefit of plants are studied far less. A range of bacteria and fungi with bioweathering capacity usually suggests the potential of releasing more than one element from the parent rock, thus making them promising for agriculture. However, it should be borne in mind that this ability is strongly influenced by available carbon and can be notable only in zones of MO accumulation, such as the rhizosphere. More data obtained under field conditions are required to recognize the real contribution of MO-driven weathering to plant nutrition (Rosenstock 2009).

To be a promising candidate for agricultural application, the MO strain should answer various demands. Besides its prominent plant-beneficial properties, the main requirements are the ability to be maintained in culture to produce inocula on a large scale and to re-colonize successfully the plant rhizosphere (Jacoby et al. 2017). Without elucidating the mechanisms of MO complex interactions within their community and with plants it is impossible to predict their effect and to design an efficient MO assembly in the agroecosystem.

As was aptly noted by Meena et al. (2014), "There is a very big gap between researchers and farmers." Practitioners should not expect MO, even with prominent plant-growth promoting capacities, to be a "magic wand" able to resolve all problems in agriculture at once. The absence of instant action can undermine confidence in the very idea of sustainable agriculture, especially with the underestimation of the enormous complexity, dynamism, and flexibility of soil biotic interactions. Further optimization, refinement, and verification of MO-based production for plant growth promotion is demanded in order not to promise "too much, too soon" (Jones and Oburger 2011). However, there is a large body of data indicating the fundamental role of MO in plant nutrient fortification and the further research and harnessing of microbes for agricultural needs is a global future challenge.

11.6 Conclusion and Perspectives

The significant role of root-associated microbiota in plant health and nutrition has been known since the nineteenth century, when A.B. Frank's keystone works on mycorrhizal and rhizobial symbioses appeared (Trappe 2005). During the twentieth century a vast mass of data on MO cultivable species diversity, biochemistry, physiology, and interactions was obtained on a global scale via culture-dependent *in vitro* research (Uroz and Frey-Klett 2011). With the molecular approach implication in the 1990s, the research focus shifted from single species/strains with its properties to biome level, thus revealing an overwhelming MO diversity in the root zone, drastically expanded with direct molecular detection of unculturable bacteria and fungi (Cardon and Whitbeck 2007). Recently the "next-generation" tools, such as metagenomics, metatranscriptomics, and metaproteomics applied to plant microbiome allowed *in situ* exploration of MO possible functionality and interrelations (Philippot et al. 2013; Turner et al. 2013). These aspects are of extreme importance for agriculture for synergism or antagonism of beneficial MO in the (mycor-)rhizosphere zone, and are considered to be crucial for plant nutrition and health (Raaijmakers et al. 2009). But despite the evolution of research tools and other advances, there are lots of challenges for future researchers to face. As is pointed out by Jacoby et al. (2017), the main goal in the field of plant growth promotion is gaining a mechanistic understanding of MO ways of action to design microbial communities suitable for application.

To achieve this goal the following questions should be addressed in future research:

- better understanding of rhizosphere signaling between plant and MO community;
- understanding of the dynamics of rhizosphere competition, especially between indigenous and introduced MO, for successful inoculation;
- the ability of certain plants to attract beneficial MO under different limiting conditions and levels of nutrient supply elucidation;
- the screening of promising MO in natural self-sustained soils with the focus on (mycor-)rhizosphere communities;
- the development of an integrative approach concerning biological aspects of plant nutrition along with physical and chemical properties and accurate quantification;
- the fortification of research of plant-MO interactions and MO beneficial properties under realistic soil conditions;
- the responsible application of recombinant DNA technology for creating more efficient plant growth promoters.

Acknowledgments

Financial support by the Russian Science Foundation (RSCF) to Elena Voronina (programme 14-50-00029) is gratefully acknowledged.

References

Agerer, R. (2001). Exploration types of ectomycorrhizae. *Mycorrhiza* 11: 107–114. https://doi.org/10.1007/s005720100108.

Allen, J.W. and Shachar-Hill, Y. (2009). Sulfur transfer through an arbuscular mycorrhiza. *Plant Physiology* 149: 549–560. https://doi.org/10.1104/pp.108.129866.

Amora-Lazcano, E., Vázquez, M.M., and Azcón, R. (1998). Response of nitrogen-transforming microorganisms to arbuscular mycorrhizal fungi. *Biology and Fertility of Soils* 27: 65–70. https://doi.org/10.1007/s003740050401.

Bago, B., Azcon-Aguilar, C., and Piché, Y. (1998). Architecture and developmental dynamics of the external mycelium of the arbuscular mycorrhizal fungus *Glomus intraradices* grown under monoxenic conditions. *Mycologia* 90 (1): 52–62.

Barreto, T.R., da Silva, A.C.M., Soares, A.C.F., and de Souza, J.T. (2008). Population densities and genetic diversity of actinomycetes associated to the rhizosphere of *Theobroma cacao*. *Brazilian Journal of Microbiology* 39: 464–470. https://doi.org/10.1590/S1517-838220080003000010.

Ben Farhat, M., Farhat, A., Bejar, W. et al. (2009). Characterization of the mineral phosphate solubilizing activity of *Serratia marcescens* CTM 50650 isolated from the phosphate mine of Gafsa. *Archives of Microbiology* 191: 815–824. https://doi.org/10.1007/s00203-009-0513-8.

Bitterlich, M., Graefe, J., and Franken, P. (2017). Primary metabolism in arbuscular mycorrhizal symbiosis: carbon, nitrogen and sulfur. In: *Molecular Mycorrhizal Symbiosis* (ed. F. Martin), 217–238. Hoboken: Wiley Blackwell.

Bloemberg, G.V. and Lugtenberg, B.J.J. (2001). Molecular basis of plant growth promotion and biocontrol by rhizobacteria. *Current Opinion in Plant Biology* 4: 343–350.

Buscot, F. and Varma, A. (eds.) (2005). *Microorganisms in Soils: Roles in Genesis and Functions*, Soil Biology, vol. 3. Heidelberg: Springer.

Cardon, Z.G. and Whitbeck, J.L. (eds.) (2007). *The Rhizosphere: An Ecological Perspective*. Burlington: Academic Press.

Cavagnaro, T.R., Bender, S.F., Ashgari, H.R. et al. (2015). The role of arbuscular mycorrhizas in reducing soil nutrient loss. *Trends in Plant Science* 20: 283–290. https://doi.org/10.1016/j.tplants.2015.03.004.

Chen, W.-M., James, E.K., Prescott, A.R. et al. (2003). Nodulation of *Mimosa* spp. by the betaproteobacterium *Ralstonia taiwanensis*. *Molecular Plant-Microbe Interactions* 16: 1051–1061. https://doi.org/10.1094/MPMI.2003.16.12.1051.

Chotte, J.-L. (2005). Importance of microorganisms for soil aggregation. In: *Microorganisms in Soils: Roles in Genesis and Functions*, Soil Biology, vol. 3 (eds. F. Buscot and A. Varma), 107–119. Heidelberg: Springer.

Criquet, S., Ferre, E., Farner, A.M., and Le Petit, J. (2004). Annual dynamics of phosphatase activities in an evergreen oak litter—influence of biotic and abiotic factors. *Soil Biology & Biochemistry* 36: 1111–1118. https://doi.org/10.1016/j.soilbio.2004.02.021.

Crowley, D.E. (2006). Microbial siderophores in the plant rhizosphere. In: *Iron Nutrition in Plants and Rhizospheric Microorganisms* (eds. L.L. Barton and J. Abadia), 169–198. Dordrecht: Springer https://doi.org/10.1007/1-4020-4743-6.

Dawson, J.O. (2008). Ecology of actinorhizal plants. In: *Nitrogen-Fixing Actinorhizal Symbioses* (eds. K. Pawlowski and W.E. Newton), 199–234. Dordrecht: Springer https://doi.org/10.1007/978-1-4020-3547-0.

Doty, S.F. (ed.) (2017). *Functional Importance of the Plant Microbiome: Implications for Agriculture, Forestry and Bioenergy*. Cham: Springer https://doi.org/10.1007/978-3-319-65897-1.

Duponnois, R., Galiana, A., and Prin, Y. (2008). The mycorrhizosphere effect: a multitrophic interaction complex improves Mycorrhizal Symbiosis and plant growth. In: *Mycorrhizae: Sustainable Agriculture and Forestry* (eds. Z.A. Siddiqui et al.), 227–240. Dordrecht: Springer https://doi.org/10.1007/978-1-4020-8770-7_10.

Finlay, R. (1992). Uptake and translocation of nutrients by ectomycorrhizal fungal mycelia. In: *Mycorrhizas in Ecosystems* (eds. D.J. Read, D.H. Lewis, A.H. Fitter and I.J. Alexander), 91–97. Wallingford: Centre for Agriculture and Bioscience International.

Fomina, M., Burford, E.P., and Gadd, G.M. (2006). Fungal dissolution and transformation of minerals: significance for nutrient and metal mobility. In: *Fungi in Biogeochemical Cycles* (ed. G.M. Gadd), 236–266. New York: Cambridge University Press.

Frey-Klett, P., Chavatte, M., Clausse, M.-L. et al. (2005). Ectomycorrhizal symbiosis affects functional diversity of rhizosphere fluorescent pseudomonads. *New Phytologist* 165: 317–328. https://doi.org/10.1111/j.1469-8137.2004.01212.x.

Galloway, J.N., Schlesinger, W., Levy II, H. et al. (1995). Nitrogen fixation: anthropogenic enhancement-environmental response. *Global Biogeochemical Cycles* 9: 235–252. https://doi.org/10.1029/95GB00158.

Gibson, B.R. and Mitchell, D.T. (2004). Nutritional influences on the solubilization of metal phosphate by ericoid mycorrhizal fungi. *Mycological Research* 108: 947–954.

Giovannetti, M., Volpe, V., Salvioli, A., and Bonfante, P. (2017). Fungal and plant tools for the uptake of nutrients in arbuscular mycorrhizas: a molecular view. In: *Mycorrhizal Mediation of Soils: Fertility, Structure and Carbon Storage* (eds. N.C. Johnson, C. Gehring and J. Jansa), 107–128. Amsterdam: Elsevier https://doi.org/10.1016/B978-0-12-804312-7.00007-3.

Graham, P.H. (2008). Ecology of the root-nodule bacteria of legumes. In: *Nitrogen-Fixing Leguminous Symbioses* (eds. M.J. Dilworth et al.), 23–58. Dordrecht: Springer https://doi.org/10.1007/978-1-4020-3548-7.

Haselwandter, K. (2008). Structure and function of siderophores produced by mycorrhizal fungi. *Mineralogical Magazine* 72: 61–64. https://doi.org/10.1180/minmag.2008.072.1.61.

Hodge, A. (2017). Accessibility of inorganic and organic nutrients for Mycorrhizas. In: *Mycorrhizal Mediation of Soils: Fertility, Structure and Carbon Storage* (eds. N.C. Johnson et al.), 129–148. Amsterdam: Elsevier https://doi.org/10.1016/B978-0-12-804312-7.00008-5.

Jacoby, R., Peukert, M., Succurro, A. et al. (2017). The role of soil microorganisms in plant mineral nutrition—current knowledge and future directions. *Frontiers in Plant Science* 8: 1617. https://doi.org/10.3389/fpls.2017.01617.

James, E.K. (2000). Nitrogen fixation in endophytic and associative symbiosis. *Field Crops Research* 65: 197–209. https://doi.org/10.1016/S0378-4290(99)00087-8.

Jansa, J., Finlay, R., Wallander, H. et al. (2011). Role of mycorrhizal symbioses in phosphorus cycling. In: *Phosphorus in Action. Biological Processes in Soil Phosphorus Cycling*, Soil Biology, vol. 26 (eds. E.K. Bünemann et al.), 137–168. Heidelberg: Springer https://doi.org/10.1007/978-3-642-15271-9_6.

Jenkinson, D.S. (1990). The turnover of organic carbon and nitrogen in soil. *Philosophical Transactions of the Royal Society, B: Biological Sciences* 329: 361–368. https://doi.org/10.1098/rstb.1990.0177.

Jones, D.L. and Oburger, E. (2011). Solubilization of phosphorus by soil microorganisms. In: *Phosphorus in Action. Biological Processes in Soil Phosphorus Cycling*, Soil Biology, vol. 26 (eds. E.K. Bünemann et al.), 169–198. Heidelberg: Springer https://doi.org/10.1007/978-3-642-15271-9_7.

Jongmans, A.G., Van Breemen, N., Lundström, U.S. et al. (1997). Rock-eating fungi. *Nature* 389: 682–683. https://doi.org/10.1038/39493.

Kashyap, P.L., Srivastava, A.K., Tiwari, S.P., and Kumar, S. (eds.) (2018). *Microbes for Climate Resilient Agriculture*. Hoboken, NJ: Wiley Blackwell.

Kaur, G. and Reddy, M.S. (2017). Role of phosphate-solubilizing fungi in sustainable agriculture. In: *Developments in Fungal Biology and Applied Mycology* (eds. T. Satyanarayana et al.), 391–412. Singapore: Springer https://doi.org/10.1007/978-981-10-4768-8_20.

Kertesz, M.A. and Mirleau, P. (2004). The role of soil microbes in plant sulphur nutrition. *Journal of Experimental Botany* 55: 1939–1945. https://doi.org/10.1093/jxb/erh176.

Khan, M.S., Zaidi, A., and Wani, P.A. (2007). Role of phosphate-solubilizing microorganisms in sustainable agriculture – a review. *Agronomy for Sustainable Development* 27: 29–43. https://doi.org/10.1051/agro:2006011.

Koele, N., Turpault, M.-P., Hildebrand, E.E. et al. (2009). Interactions between mycorrhizal fungi and mycorrhizosphere bacteria during mineral weathering: budget analysis and bacterial quantification. *Soil Biology & Biochemistry* 41: 1935–1942. https://doi.org/10.1016/j.soilbio.2009.06.017.

Lehmann, A., Leifheit, E.F., and Rillig, M.C. (2017). Mycorrhizas and soil aggregation. In: *Mycorrhizal Mediation of Soils: Fertility, Structure and Carbon Storage* (eds. N.C. Johnson et al.), 241–262. Amsterdam: Elsevier https://doi.org/10.1016/B978-0-12-804312-7.00014-0.

Li, C.-Y., Massicotte, H.B., and Moore, L.V.H. (1992). Nitrogen-fixing *Bacillus* sp. associated with Douglas-fir tuberculate ectomycorrhizae. *Plant and Soil* 140: 35–40. https://doi.org/10.1007/BF00012804.

Li, R.-X., Cai, F., Pang, G. et al. (2015). Solubilisation of phosphate and micronutrients by *Trichoderma harzianum* and its relationship with the promotion of tomato plant growth. *PLoS One* 10 (6): e0130081. https://doi.org/10.1371/journal.pone.0130081.

Martino, E., Perotto, S., Parsons, R., and Gadd, G.M. (2003). Solubilization of insoluble inorganic zinc compounds by ericoid mycorrhizal fungi derived from heavy metal polluted sites. *Soil Biology & Biochemistry* 35: 133–141. https://doi.org/10.1016/S0038-0717(02)00247-X.

McGill, W.B. and Cole, C.V. (1981). Comparative aspects of cycling of organic C, N, S and P through soil organic matter. *Geoderma* 26: 267–268. https://doi.org/10.1016/0016-7061(81)90024-0.

McNeill, A. and Unkovich, M. (2007). The nitrogen cycle in terrestrial ecosystems. In: *Nutrient Cycling in Terrestrial Ecosystems*, Soil Biology, vol. 10 (eds. P. Marschner et al.), 37–64. Heidelberg: Springer https://doi.org/10.1007/978-3-540-68027-7_2.

Meena, V.S., Maurya, B.R., and Verma, J.P. (2014). Does a rhizospheric microorganism enhance K^+ availability in agricultural soils? *Microbiological Research* 169: 337–347. https://doi.org/10.1016/j.micres.2013.09.003.

Mickan, B. (2017). Mechanisms for alleviation of plant water stress involving arbuscular mycorrhizas. In: *Mycorrhizal Fungi: Use in Sustainable Agriculture and Land Restoration*, Soil Biology, vol. 41 (eds. M. Zakaria et al.), 225–239. Heidelberg: Springer https://doi.org/10.1007/978-3-662-45370-4_14.

Pena, R. (2017). Nitrogen acquisition in ectomycorrhizal symbiosis. In: *Molecular Mycorrhizal Symbiosis* (ed. F. Martin), 179–196. Hoboken: Wiley Blackwell.

Philippot, L., Raaijmakers, J.M., Lemanceau, P., and van der Putten, W.H. (2013). Going back to the roots: the microbial ecology of the rhizosphere. *Nature Reviews Microbiology* 11: 789–799. https://doi.org/10.1038/nrmicro3109.

Raaijmakers, J.M., Paulitz, T.C., Steinberg, C. et al. (2009). The rhizosphere: a playground and battlefield for soil borne pathogens and beneficial microorganisms. *Plant and Soil* 321: 341–361. https://doi.org/10.1007/s11104-008-9568-6.

Rai, A.N., Bergman, B., and Rasmussen, U. (eds.) (2002). *Cyanobacteria in Symbiosis*. Dordrecht: Kluwer Academic Publishers https://doi.org/10.1007/0-306-48005-0.

Read, D.J., Leake, J.R., and Perez-Moreno, J. (2004). Mycorrhizal fungi as drivers of ecosystem processes in heathland and boreal forest biomes. *Canadian Journal of Botany* 82: 1243–1263. https://doi.org/10.1139/b04-123.

Richardson, A.E. and Simpson, R.J. (2011). Soil microorganisms mediating phosphorus availability. *Plant Physiology* 156: 989–996. https://doi.org/10.1104/pp.111.175448.

Richardson, A.E., Barea, J.-M., McNeill, A.M., and Prigent-Combaret, C. (2009). Acquisition of phosphorus and nitrogen in the rhizosphere and plant growth promotion by microorganisms. *Plant and Soil* 321: 305–339. https://doi.org/10.1007/s11104-009-9895-2.

Rillig, M.C. and Mummey, D.L. (2006). Mycorrhizas and soil structure. *New Phytologist* 171: 41–53. https://doi.org/10.1111/j.1469-8137.2006.01750.x.

Rineau, F., Shah, F., Smits, M.M. et al. (2013). Carbon availability triggers the decomposition of plant litter and assimilation of nitrogen by an ectomycorrhizal fungus. *ISME Journal* (10): 2010–2022. https://doi.org/10.1038/ismej.2013.91.

Römheld, V. and Kirkby, E.A. (2010). Research on potassium in agriculture: needs and prospects. *Plant and Soil* 335: 155–180. https://doi.org/10.1007/s11104-010-0520-1.

Rosenstock, N.P. (2009). Can ectomycorrhizal weathering activity respond to host nutrient demands? *Fungal Biology Reviews* 23: 107–114. https://doi.org/10.1016/j.fbr.2009.11.003.

Rosling, A., Lindahl, B.D., Taylor, A.F.S., and Finlay, R.D. (2004). Mycelial growth and substrate acidification of ectomycorrhizal fungi in response to different minerals. *FEMS Microbiology Ecology* 47: 31–37. https://doi.org/10.1016/S0168-6496(03)00222-8.

Saito, K. and Ezawa, T. (2017). Phosphorus metabolism and transport in arbuscular mycorrhizal symbiosis. In: *Molecular Mycorrhizal Symbiosis* (ed. F. Martin), 197–216. Hoboken: Wiley Blackwell.

Santi, C., Bogusz, D., and Franche, C. (2013). Biological nitrogen fixation in non-legume plants. *Annals of Botany* 111: 743–767. https://doi.org/10.1093/aob/mct048.

Sidorova, I.I., Alexandrova, A.V., and Voronina, E.Y. (2014). Inorganic phosphate solubilising bacteria isolated from agaricomycete hyphosphere. In: *Biogenic-Abiogenic Interactions in Natural and Anthropogenic Systems*. V International Symposium, 112–113. Saint-Petersburg: VVM Publishing Ltd. https://doi.org/10.13140/RG.2.1.2963.3040.

Sindhu, S.S., Parmar, P., and Phour, M. (2014a). Nutrient cycling: potassium solubilization by microorganisms and improvement of crop growth. In: *Geomicrobiology and Biogeochemistry*, Soil Biology, vol. 39 (eds. N. Parmar and A. Singh), 175–198. Heidelberg: Springer https://doi.org/10.1007/978-3-642-41837-2_10.

Sindhu, S.S., Phour, M., Choudhary, S.R., and Chaudhary, D. (2014b). Phosphorus cycling: prospects of using rhizosphere microorganisms for improving phosphorus nutrition of plants. In: *Geomicrobiology and Biogeochemistry*, Soil Biology, vol. 39 (eds. N. Parmar and A. Singh), 199–237. Heidelberg: Springer https://doi.org/10.1007/978-3-642-41837-2_11.

Singh, H. and Reddy, M.S. (2012). Improvement of wheat and maize crops by inoculating *Aspergillus* spp. in alkaline soil fertilized with rock phosphate. *Archives of Agronomy and Soil Science* 58: 535–546. https://doi.org/10.1080/03650340.2010.532125.

Smith, S.E. and Read, D.J. (2008). *Mycorrhizal Symbiosis*, 3e. New York: Academic Press.

Sprent, J.I. (2001). *Nodulation in Legumes*. London: Royal Botanic Gardens, Kew.

Sprent, J.I. and James, E.K. (2007). Legume evolution: where do nodules and mycorrhizas fit in? *Plant Physiology* 144: 575–581. https://doi.org/10.1104/pp.107.096156.

Summerbell, R.C. (2005). From Lamarckian fertilizers to fungal castles: recapturing the pre-1985 literature on endophytic and saprotrophic fungi associated with ectomycorrhizal root systems. *Studies in Mycology* 53: 191–256.

Sverdrup, H. (2009). Chemical weathering of soil minerals and the role of biological processes. *Fungal Biology Reviews* 23: 94–100. https://doi.org/10.1016/j.fbr.2009.12.001.

Thirkell, T.J., Cameron, D.D., and Hodge, A. (2016). Resolving the "nitrogen paradox" of arbuscular mycorrhizas: fertilization with organic matter brings considerable benefits for plant nutrition and growth. *Plant, Cell and Environment* 39: 1683–1690. https://doi.org/10.1111/pce.12667.

Timonen, S. and Marschner, P. (2006). Mycorrhizosphere concept. In: *Microbial Activity in the Rhizosphere*, Soil Biology, vol. 7 (eds. K.G. Mukerji et al.), 155–172. Heidelberg: Springer.

Trappe, J.M. (2005). A.B. Frank and mycorrhizae: the challenge to evolutionary and ecologic theory. *Mycorrhiza* 15: 277–281. https://doi.org/10.1007/s00572-004-0330-5.

Tunlid, A., Floudas, D., Koide, R., and Rineau, F. (2017). Soil organic matter decomposition mechanisms in ectomycorrhizal fungi. In: *Molecular Mycorrhizal Symbiosis* (ed. F. Martin), 257–275. Hoboken: Wiley Blackwell.

Turner, T.R., James, E.K., and Poole, P.S. (2013). The plant microbiome. *Genome Biology* 14: 209. https://doi.org/10.1186/gb-2013-14-6-209.

Uroz, S. and Frey-Klett, P. (2011). Linking diversity to function: highlight on the mineral weathering bacteria. *Central European Journal of Biology* 6: 817–820. https://doi.org/10.2478/s11535-011-0053-5.

Uroz, S., Calvaruso, C., Turpault, M.-P., and Frey-Klett, P. (2009). Mineral weathering by bacteria: ecology, actors and mechanisms. *Trends in Microbiology* 17: 378–387. https://doi.org/10.1016/j.tim.2009.05.004.

Vansuyt, G., Robin, A., Briat, J.F. et al. (2007). Iron acquisition from Fe-pyoverdine by *Arabidopsis thaliana*. *Molecular Plant-Microbe Interactions* 20: 441–447. https://doi.org/10.1094/MPMI-20-4-0441.

Vazquez, P., Holguin, G., Puente, M.E. et al. (2000). Phosphate solubilizing microorganisms associated with the rhizosphere of mangroves in a semiarid coastal lagoon. *Biology and Fertility of Soils* 30: 460–468. https://doi.org/10.1007/s003740050024.

Voronina, E. and Sidorova, I. (2017). Rhizosphere, mycorrhizosphere and hyphosphere as unique niches for soil-inhabiting bacteria and micromycetes. In: *Advances in PGPR Research* (eds. H.B. Singh et al.), 165–186. Boston: Centre for Agriculture and Bioscience International https://doi.org/10.1079/9781786390325.0165.

Wakelin, S.A., Warren, R.A., Harvey, P.R., and Ryder, M.H. (2004). Phosphate solubilization by *Penicillium* spp. closely associated with wheat roots. *Biology and Fertility of Soils* 40: 36–43. https://doi.org/10.1007/s00374-004-0750-6.

Whitelaw, M.A. (1999). Growth promotion of plants inoculated with phosphate-solubilizing fungi. *Advances in Agronomy* 69: 99–151. https://doi.org/10.1016/S0065-2113(08)60948-7.

Winkelmann, G. (2007). Ecology of siderophores with special reference to the fungi. *Biology of Metals* 20: 379–392. https://doi.org/10.1007/s10534-006-9076-1.

Winkelmann, G. (2017). A search for glomuferrin: a potential siderophore of arbuscular mycorrhizal fungi of the genus *Glomus*. *Biology of Metals* 30: 559–564. https://doi.org/10.1007/s10534-017-0026-x.

12

Engineering Microbes to Improve Crop Health

A New Dimension for Sustainable Agricultural Productivity

P. Veera Bramhachari, A.M.V.N. Prathyusha, and Ganugula Mohana Sheela

Department of Biotechnology, Krishna University, Machilipatnam, Andhra Pradesh, India

12.1 Introduction

Food crisis is one of the major challenges for the present day generations due the continuous increase of worldwide populations. Agriculture is posing serious problems due to soil fertility loss, pesticide contamination, salinization, and drought. Another major difficulty is the green revolution which poses problems to soil fertility and the environment, with pesticide contamination along with nutritional quality of food. Therefore, it is essential to enhance crop productivity in an environmentally sustainable way within the cultivable land and extreme climate conditions. Currently, the plant microbiome is one alternative to increase agriculture productivity, soil fertility and vigor, reduce environmental issues, and to provide healthy and nutrient rich food in a sustainable way (Singh 2017).

The significance of soil, plant, and microbiome interaction in agriculture productivity has been acknowledged for more than a century, which describes the influence of microbial inhabitance in the soil and their metabolic capabilities on soil fertility (del Carmen Orozco-Mosqueda et al. 2018). Furthermore, many breakthroughs made by microbiologists chiefly originated from work with soil microbes, *viz.*, discovery of antibiotics, and unique microbial metabolic pathways (fixation of N_2 gas and ammonia oxidation). Interestingly, perturbation to microbiomes may lead to vast changes in hosts and their environment such as decreased host vigor and soil fertility (Round and Mazmanian 2009; Navarrete et al. 2015). Plant microbiomes facilitate plant growth, increase crop production, decrease global use of synthetic fertilizers, enhance nutrient acquisition, decrease greenhouse gas emissions and combat plant stress. Therefore, the plant microbiome may help to cater for cleaner and sustainable production of agriculture under extreme climatic conditions.

Microbiomes are crucial for plant survival, which describe more than phenotypic variations as plant genotypes. Alteration of the plant microbiome plays a prospective role in reducing plant disease, chemical inputs and emissions of greenhouse gases, increases crop production, and enhances water cycling, which reflects biogeochemical functions of the rhizosphere resulting in more sustainable agricultural practices (Mueller and Sachs 2015).

The Plant Microbiome in Sustainable Agriculture, First Edition. Edited by Alok Kumar Srivastava, Prem Lal Kashyap and Madhumita Srivastava.

Through a better knowledge of beneficial plant–microbiome interactions, developments in economic and ecological sustainability of agriculture, forestry, and bioenergy may succeed.

With advances in modern techniques, *viz.*, culture-independent microbiome analyses, "metagenomic" tools and genome sequencing, microbiome research has accelerated. These techniques also help the researchers to chart soil microbe diversity, and build a more inclusive perception of specific microbial populations on soil fertility (Schweitzer et al. 2016). It gradually became clearer how composition of the microbiome may effect ecosystems, and interest is significantly evolving in microbiome engineering for modeling microbiota to amend the ecosystem of importance (Mueller and Sachs 2015; Moran 2015). Particularly, DNA and RNA-based analysis of the rhizomicrobiome are widespread and have significantly expanded what is known about phylogenetic and taxonomic nature of soil microbial populations (Busby et al. 2017). It is now documented that culture-based methodologies have substantively underestimated microbial diversity in soil, the majority of which remain uncharacterized (Mueller and Sachs 2015).

Employing microbiomes is one possible way to accomplish these preferred objectives through microbiome engineering (Vejan et al. 2016). The advancements in microbiome engineering may appreciably accelerate innovations of successful microbial products for sustainable farming. This updated understanding is imperative for prediction of the impacts of environmental perturbations on vital functions carried out by microbiome and will facilitate development of novel approaches for optimizing soil carbon cycling, managing nutrient transport, and sustainable crop production. Hence, due to the extensive impacts of the microbiome over varied soil ecosystems, there is an outcry of concern currently to comprehend the microbiome and obtain insight on factors that contribute to the diversity of microbiomes. The current review primarily recapitulates the challenges and opportunities of evolving novel microbiome technologies for sustainable agriculture and productivity.

12.2 Complexity of Soil and Its Ecosystem Functions

Current farming practices that use chemical fertilizers and pesticides have contributed radically to boost farm productivity, but also added to chemical contamination, soil degradation, loss of biodiversity and compromised soil and water quality, which in turn impacts largely on environmental sustainability and may also affect human health. Agriculture productivity is facing several challenges, together with a structural turn down in soil fertility and the impact of extreme weather events (Singh and Trivedi 2017). Increasing the area of cultivable land is virtually impossible; exceptional demand has meant severe pressure on the terrestrial ecosystem for increased agriculture production. Consequently, additional scientific and enhanced farming techniques are indispensable for satisfying the escalating demands, and furthermore to sustain the fertility of soil. Few existing techniques implicated in sustainable agriculture (Ubertino et al. 2016), include: agricultural spiraling; GMO (genetically modified crops) crops to form N_2-fixing symbioses; N_2-fixing devoid of microbial symbionts (Passari et al. 2016); use of GMO microbes to stimulate plant growth (Perez et al. 2016); and using biofertilizers (Kamkar 2016). Additionally, numerous other scientific

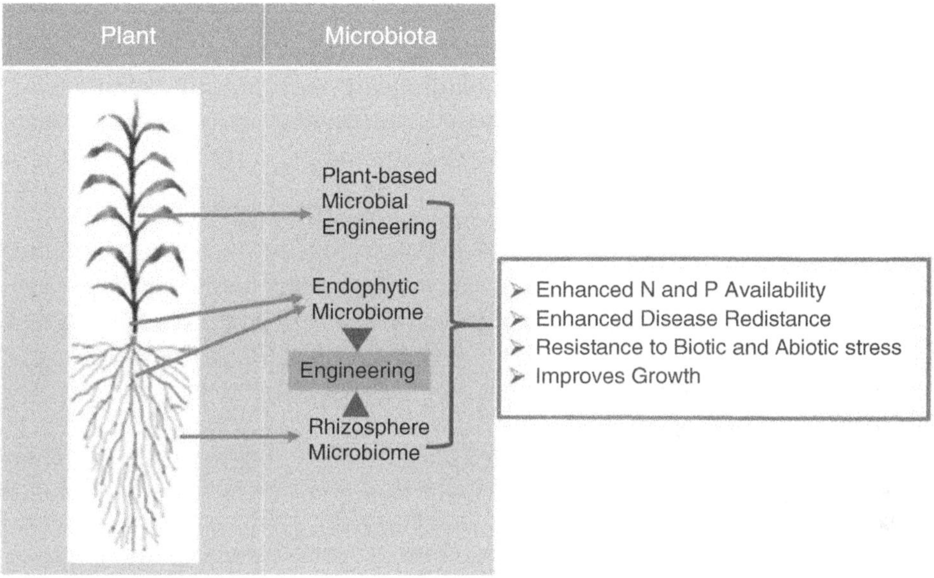

Figure 12.1 Applications of microbiome engineering to improve crop health and sustainable agricultural productivity.

techniques commenced towards sustainable development of novel agriculture practices primarily include the following traits: disease resistance, drought tolerance, salt tolerance, metal stress tolerance and superior nutritional value (Figure 12.1).

Nonetheless the microbes are the sole drivers of universal biogeochemical cycles and therefore key to sustainable agriculture. Microbes not only facilitate an array of ecosystem functions, but also interrelate with each other directly or indirectly in the course of competitive, mutualistic, facilitative and pathogenic effects (Bardgett and Wardle 2010). To comprehend these multitude interactions is an overwhelming task. A plethora of studies exploring the effects of single microbe groups to explain different plant functions is already in progress. On the contrary, these studies have not aimed out to address the intricate interplay between diverse organisms and ecosystem functions. The microbiome is an entity of microbes inhabiting a definite habitat, such as the soil or rhizosphere. The emerging microbiome approaches have enormous potential to deal with unexploited challenges as a complementary approach to conventional farming. There are increasing evidences that with development of suitable technologies, the plant microbiome may be exploited to potentially reduce the occurrence of plant diseases, increase resource use efficiencies and ultimately improve agricultural productivity. Reducing the input of chemical fertilizers and pesticides will result in reduced greenhouse gas emissions and promote environmental sustainability. Most microbial communities exist in complex, dynamic consortia with extremely interconnected networks of ecological and metabolic interactions which are yet to be deciphered. Foley et al. (2011) opined that global agriculture necessitates radical and ground-breaking approaches to congregate the challenges of food security and environmental sustainability in the years to come.

12.3 Plant Microbiome Interactions

Virtually all plants are inhabited by diverse microorganisms including bacteria, archaea, and fungi, collectively considered as the plant–microbiome. Phytomicrobiomes are classified as endophytic, phyllospheric, or rhizospheric depending on their interaction with plant components. Microbes colonize different plant parts and their interactions may be symbiotic or parasitic. Interestingly the plant rhizosphere harbors a largely augmented microbial abundance and activity. Approximately a range between 5 and 20 per cent of a plant's photosynthetic yield is transferred to its microbiome, and this occurs chiefly through the roots. Root exudates comprise different amino acids, sugars, proteins, fatty acids, and flavonoids (Badri and Vivanco 2009) which serve as growth substrates/signals for some microbial partners and also antimicrobials or growth deterrents for other microbes (Bais et al. 2006).

Microbe interaction with plants is complex, often produce synergistic effects, and are inextricable from their host which is thought to function as a metaorganism or holobiont. The close symbiotic association of plants and microbes as an integrated ecological unit is known as the holobiont (Quiza et al. 2015; Yergeau et al. 2015). There are ample evidences reported that plant–microbe relationships are decisive for health, crop productivity and overall condition of the plant (Santhanam et al. 2015; Wei et al. 2015; Goyal et al. 2019). However in analogy with humans, plants may not move and face with new challenges to attain sufficient nutrients from the experimental area, protecting from fungal, insecticidal, bacterial pathogens and bearing multiple stresses, i.e. drought, temperature, pesticides, salinity, and other pollutants. The phytomicrobiome protects plants from an inundation of environmental challenges. The colonization ability or resilience of the phytomicrobiome depends on biotic factors (predation, competition, mutualism and parasitism, within the microbiome) and abiotic factors (light, temperature, water, acidity, and nutrient availability) (Busby et al. 2017).

12.4 Engineering Microbiomes for Plant-Beneficial Phenotypes

The association among plants and microorganisms is well-known. Nonetheless, scientists are still working to comprehend the degree of collaboration and full complexity of plant associated microbes. It has been reported that microbes are significant for plant nutrient acquisition, growth, and protection against diseases. Certain types of bacteria *viz.*, *Azospirillum*, *Enterobacter* sp. Strain 638, *Herbaspirillum*, *Azoarcus*, *Bacillus*, and *Klebsiella* are commercially available and used to augment yields and decrease fertilizer use (Farrar et al. 2014). Strikingly, now there is clear evidence that plants engineer the rhizomicrobiome (Chaparro et al. 2014). However, most largely ancient plant descendants exhibit a strong capability to amend the relative abundance of microbial clusters surrounding the rhizosphere (Vandenkoornhuyse et al. 2015; Valverde et al. 2016). Diverse plant species assist the unique microbiomes attributed to diversity in root exudate chemistry and rates of nutrient uptake (Bais et al. 2006). Plants can modify the rhizomicrobiome by numerous pathways which can affect plant health (Berendsen et al. 2012; Mueller and Sachs 2015). For instance,

Castrillo et al. (2017) recently studied and established the molecular mechanism regulating microbiome-mediated phosphate uptake and immunity. A hypothesis emerges that phenotypic and genotypic variation of plant traits which support microbiomes will enhance plant nutrient availability, plant health, growth, performance, resistance to pathogens and incur a fitness advantage. This fitness advantage incurred by the microbiome must outweigh cost to the plant in diverted carbon and energy (Vandenkoornhuyse et al. 2015).

Genetic engineering has the potential to improve or redesign microorganisms and provides a unique possibility to alter plant–microbe signaling, to encourage beneficial microbial associations and functions in the rhizosphere, diversify root exudation, promote diverse and microbiomes (Quiza et al. 2015). Therefore, through engineering microbiomes, microbial diversity can be significantly transformed to mend plant phenotypes and serve ecosystems. Nocera (2017) reported on engineering of *Ralstonia eutropha* to maintain H_2 and CO_2 obtained from air and to make hydrocarbon fuels. They also expanded the scope of their work by engineering an additional type of bacterium (*Xanthobacter autotrophicus*) to take nitrogen out of the air to make fertilizer. Liu et al. (2017) designed a genetically engineered *X. autotrophicus* that can derive hydrogen from electrolysis of water which combine with N_2 from air to produce NH_3. This was achieved by providing the plant with an enzyme called a hydrogenase, which naturally harbors a nitrogenase enzyme, and allowing them to grow on H_2 to make cellular energy. These plants utilize ATP, H_2, and CO_2 from air to produce a kind of bioplastic called polyhydroxy butyrate, which can be deposited in their bodies. Interestingly, the bacterium harvested H_2 from the PHB store and utilized their nitrogenase enzymes to combine with N_2 from air to make NH_3, which is a primary need for fertilizer. They also confirmed the practical application of this method, by adding these engineered *Xanthobacter microbe* in water solution and used it to nourish radish crops, which grew 150 per cent larger than controls (Liu et al. 2017).

The key to mapping genomes of every microbe in complex soil community is a comparatively novel method identified as genome binning. To meet the challenge to unravel every microbe in the laboratory, researchers are amalgamating all microbes in 1 g soil and sequencing the entire DNA that spills out at same time for its molecular-scale resemblance. However, genome binning has been successful in simple microbial communities with only a few hundred diverse species. Notably, several reports provided insight into taxonomic and functional basis of microbiome-associated phenotypes (MAPs) through isolation and phenotypic screening; amplicon sequencing, and shotgun'omics approaches (Falony et al. 2016; Lemanceau et al. 2017). Nevertheless, widespread practical application of this elemental knowledge is still overdue, until there is integration of complex multivariate interactions to achieve consistent, effective, and cost-effective strategies for engineering microbiomes. However, this basic knowledge will help to tackle challenges in engineering complex microbial communities with anticipated results and provide a guide for missing information that needs to be integrated when improving models. Identified root exudates, architecture or other plant traits that promote a livelihood of beneficial microbiomes, can be engineered to crops through CRISPR-CAS9 technology and other gene editing tools. These approaches seem promising as they emulate interactions that maintain beneficial microbes in natural systems or those selected during evolution of holobionts. Strategies for enhanced agriculture production might comprise breeding plants for selection of beneficial microbes (Kroll et al. 2017) or employing targeted genetic engineering techniques directly (Borel 2017).

12.5 Engineering of Plant Microbiome for Crop Productivity

The indigenous microbes affect plant properties *viz.*, resistance to biotic or abiotic stresses, nutrient status and biomass production (Vandenkoornhuyse et al. 2015; Mueller et al. 2016; Busby et al. 2017). Plant microbial communities with superior properties can be identified through artificial microbiome selection (Mueller and Sachs 2015). Swenson et al. (2000) first employed the principle for engineering the rhizomicrobiome for selection of plant traits with difference in plant biomass and pH. Recently, this method was used to select for microbiota that affect flowering time and thus plant development (Panke-Buisse et al. 2015). Therefore, engineering rhizomicrobiome was successfully employed to modulate microbiome networking and composition for improved plant phenotypes, *viz.*, detoxification of toxic compounds, flowering, plant growth, health, fitness, and biomass (Swenson et al. 2000; Edwards et al. 2015).

Notably, microbiome transfer, synthetic microbiomes and host-mediated artificial selection have been employed to engineer the rhizomicrobiome. Microbial strains with desired functions can be identified and combined to form synthetic microbiomes for engineering the plant microbiome. For instance, five rhizobacterial strains (*Arthrobacter nitroguajacolicus* E46, *Bacillus megaterium* B55, *Bacillus mojavensis* K1, *Pseudomonas azotoformans* A70, and *Pseudomonas frederiksbergensis* A176) were reported to defend *Nicotiana attenuata* from sudden-wilt disease (Santhanam et al. 2015). Nonetheless, most of the current studies on the plant microbiome were presented in model plants, principally *Arabidopsis thaliana*, where insights gained are extrapolated to other plants. Therefore, more efforts are needed to be directed regarding studies on engineering the microbiome for enhancing crop characteristics, *viz.*, to improve drought and pathogen resistance, nutrient efficiency, agricultural productivity and sustainability. Moreover, facilitating engineering of agricultural plant–microbiome associations for agriculture sustainability (Schlaeppi and Bulgarelli 2014) requires substantial developments in experimental approaches, advanced characterization procedures, modeling, and theory which integrate, but also move beyond, the current dominant consensus-based descriptive methods (Lebeis 2015; Widder et al. 2016).

12.6 CRISPR/Cas9 Technology

CRISPR/Cas9 is the latest synthetic endonuclease technology which transformed molecular research, via permitting facile genome editing in prokaryotes and can operate only with Cas9-sgRNA complex. This programmable technique was first implemented to eukaryotic genomes (animals), along with numerous effective applications in plants as well as crops (Bortesi and Fischer 2015; Butler et al. 2015). Interestingly, CRISPR–Cas9 technology can be applied in plants to knock out all/single multigene family members and multiple unrelated genes (Endo et al. 2015; Lowder et al. 2016). The CRISPR–Cas9 technique has now made gene editing feasible in fungi (Matsu-ura et al. 2015; Nødvig et al. 2015). Fang and Tyler (2016) reported knock out of effector Avr4/6 in soybean pathogen *Phytophthora sojae* and reinstatement by selectable marker nptII, revealing supplementary roles for corresponding R gene loci RPS4 and RPS6 (Fang and Tyler 2016).

Notably, three different ways are currently in practice to introduce *Cas9* and *sgRNA* genes into plant cells:

1) ***Agrobacterium*-mediated transformation** – conventional and established technique for transformation in plants. Integration of T-DNA into plant genome by *Agrobacterium* permits constant expression of *sgRNA* and *Cas9* to yield extremely effective genome editing. *Agrobacterium* can transfer T-DNA into several plant spp..
2) **Polyethylene glycol (PEG)-based transformation method** – uses plant protoplasts, this technique has advantage where numerous substances, *viz.*, ribonucleoproteins (RNPs), can be presented. Cas9-sgRNA complex can be transformed into protoplasts through PEG method in *Arabidopsis*, lettuce, rice and tobacco as protein/RNA complex to attain DNA-free genome editing (Woo et al. 2015). Instead, experimentally challenging tissue regeneration is needed to attain individual plants from protoplasts.
3) **Particle gun method** – it can introduce several molecules unconventionally into plant spp. Genome editing in maize, barley (Lawrenson et al. 2015), soybeans (Du et al. 2016), wheat (Liang et al. 2017), and banana (Kaur et al. 2018) have been reported by particle gun transformations.

Interestingly, the complexity of plant microbial associations with beneficial and harmful microbes is beginning to be unraveled (Bai et al. 2015). The CRISPR–Cas9 technique may facilitate forthcoming research on plant–pathogen interactions to surpass individual genes or pathogens. This technique will become a more holistic approach for elucidation of plant microbiome ecosystems. However, the multitude functions that might be accomplished with CRISPR–Cas9 and its derivatives (Sander and Joung 2014) make it a molecular tool that will open new opportunities in the complicated world of plant–pathogen interactions and assist in designing durable crop resistance to pathogens.

12.7 Glimpses of Successful Microbiome Based Products

Interestingly, it is anticipated that microbiome-based products may have a similar market size to chemical agrochemicals in subsequent years. For instance, the Indigo Company has launched a new cotton seed product coated with microbes which provided ~10 per cent yield benefits under water stress (Singh 2017). Likewise, the Bioag Alliance Company broadcast several novel microbiome products that can considerably augment corn yield by improving nutrient uptake and nematicidal activities. These plant microbiome products fuel expectation for novel and effective biopesticides by progressing scientific knowledge of the soil. A transformed focus on microbiomes may unearth a few more efficient biopesticides and biofertilizers (Timmis et al. 2014). Additionally, microbes offer essential genes for GM plants (James 2015), in the face of legal discernment hurdles. These outcomes will unswervingly add to food and nutrient security and curtailing the negative impact on the environment and non-target organisms by decreasing chemical usage, controlling the rate of climate change by limiting the emission of greenhouse gas and stimulating social stability. Application of microbiome techniques in agriculture will undoubtedly enhance innovation and commercialization of allied companies, formulation and precision tools industries, and plant breeding. All these activities collectively make a major contribution toward environmental sustainability and global economic growth (Singh 2017).

12.8 Implications for Agriculture: A Negative Note

However increasing agriculture productivity is a centuries-old agrarian goal, current concerns include globalization, high energy prices, climate change and changing landscape in quest of solutions (Godfray et al. 2010). However agriculture should consider maximizing the co-adaptation between microbes and plants in an attempt to support microbial diversity in the soils, while, even though this may perhaps temporarily lessen plant productivity. Evidence signifies the importance of co-adaptation of plant microbial communities in plant health and persistent crop yield (Badri and Vivanco 2009). What connotations do the co-adapted plant–microbial association hold for agriculture? The lack of plants' ability to uphold diversity and constancy of microbiome which is not co-adaptive have been described previously (Broeckling et al. 2008). Nonetheless the loss of diversity and evenness is detrimental to plant productivity and ecosystem functioning. Therefore, molecular communication oscillates in accordance with the plant development stage, proximity to neighboring species and several other factors. Finally the experimental manipulations may be disruptive to host fitness (e.g. antibiotics may harm the host) and similarly experimental inoculation with single strains is normally limited to microbes that might be cultured.

12.9 Future Perspectives and Outlook

Microbiome engineering in plants has made great advances in the past decades. Therefore, despite being a relatively new field, microbiome engineering exhibits promising prospects for agriculture to improve human health and augment crop productivity. The importance of microbiomes for functioning of plant agro-ecosystems has been broadly familiar, but the majority of studies lack a quantitative assessment of virtual contributions of host genotype, microbiome and environmental conditions for any given phenotype. Exploring the vast and yet unknown functional potential in host-associated microbiomes and unraveling the dynamic chemical interplay between host and microbiome will be indispensable to explain to what extent and how hosts recruit or activate specific members in their microbiomes for their own benefit. Furthermore, with regard to food and nutrient security, microbiome tools may significantly enhance economic performance by commercializing novel products, and improving environmental health by reducing chemical contamination. In summary, development of technologies will allow manipulation of crop microbiome *in situ*. These technologies turn out to be an integral part of sustainable increase in agricultural productivity, ensuring food and nutrient security for future global populations. Capitalizing on pristine understanding, microbiome engineering has demonstrated success in application for treatment of plant diseases and improving agricultural productivity. Realization of anticipated benefits of the plant microbiomes in agriculture will require a thorough understanding of functional and mechanistic aspects of interactions between microbes, plants, the farm environment, and agricultural management practices. Molecular biology and physiology of the vast majority of microbes remains uncharacterized. If genomes of every microbe in the complex soil community are mapped, one can figure out which microbes are implicated in plant growth and could unravel the intricate web of complex soil communications. In addition to that, a few outstanding questions arise on which mechanisms plants recruit in their root microbiome. How can this knowledge be used to manipulate ecosystem processes? However, microbiome engineering is still in its infancy, and has immense potential for the future.

Acknowledgment

The authors gratefully acknowledges UGC, Government of India, for financial support in the form of UGC-CSIR-SRF under the Grant No: F.No. 19-1/2015(SA-1).

Conflict of Interest

We declare we do not have any conflicts of interest.

References

Badri, D.V. and Vivanco, J.M. (2009). Regulation and function of root exudates. *Plant, Cell & Environment* 32: 666–681.

Bai, Y., Müller, D.B., Srinivas, G. et al. (2015). Functional overlap of the *Arabidopsis* leaf and root microbiota. *Nature* 528 (7582): 364.

Bais, H.P., Weir, T.L., Perry, L.G. et al. (2006). The role of root exudates in rhizosphere interactions with plants and other organisms. *Annual Review of Plant Biology* 57: 233–266.

Bardgett, R.D. and Wardle, D.A. (2010). *Above Ground–Below-Ground Linkages: Biotic Interactions, Ecosystem Processes, and Global Change*. Oxford University Press.

Berendsen, R.L., Pieterse, C.M.J., and Bakker, P.A.H.M. (2012). The rhizosphere microbiome and plant health. *Trends in Plant Science* 17: 478–486.

Borel, B. (2017). When the pesticides run out. *Nature* 543: 302–304.

Bortesi, L. and Fischer, R. (2015). The CRISPR/Cas9 system for plant genome editing and beyond. *Biotechnology Advances* 33 (1): 41–52.

Broeckling, C.D., Broz, A.K., Bergelson, J. et al. (2008). Root exudates regulate soil fungal community composition and diversity. *Applied and Environmental Microbiology* 74: 738–744.

Busby, P.E., Soman, C., Wagner, M.R. et al. (2017). Research priorities for harnessing plant microbiomes in sustainable agriculture. *PLoS Biology* 15 (3): e2001793.

Butler, N.M., Atkins, P.A., Voytas, D.F., and Douches, D.S. (2015). Generation and inheritance of targeted mutations in potato (*Solanum tuberosum* L.) using the CRISPR/Cas system. *PLoS One* 10 (12): e0144591.

del Carmen Orozco-Mosqueda, M., del Carmen Rocha-Granados, M., Glick, B.R., and Santoyo, G. (2018). Microbiome engineering to improve biocontrol and plant growth-promoting mechanisms. *Microbiological Research* 208: 25–31.

Castrillo, G., Teixeira, P.J., Paredes, S.H. et al. (2017). Root microbiota drive direct integration of phosphate stress and immunity. *Nature* 543: 513–518.

Chaparro, J.M., Badri, D.V., and Vivanco, J.M. (2014). Rhizosphere microbiome assemblage is affected by plant development. *ISME Journal* 8: 790–803.

Du, H., Zeng, X., Zhao, M. et al. (2016). Efficient targeted mutagenesis in soybean by TALENs and CRISPR/Cas9. *Journal of Biotechnology* 217: 90–97.

Edwards, J., Johnson, C., Santos-Medellin, C. et al. (2015). Structure, variation, and assembly of the root-associated microbiomes of rice. *Proceedings of the National Academy of Sciences of the United States of America* 112: E911–E920.

Endo, M., Mikami, M., and Toki, S. (2015). Multigene knockout utilizing off-target mutations of the CRISPR/Cas9 system in rice. *Plant and Cell Physiology* 56 (1): 41–47.

Falony, G., Joossens, M., Vieira-Silva, S. et al. (2016). Population-level analysis of gut microbiome variation. *Science* 352: 560–564.

Fang, Y. and Tyler, B.M. (2016). Efficient disruption and replacement of an effector gene in the oomycete *Phytophthora sojae* using CRISPR/C as9. *Molecular Plant Pathology* 17 (1): 127–139.

Farrar, K., Bryant, D., and Cope-Selby, N. (2014). Understanding and engineering beneficial plant–microbe interactions: plant growth promotion in energy crops. *Plant Biotechnology Journal* 12 (9): 1193–1206.

Foley, J.A., Ramankutty, N., Brauman, K.A. et al. (2011). Solutions for a cultivated planet. *Nature* 478: 337–342.

Godfray, H.C., Beddington, J.R., Crute, I.R. et al. (2010). Food security: the challenge of feeding 9 billion people. *Science* 327 (5967): 812–818. https://doi.org/10.1126/science.1185383.

Goyal, D., Prakash, O., and Pandey, J. (2019). Rhizospheric microbial diversity: an important component for abiotic stress management in crop plants toward sustainable agriculture. In: *New and Future Developments in Microbial Biotechnology and Bioengineering* (ed. V. Gupta), 115–134. Elsevier.

James, C. 2015. 20th Anniversary (1996 to 2015) of the Global Commercialization of Biotech Crops and Biotech Crop Highlights in 2015. ISAAA Brief No. 51. Ithaca, NY: ISAAA.

Kamkar, B. (2016). Sustainable development principles for agricultural activities. *Advances in Plants and Agricultural Research* 3 (5): 1–2.

Kaur, N., Alok, A., Kaur, N. et al. (2018). CRISPR/Cas9-mediated efficient editing in phytoene desaturase (PDS) demonstrates precise manipulation in banana cv. Rasthali genome. *Functional and Integrative Genomics* 18 (1): 89–99.

Kroll, S., Agler, M.T., and Kemen, E. (2017). Genomic dissection of hostmicrobe and microbe-microbe interactions for advanced plant breeding. *Current Opinion in Plant Biology* 36: 71–78.

Lawrenson, T., Shorinola, O., Stacey, N. et al. (2015). Induction of targeted, heritable mutations in barley and *Brassica oleracea* using RNA-guided Cas9 nuclease. *Genome Biology* 16 (1): 258.

Lebeis, S.L. (2015). Greater than the sum of their parts: characterizing plant microbiomes at the community-level. *Current Opinion in Plant Biology* 24: 82–86.

Lemanceau, P., Blouin, M., Muller, D., and Moënne-Loccoz, Y. (2017). Let the core microbiota be functional. *Trends in Plant Science* 22: 583–595.

Liang, Z., Chen, K., Li, T. et al. (2017). Efficient DNA-free genome editing of bread wheat using CRISPR/Cas9 ribonucleoprotein complexes. *Nature Communications* 8: 14261.

Liu, C., Sakimoto, K.K., Colón, B.C. et al. (2017). Ambient nitrogen reduction cycle using a hybrid inorganic–biological system. *Proceedings of the National Academy of Sciences of the United States of America* 114 (25): 6450–6455.

Lowder, L., Malzahn, A., and Qi, Y. (2016). Rapid evolution of manifold CRISPR systems for plant genome editing. *Frontiers in Plant Science* 7: 1683.

Matsu-ura, T., Baek, M., Kwon, J., and Hong, C. (2015). Efficient gene editing in *Neurospora crassa* with CRISPR technology. *Fungal Biology and Biotechnology* 2 (1): 4.

Moran, M.A. (2015). The global ocean microbiome. *Science* 350: 8455.

Mueller, U.G. and Sachs, J.L. (2015). Engineering microbiomes to improve plant and animal health. *Trends in Microbiology* 23: 606–617.

Navarrete, A.A., Tsai, S.M., Mendes, L.W. et al. (2015). Soil microbiome responses to the short-term effects of Amazonian deforestation. *Molecular Ecology* 24: 2433–2448.

Nocera, D.G. (2017). Solar fuels and solar chemicals industry. *Accounts of Chemical Research* 50 (3): 616–619.

Nødvig, C.S., Nielsen, J.B., Kogle, M.E., and Mortensen, U.H. (2015). A CRISPR-Cas9 system for genetic engineering of filamentous fungi. *PLoS One* 10 (7): e0133085.

Panke-Buisse, K., Poole, A.C., Goodrich, J.K. et al. (2015). Selection on soil microbiomes reveals reproducible impacts on plant function. *The ISME Journal* 9: 980–989.

Passari, A.K., Chandra, P., Zothanpuia et al. (2016). Detection of biosynthetic gene and phytohormone production by endophytic actinobacteria associated with *Solanum lycopersicum* and their plant growth-promoting effect. *Research in Microbiology* 167 (8): 692–705.

Perez, Y.M., Charest, C., Dalpe, Y. et al. (2016). Effect of inoculation with Arbuscular Mycorrhizal fungi on selected spring wheat lines. *Sustainable Agricultural Research* 5 (4): 24–29.

Round, J.L. and Mazmanian, S.K. (2009). The gut microbiota shapes intestinal immune responses during health and disease. *Nature Reviews Immunology* 9: 313–323.

Sander, J.D. and Keith, J. (2014). CRISPR-Cas systems for editing, regulating and targeting genomes. *Nature biotechnology vol.* 32 (4): 347–355. https://doi.org/10.1038/nbt.2842.

Santhanam, R., Luu, V.T., Weinhold, A. et al. (2015). Native root-associated bacteria rescue a plant from a sudden-wilt disease that emerged during continuous cropping. *Proceedings of the National Academy of Sciences of the United States of America* 112: E5013–E5020.

Schlaeppi, K. and Bulgarelli, D. (2014). The plant microbiome at work. *Molecular Plant-Microbe Interactions* 28 (3): 212–217.

Schweitzer, J.A., Bailey, J.K., Fischer, D.G. et al. (2016). A review on sustainable agricultural intensification in Nepal. *International Journal of Business, Social and Scientific Research* 4 (3): 152–156.

Singh, B.K. (2017). Creating new business, economic growth and regional prosperity through microbiome-based products in the agriculture industry. *Microbial Biotechnology* 10 (2): 224–227.

Singh, B.K. and Trivedi, P. (2017). Microbiome and the future for food and nutrient security. *Microbial Biotechnology* 10: 50–53.

Singh, U.B., Malviya, D., Singh, S. et al. (2016). Bio-protective microbial agents from rhizosphere eco-systems trigger plant defense responses provide protection against sheath blight disease in rice (*Oryza sativa* L.). *Microbiological Research* 192: 300–312.

Swenson, W., Wilson, D.S., and Elias, R. (2000). Artificial ecosystem selection. *Proceedings of the National Academy of Sciences of the United States of America* 97: 9110–9114.

Timmis, K., de Lorenzo, V., Verstraete, W. et al. (2014). Pipelines for New Chemicals: a strategy to create new value chains and stimulate innovation-based economic revival in Southern European countries. *Environmental Microbiology* 16: 9–18.

Ubertino, S., Mundler, P., and Tamini, L.D. (2016). The adoption of sustainable management practices by Mexican coffee producers. *Sustainable Agricultural Research* 5 (4): 1–12.

Valverde, A., De Maayer, P., Oberholster, T. et al. (2016). Specific microbial communities associate with the rhizosphere of *Welwitschia mirabilis*, a living fossil. *PLoS One* 11: e0153353.

Vandenkoornhuyse, P., Quaiser, A., Duhamel, M. et al. (2015). The importance of the microbiome of the plant holobiont. *The New Phytologist* 206: 1196–1206.

Vejan, P., Abdullah, R., Khadiran, T. et al. (2016). Role of plant growth promoting Rhizobacteria in agricultural sustainability – a review. *Molecules* 21 (573): 1–17.

Wei, Z., Yang, T., Friman, V.P. et al. (2015). Trophic network architecture of root-associated bacterial communities determines pathogen invasion and plant health. *Nature Communications* 6: 8413.

Widder, S., Allen, R.J., Pfeiffer, T. et al. (2016). Challenges in microbial ecology: building predictive understanding of community function and dynamics. *The ISME Journal* 10: 2557–2568.

Woo, J.W., Kim, J., Kwon, S.I. et al. (2015). DNA-free genome editing in plants with preassembled CRISPR-Cas9 ribonucleoproteins. *Nature Biotechnology* 33 (11): 1162.

13

Biotechnology of Plant-Associated Microbiomes

Son Truong Dinh[1], Van T. Luu[2], Long Hoa Hoang[3], Xuan Canh Nguyen[1], and Cuong Tu Ho[4]

[1] *Vietnam National University of Agriculture, Hanoi, Vietnam*
[2] *Institute for Molecular Physiology, Heinrich-Heine-Universität, Düsseldorf, Germany and Max Planck Institute for plant breeding research, Cologne, Germany*
[3] *Agricultural Genetics Institute – Vietnam Academy of Agricultural Sciences, Hanoi, Vietnam*
[4] *Institute of Environmental Technology, Vietnam Academy of Science and Technology, Hanoi, Vietnam*

13.1 What Are Plant-Associated Microbiomes

Plants do not live alone; however, a vast diversity of microorganisms colonize their surfaces and inner tissues which make up the plant-associated microbiomes. The term "microbiome" was first mentioned by Joshua Lederberg as the "ecological community of commensal microorganisms, symbionts or pathogens, that literally occupy a space in our body" (Lederberg and McCray 2001). Later on, this term has been widely used in two different ways to indicate (i) a set of microbes which can interact with different issues in specific hosts or (ii) microbial communities of a particular habitat (Boon et al. 2014; Ofek et al. 2014; Orozco-Mosqueda et al. 2018). In this chapter, the term plant microbiome, is reflecting the first definition, indicates the reciprocal interactions between microbiome and its host which can occur, and it is recognized as parts of a whole meta-organism. Most microbiota individuals are horizontally obtained from the encompassing environment where soil is the fundamental source of the plant's microbiome, whereas a small fragment of microbiota is originally from the air (Sánchez-Cañizares et al. 2017). There are many descriptions of the core plant microbiome in many model plants, for example, *Arabidopsis thaliana, Brachypodium distachyon, Medicago truncatula, Lotus japonicus, Marchantia polymorpha*, and *Nicotiana attenuata* (Lundberg et al. 2012; Tkacz et al. 2015; Kawasaki et al. 2016; Zgadzaj et al. 2016; Alcaraz et al. 2017; Santhanam et al. 2017). Many crops like sugar beet, rice, maize, sugarcane, barley, soybean and cowpea, potato, seven different *Brassica* vegetables including: broccoli, white cabbage, arugula turnip, cabbage, cauliflower, radish, and rucola have also been documented (Mendes et al. 2011; Peiffer et al. 2013; Sugiyama et al. 2014; Bulgarelli et al. 2015; Edwards et al. 2015; de Souza et al. 2016; Leite et al. 2016; Pfeiffer et al. 2017; Santos-Medellin et al. 2017; Wassermann et al. 2017).

The Plant Microbiome in Sustainable Agriculture, First Edition. Edited by Alok Kumar Srivastava, Prem Lal Kashyap and Madhumita Srivastava.

Plants, like humans and other eukaryotic hosts, endure a compact symbiotic relationship with their microbiome, making plant-associated microbiomes a potential field of study. The issue addressed here is not just describing the place where the microbes live (rhizosphere, endosphere, and phyllosphere) but also reviewing the consequence of interactions between microorganisms and plants, as well as providing selected studies on the functional roles of microorganism communities in plant health and fitness.

Plant-associated microbiomes are understood as microbial organisms which can colonize internal tissues or live on external surfaces and play an appreciable role in regulating the physiological processes of their hosts. Though there is no clear picture of the overall function of the plant microbiome, there is a substantial evidence that these communities are involved in enhancing nutrient acquisition, disease control, and stress tolerance under adverse environmental conditions (Andreote et al. 2014). They even serve as an extension of the plant's genetic compendium (Berendsen et al. 2012). Particularly, in supplying nutrients, these microorganisms have a function in phosphorus solubilization and nitrogen fixation; or in promoting plant protection by hindering agents of plant stresses such as infection by pathogens and pests (Andreote et al. 2014).

Most microbes have neutral or beneficial roles on plant health and fitness (Andreote et al. 2014). There are three major compartments, namely, rhizosphere, endosphere, and phyllosphere where microbes can establish and develop (Hardoim et al. 2008; Hirsch and Mauchline 2012). The rhizosphere generally is the portion of soil surrounding the roots of living plants and influenced by the plant (Bulgarelli et al. 2013; Philippot et al. 2013); The endosphere consists of all the endophytes of a plant that intimately interact with the host (Hallmann et al. 1997; Hardoim et al. 2008) and the phyllosphere is usually defined by microbial cells that are able to colonize above-ground surfaces (Lindow and Brandl 2003; Lambais et al. 2006).

13.1.1 The Rhizosphere Microbiomes

The rhizosphere is described as the soil region under the influence of the roots (Hartmann et al. 2008; Philippot et al. 2013). This is a hot spot for diverse organisms and is recognized as one of the most complicated ecosystems on Earth (Hiltner and Hinsinger 2006; Pierret et al. 2007; Jones and Hinsinger 2008; Hinsinger et al. 2009; Raaijmakers et al. 2009). The rhizodeposits (e.g. exudates, border cells, mucilage) have been known to be major players in modulating microbial diversity and activity on plant roots. It is postulated that plants might shape the rhizosphere microbiome by affecting microorganisms with traits that are beneficial to their growth and fitness (Cook et al. 1995). Organisms known in the rhizosphere include bacteria, fungi, oomycetes, nematodes, protozoa, algae, viruses, archaea, and arthropods. The soil microbiome contributes to many substantial processes occurring in the rhizosphere, which could directly or indirectly relate to plant health. For instance, the soil microbiome inhibits plant diseases by imposing physiological restrictions on pathogens establishing and infecting plant tissues. In addition, exudates can effectively be a defensive barrier of the below-ground plant–microbe interface (Rout 2014). Various researches have been focusing on the crucial effect of local environmental conditions such as soil properties on soil microbial community composition and root microbiota.

The relationship between microorganisms and environmental conditions affects the regulation of plant activities. Also, the microbial community structure is differentially established, depending on types of soil and driven by two types of selection: the root exudates that leads to the high availability of nutrients; and the increase of microbial biomass, which alters the environmental conditions in the rhizosphere (Andreote et al. 2014). Table 13.1 enlists some specific rhizosphere microbes and their effect on hosts.

13.1.2 The Endosphere Microbiomes

Rhizospheric microorganisms have been extensively explored for decades, however, endophytes, the microorganisms living fully within plants and more intimate associations of plants, have been only recently studied. Endophytes are defined as microorganisms which occupy the plant tissue without causing any symptoms of disease. Endophytes are typical for different plant organs, such as roots, leaves and seeds, and can be found on the surface or within the plant tissue. Endophytes can be divided into groups, which perform different functions such as N-fixation, P and Fe nutrient acquisition, and even plant growth promotion (Cankar et al. 2005).

Comprising an ecosystem within plants, endophytes are interacting with each other and involved in nutrient acquisition and cycling. The specific members of the microbiome are influenced by the environment, abiotic (including temperature, drought, and salinity) or biotic stresses and plant genotype. Endophytes can improve nutrient availability for plants via many activities such as nitrogen fixation, phosphate solubilization, and siderophore production. The phytobiome can improve photosynthetic and water use efficiency (Table 13.1). Many endophyte strains can produce hormones or adjust the host phytohormones, and improve both plant growth and stress tolerance (Doty 2017).

13.1.3 The Phyllosphere Microbiomes

Beside the roots zone, the above-ground parts of plants, termed the phyllosphere, are normally colonized by a variety of microorganisms and the inhabitants are called epiphytes. While a few microbial species can be isolated from the inner part of plant tissues, many more are collected from the surfaces of plants; especially on the leaf, a dominant aerial plant structure (Lindow and Brandl 2003). Typically, both the age and size of the hosts positively influence the abundance of epiphytes (Andersohn 2004).

The leaf microbial communities comprise many different bacterial genera, filamentous fungi, yeasts, algae, and, less frequently, protozoa and nematodes. Among these, the largest group is the bacterial composition, presented in the phyllosphere at concentration between 10^5 and 10^7 cells cm^{-2} (Beattie and Lindow 1995; Andrews and Harris 2000). Phyllosphere-living microorganisms exert essential roles related to development of plant growth, for example, fixing nitrogen, phytopathogenic biocontrol and biosynthesizing phytohormones (Table 13.1). In addition, these microbial communities are able to thrive and live under stressful environmental conditions such as humidity, UV radiation, pH and temperature (Andrews and Harris 2000).

Table 13.1 Diversity of plant-associated microorganism communities.

Host plant (Plant part)	Isolated strains/plant-associated microorganisms	Functional traits	References
ENDOSPHERE			
Theobroma cacao (Roots)	*Trichoderma hamatum*	Drought tolerance, Enhancement of nutrient availability	Bae et al. (2009)
Triticum spp. (Seeds)	*Paenibacillus* sp. *Pantoea* sp. *Bacillus* sp. *Fictibacillus* sp.	Indoleacetic acid and siderophore production Phosphate solubilization Antifungal (*Fusarium graminearum)* activity Biofilm formation, Antibiotic resistance	Herrera et al. (2016)
Panicum virgatum L. (leaves/seeds)	*Microbacterium testaceum* *Curtobacterium flaccumfaciens* *Bacillus subtilis/B.pumilus* *Pseudomonas fluorescens* *Sphingomonas parapaucimobilis* *Serratia* sp. *Pantoea ananatis*	Solubilizing inorganic phosphorus capability Lipopeptide toxins and phytohormones production Antifungal activity	Gagne-Bourgue et al. (2013)
Fragaria ananassa (tissues)	*Bacillus* sp. *Sphingopyxis* sp. *Virgibacillus* sp.	Auxin production Phosphate solubilization Root and shoot development	Dias et al. (2009)
Oryza sativa L. (seeds/roots)	*Bacillus* sp. *Caulobacter crescentus* *Kocuria palustris* *Micrococcus luteus* *Methylobacterium fujisawaense* *Me. radiotolerans* *Pantoea ananatis*	Tolerance to high osmotic pressure Enhanced nutrient availability	Kaga et al. (2009)
RHIZOSPHERE			
Calendula officinalis	*Bacillus* spp.	Antifungal activity, Indole 3 acetic (IAA) and siderophore production, Lipopeptides and cell-wall degrading enzyme production.	Kaki et al. (2013)
Helianthus annuus	*Pseudomonas gessardii* *P. fluorescens*	Lead (Pb) tolerance Enhancing the activity of antioxidant enzymes, growth, physiology and yield Phytoremediation potential in lead stressed conditions	Saleem et al. (2018)

Table 13.1 (Continued)

Host plant (Plant part)	Isolated strains/plant-associated microorganisms	Functional traits	References
Hordeum vulgare L.	*B. amyloliquefaciens*, *B. circulans*, *P. fluorescens*, *Paenibacillus polymyxa*, *B. megatherium*, *Azotobacter chroococcum*, *Azospirillum brasilense*	Biofilm formation Salinity stress tolerance	Kasim et al. (2016)
Solanum lycopersicum L.	*B. pumilus Pseudomonas fluorescens* *Piriformospora indica* (endophytic fungi)	Antagonism against the root-knot nematode infection (*Meloidogyne incognita*)	Varkey et al. (2018)
PHYLLOSPHERE			
Oryza sativa L.	*Bacillus* sp. *Pseudomonas* sp.	Phosphate solubilization IAA and siderophores production, N fixation Antagonistic activity against the phytopathogenic fungi (*Rhizoctonia solani, Drechslera oryzae* and *Magnaporthe oryzae*) Leaf nutrient enhancement	Thapa et al. (2017)
Vitis vinifera	*Aureobasidium pullulans* *Epicoccum nigrum* *Pseudomonas* sp.	Antibacterial activity	Grube et al. (2011)
Lactuca sativa L.	*Proteobacteria, Bacteroidetes* *Actinobacteria, Firmicutes,* *Gammaproteobacteria* *Betaproteobacteria*	Solar radiation resistance	Truchado et al. (2017)

13.2 Factors Influencing the Diversity and Structure of Plant Microbiomes

The association of plants with their microbiomes plays an important role in all aspects of their growth, development, defense, and reproduction (Berendsen et al. 2012; Gaiero et al. 2013; Chaparro et al. 2014; Antoniou et al. 2017). Therefore, the ability of a plant to recruit and maintain its microbiome is crucial for the plant's fitness and evolution in nature (Bouffaud et al. 2014). In contrast to mammals whose gut microbiome is assembled internally and transmissible through birth, when plants germinate and grow in soil they typically acquire a selection of the bulk soil microbial community and a subset of this community develops into a root associated microbiome (Vaishampayan et al. 2010; Koenig et al. 2011; Santhanam et al. 2015; Sasse et al. 2017). For rice, the core endospheric microbiome attaches to the rhizosphere, relocates to the rhizoplane and then migrates to the root interior. This assembly process begins early – within a day – after transplanting into the soil, and approaches steady state within two weeks (Edwards et al. 2015). A two-step model for root microbiota assembly was proposed: the first step is the enrichment process which is largely driven by root exudates and begins at a distance in the rhizosphere and then on the rhizoplane. The second step is the exclusion process that occurs first on the rhizoplane and then continues in the root endosphere (Edwards et al. 2015). While the rhizoplane selects for biofilm-forming bacteria that can compete for the elevated nutrient levels, the root endosphere selects microbes which could evade plant recognition or manipulate host defense responses to enable their colonization (van der Heijden and Schlaeppi 2015).

The association of plants with their microbiome is strongly affected by plant evolutionary history and the plant's species, genotype, phenological and physiological characteristics, as well as both biotic and abiotic factors in the environment where the plant grows.

13.2.1 Plant Evolutionary History and Plant Species

The evolution of a plant involves traits that are important for association with microbiomes; therefore, the more phylogenetically distant the host plants, the more distinct their microbiomes are (Bouffaud et al. 2014; Emmett et al. 2017). It has been shown that phylogenetic variation in different angiosperm species which diverged 140 million years ago affect diversity and composition of the root-associated bacterial communities (Fitzpatrick et al. 2018). For Poaceae species, their rhizobiome clusters were significantly correlated with their phylogenetic distances (Bouffaud et al. 2014). Similarly, beta-diversity of the rhizosphere was positively correlated with the phylogenetic distance of 11 annual plant species (Emmett et al. 2017). This indicates that plant evolutionary history influences its associated microbial community and that incorporating phylogenetic diversity into crop rotation to manipulate plant microbiome is important.

Plant domestication causes a decrease in the genetic diversity of crop plants, hence, affecting the ability of these plants to recruit beneficial microbial communities (Perez-Jaramillo et al. 2016). Domesticated crop plants such as barley, maize, sugar beet, and lettuce have distinct root-associated microbiomes compared to their wild relatives (Zachow et al. 2014; Bulgarelli et al. 2015; Cardinale et al. 2015; Szoboszlay et al. 2015). Interestingly, crop wild relatives, for example, sugar beet and *Agave*, harbor more diverse bacterial communities that differ from those in domesticated plants at functional levels, leading to a stronger ability to cope with environmental stress (Zachow et al. 2014; Coleman-Derr et al. 2016). It was suggested that native plants selected soil microbial communities throughout their evolutionary history to cope with context-specific problems (Santhanam et al. 2015). Hence, understanding the functional differences between crop-associated and wild relative-associated microbiomes is important for engineering crop microbial communities. In addition, wild relatives of crops, as well as soils from the area where the plant was originally distributed, are important sources for isolating crop beneficial microbes.

13.2.2 Plant Genotype

Plant genotypes are important for shaping plant microbiomes but are not always necessary to do so. Root associated microbial communities, and especially the rhizospheric community, are significantly affected by different rice genotypes (Edwards et al. 2015). In *Arabidopsis*, some genotype-dependent variations within the species were observed for endophytic microbiomes (Bulgarelli et al. 2012; Lundberg et al. 2012). For different maize inbred lines, the combination of root rhizospheric and endospheric microbiomes have relatively small genotype dependence (Peiffer et al. 2013; Emmett et al. 2017). Fungal communities within leaf niches, but not archaeal/bacterial communities, were greatly impacted by the genotypes of *Populus* trees (Cregger et al. 2018). Understanding the impact of genetic variation on the composition of the microbiota is important for identification of heritable plant–microbe interactions by genome-wide association, thus, plant microbiome selection can be included into plant breeding. Selection of plant genotypes that can specifically recruit beneficial microbes offers opportunities to enhance agricultural productivity and environmental sustainability.

13.2.3 Phenological and Physiological Characteristics of the Plant

13.2.3.1 Plant Compartments

Different plant compartments harbor different microbial communities. Archaeal/bacterial and fungal microbiomes were shown to vary significantly across different *Populus* plant compartments. Fungal diversity was the greatest in stems and soils, whereas archaeal/bacterial diversity increased from leaves to stems, then to roots, and was the highest in soil habitats (Cregger et al. 2018). The composition of prokaryotic communities was primarily determined by *Agave* plant compartments, in which the leaf endosphere, the phyllosphere, the rhizosphere and the root endosphere are clearly different from one another (Coleman-Derr et al. 2016). Proteobacteria and Ascomycota were dominant in *Populus* root microbial communities, whereas Chytridiomycota and Acidobacteria were greatly reduced in the endospheres compared to the rhizosphere (Shakya et al. 2013). Different types of Brachypodium lateral roots were associated with different microbial communities (Kawasaki et al. 2016). While root tips are often associated with a wide range of active bacteria, the root elongation zone is colonized by plant-specific microbes. Mature root zones also harbor distinct microbial communities compared to root tips (Massalha et al. 2017). Variation in microbiomes among different plant compartments suggests the different mechanisms that plant use to recruit tissue-specific microbiomes. In particular, the root surface or rhizoplane can form a critical interface between a plant and the soil (Edwards et al. 2015). Features of *Arabidopsis* cell wall supply enough cues for the assembly of 40 per cent root associated bacteria, however, the colonization of *Arabidopsis*-specific root microbiome depends on other cues from metabolically active host cells (Bulgarelli et al. 2012).

13.2.3.2 Plant Developmental Stages

Plants at different developmental stages can exhibit distinct rhizobiomes. Studies on *Arabidopsis* showed that the microbial structure at the seedling stage is different from the other stages such as bolting and flowering. Furthermore, there is a strong association of bacterial phyla such as Acidobacteria, Actinobacteria, Bacteroidetes, and Cyanobacteria with plant development (Chaparro et al. 2014). In sweet potato, the structure of the microbial community in the tuber rhizosphere is affected by the growth stage. Among the analyzed bacterial community, Alphaproteobacteria and Bacteria communities were more strongly influenced by plant stage than the *Pseudomonas* community (Marques et al. 2014). In strawberry, potato, and oilseed rape, Gram-positive bacteria in the rhizosphere such as *Nocardia* and *Streptomyces* were found to differ during the developmental stages of these plants (Smalla et al. 2001). At the early reproductive stages, soybean have more diversity in their microbial communities than later developmental stages (Xu et al. 2009). Studying the changes in microbial communities in combination with transcriptomic and metabolomic changes along plant development allows us to uncover genes and metabolites that tailor the microbial assembly. For example, *Arabidopsis* secretes more sugars and sugar alcohols at an earlier developmental stage and increases levels of amino acids and phenolics at a later developmental stage. This suggests that seedlings release sugars to recruit a wide range of microbes but later, the plant releases more specific substrates and antimicrobial compounds to select for certain microbes (Chaparro et al. 2013).

13.2.3.3 Root Exudates

Plants exude 15 per cent nitrogen and up to 20 per cent fixed carbon (Haichar et al. 2016; Venturi and Keel 2016). Root exudates contain many chemicals that act as substrates, chemotactic molecules, or signaling molecules that structure microbial communities of plants. For instance, benzoxazinoids in root exudates of maize help the plant to attract *Pseudomonas putida* to the rhizosphere (Neal et al. 2012). Petunia roots release strigolactones that trigger the attachment of the arbuscular mycorrhizal fungus *Glomeromycota* (Kretzschmar et al. 2012). In *Pseudomonas*-infected *Arabidopsis* roots, malic acid is exudated to recruit the beneficial rhizobacterium *Bacillus subtilis* (Rudrappa et al. 2008). The identification of phytochemicals derived from root exudates that positively influence plant microbiome associations can help to increase crop productivity and sustainability.

Plant roots do not only exudate phytochemicals, but also a polysaccharide mucilage which provides a carbon source for microbes, and thus affects the rhizobiome structure (Knee et al. 2001). Consequently, adding the maize root mucilage to soil could alter the composition of the rhizobiome (Benizri et al. 2007). Border cells produced by root tips can also exude proteins, extracellular DNA and metabolites which influence the microbial community (Brigham et al. 1995; Vicre et al. 2005; Wen et al. 2009). For example, border cells of *Arabidopsis* exudate arabinogalactan protein that attracts a rhizobiome (Vicre et al. 2005). *Medicago* border cells accumulate a high level of flavonoid and triterpenoid-related metabolites which act as constitutive defenses as well as signals to motile rhizobia (Watson et al. 2015).

Root exudates vary with plant genotypes, plant developmental stages, and environmental stresses. Among different *Arabidopsis* accessions, the glucosinolates they produce present the most variability; flavonoids are relatively intermediate and phenylpropanoids have the lowest variability (Monchgesang et al. 2016). With the progression of *Arabidopsis* developmental stages, sugar exudation is decreased whereas amino acid and phenolic exudation are increased (Chaparro et al. 2013). When maize plants were grown in nitrogen-, iron-, phosphate-, or potassium-deficient soils, their exudations of amino acids, sugars, and organic acids changed (Carvalhais et al. 2013). Similarly, wheat plants increased phytosiderophore exudation under zinc-deficient conditions (Cakmak et al. 1994; Oburger et al. 2014). The proteoid root of plants growing on extremely nutrient-poor soils exudes high levels of organic acids to solubilize phosphate (Neumann and Martinoia 2002).

13.2.3.4 Plant Immune System

The host immune system can have a strong impact on the exclusion of bacterial communities moving from the rhizoplane to the root endosphere (van der Heijden and Schlaeppi 2015). Since plants normally initiate defense responses upon the detection of microbe-associated molecular patterns which are conserved throughout the bacterial kingdom, in order to colonize plant roots, microbes need to cope with the host immune system. It has been suggested that plants first recognize the beneficial microbes as potential invaders with an activation of immune responses, however, at later stages, mutualists can short-circuit plant defense responses (Zamioudis and Pieterse 2012).

Plant defense responses involve the activation of plant signaling pathways, such as salicylic acid (SA), jasmonic acid (JA) and ethylene (ET) signaling. The increase of the JA level in a plant during mycorrhizal symbiosis and the interaction with the plant growth promoting fungus (PGPF) *Piriformospora indica* is important for controlling colonization of such beneficial fungi (Schafer et al. 2009; Lopez-Raez et al. 2010). ET plays a crucial role in the regulation of microsymbionts as well as beneficial endophytic fungi. ET-insensitive *Medicago* plants are hyperinfected by the microsymbiont *Rhizobium meliloti* (Penmetsa and Cook 1997). Increasing ET signaling responses reduced root colonization by PGPF *P. indica* in *Arabidopsis* (Camehl et al. 2010). In *N. attenuata*, a wild tobacco plant, the abilities to produce and perceive ET are important for shaping the root-associated microbiome (Long et al. 2010). In *Arabidopsis*, the phytohormone SA plays a role in the colonization of the root microbiome at specific bacterial families (Lebeis et al. 2015).

The influence of phytohormone signaling on the plant microbiome occurs not only locally, but also systemically. It is known that SA promotes systemic acquired resistance (SAR), while JA and ET mediate induced systemic resistance (ISR). Pathogen-induced SAR depends on SA signaling and is associated with enhanced expression of PR genes, whereas PGPF–triggered ISR is often SA-independent (Van der Ent et al. 2009). Studies have shown that plants utilize these systemic regulation mechanisms to control the interaction with beneficial microbes. Hence, deficiency of SAR in *Arabidopsis* leads to differences in the rhizosphere bacterial community (Hein et al. 2008). The colonization of one part of the tomato root system with PGPF enhanced the ISR in another part of the plant, which in turn reduced the colonization by *Meloidogyne javanica* (Siddiqui and Shaukat 2002). These studies indicate that defense phytohormones are crucial for structuring plant microbiome diversity.

Plant defenses via the production of antimicrobial compounds such as phenolics, terpenoids, and alkaloids play a role in structuring the plant microbiome. A study showed that Rhizobiaceae and fungal communities in the rhizosphere and root tissues of *Arabidopsis* changed significantly when transgenic plants produced exogenous glucosinolate (Bressan et al. 2009). Similarly, lacking avenacins, which are triterpenoid saponins with a broad-spectrum antifungal activity, led to a significant change in culturable root-associated fungi in oat (Carter et al. 1999). Therefore, molecular breeding for disease resistance via the production of antimicrobial compounds should also consider the changes in the plant microbiome.

13.2.4 Biotic Factors

The association of a plant with pioneer microbes strongly influences the secondary colonization of the plant by other microbes, thus affecting the microbial community structure. For instance, pre-inoculation of barley with *Pseudomonas* sp. DSMZ13134 affects the dominant bacterial community in the root (Buddrus-Schiemann et al. 2010). The presence of ectomycorrhizal species influences the root bacterial communities of *Pinus sylvestris* (Deveau 2016). The presence of potential fungal pathogens leads to a significant change in overall microbiome composition of *Populus* trees (Cregger et al. 2018). The obligate biotrophic oomycete pathogen *Albugo* had strong effects on epiphytic and endophytic bacterial colonization in *Arabidopsis*. On the other hand, the presence of *Dioszegia*, a basidiomycete yeast fungus, strongly influences the phyllosphere microbiome (Agler et al. 2016). Furthermore, upon attacks by phytopathogens, plants can also exploit microbial communities from the soil for protection (Mendes et al. 2011). One classical example is the "take-all" disease in wheat caused by *Gaemannomyces graminis*: the disease severity is reduced by continuous monoculture that has built-up the disease antagonist *Pseudomonas fluorescens* (Weller et al. 2002). The association of a plant with beneficial microbes can suppress phytopathogens directly via production of antimicrobial compounds, biofilm formation, or founder effects. It can also indirectly affect pathogens by triggering ISR or increase the overall health of the host plant (Zamioudis and Pieterse 2012; Pieterse et al. 2014). Since pre-inoculation with beneficial microbes increases plant defense responses upon the presence of pathogens, it is an ideal method to modulate the plant microbiome.

Not only can plant–microbe interactions change the plant microbiome, but plant–insect interactions can also influence plant colonization by other microbes. For example, in response to aphid attack on pepper, the plants increased their association with the beneficial *B. subtilis*, which in turn reduced populations of the pathogen *Ralstonia solanacearum* (Lee et al. 2012). This tripartite plant–insect–microbe interaction increases the complexity of biotic factors that could tailor the plant associated microbiome and require more research.

13.2.5 Abiotic Factors

Abiotic factors such as biogeography, soil condition, agricultural practice and environmental stress are important for defining the plant associated microbiome.

13.2.5.1 Plant Biogeography

A plant species located at different biogeographies exhibits a different microbiome in each location. Different maize inbred lines exhibit large variations of plant-associated bacterial communities among different geographical locations (Peiffer et al. 2013). Similarly, biogeography of *Agave* host species plays the dominant role in structuring fungal communities (Coleman-Derr et al. 2016). Interestingly, plant-associated fungal communities are more differentiated by geographic distance than prokaryotic communities (Shakya et al. 2013). It was shown that habitat size and isolation have a strong effect on ectomycorrhizal species richness (Peay et al. 2007). Hence, trade-offs between competition and dispersal could play a role in the assembling of plant-associated fungal communities.

13.2.5.2 Soil Condition

The bacterial communities in the endophytic compartments, rhizosphere and bulk soil of *Arabidopsis* are structured by soil type (Lundberg et al. 2012). For rice growing under controlled conditions, the root microbiome composition varied by soil type. Interestingly, soil type can have more effect on the frequency of rare taxa than abundant taxa (Edwards et al. 2015). Different plant microbiomes of the same plant species grown in different soils are most likely due to the different physical and chemical characteristics of the soil. For instance, in tomato which was grown in soil that has a greater carbonate content but lower organic and nitrogen contents, the abundance of Proteobacteria in the rhizobiome was reduced (Pii et al. 2016). Root of white lupin growing in low pH and carboxylate-rich soil are dominated by *Burkholderia* species that help to metabolize oxalate and citrate (Weisskopf et al. 2011). Physical disturbances of soil such as short periods of sudden heat could also rearrange the rhizobacterial community due to the differences in heat tolerance of bacterial families (van der Voort et al. 2016). Different land use intensities, for example, natural grassland vs. intensively managed meadows, were shown to alter root endophytic communities by causing significant changes in soil structure and soil microbial community (Estendorfer et al. 2017). Other agriculture practices, and especially the nurturing of seedlings in sterilized conditions, can lead to a lack of early colonization with beneficial microbial communities, which are important for protecting the plant from pathogenic diseases (Santhanam et al. 2015). Thus, it has been suggested reconsidering using non-specific antimicrobial seed treatments that could affect the microbial recruitment process.

13.2.5.3 Environmental Stress

Environmental factors including temperature, moisture, and radiation strongly influence plant metabolism, which in turn affect the phyllosphere microbiome (Turner et al. 2013a). Interestingly, phyllosphere microbial diversity is greater in warm and humid climates than in temperate zones (Turner et al. 2013b). Lactic acid bacteria were found to be highly abundant in the phyllosphere of plants grown in the harsh conditions of the Mediterranean summer (Vokou et al. 2012). Drought, one of the most important environmental stressors that plants face, can affect plant productivity. Drought has been shown to induce a shift in the composition of the root microbiome. For instance, drought increased the relative abundance of Actinobacteria, leading to a shift in root microbiome composition in grass and other angiosperm species (Naylor et al. 2017; Fitzpatrick et al. 2018). Importantly, members of the Streptomycetaceae were highly associated with increased drought tolerance of the host plant (Fitzpatrick et al. 2018). This finding supports the hypothesis that plant-microbiome change under abiotic stress is an adaptive mechanism of the plant (Goh et al. 2013).

13.3 Roles of Plant-Associated Microbiome in Plant

The plant-associated microbiome modulates almost every aspect of the plant's life either directly or indirectly. The role of these microbes is discussed in two broad categories: plant growth and fitness and the integration of plant-associated microbiomes into sustainable agriculture. Plant growth promotion includes direct production or secretion of nutrients like nitrogen fixation, phosphate solubilization (Wakelin et al. 2004) or iron chelation (Costa and Loper 1994), production of hormones such as auxin or cytokinin (Madhaiyan et al. 2006), and plant growth-promoting (PGP) substances like the release of volatile organic compounds (VOCs) (Ryu et al. 2003). Plant-associated microbes are known to indirectly promote plant growth by protecting the host plant from microbial pathogens and herbivores. They also protect their host plants from abiotic stresses including drought, salinity, and pollutants. The microbiome thereby increases plant fitness by protecting plants from pathogens and limiting the effects of abiotic stresses on plants.

The plant microbiome has been intensively studied in agriculture in order to increase crop productivity (Maheshwari 2011). However, their role in an ecological context, especially if they may increase the growth or fitness of some native plant species remain unknown. A few studies have reported on the root-associated bacterial communities of native plants (Zinniel et al. 2002; Long et al. 2008; Long et al. 2010), and their PGP effects which occur in nature are not known. In this review, we discuss the main functions of the plant-associated microbiome and their potential impacts on plant growth and fitness. The mechanisms of the multitrophic interactions and chemical dialogs in native plant species are also discussed.

13.3.1 Impact of the Microbiome on Plant Growth

Almost all plants in nature possess a diverse community of endophytic bacteria which can promote host plant growth. Changes in phytohormone homoeostasis are contributed by PGP rhizobacteria producing 1-aminocyclopropane-1-carboxylate (ACC) deaminase or indole acetic acid (IAA). The growth of two Solanaceous native plant species, *Solanum nigrum* and *N. attenuata,* has been significantly impacted by bacterial endophytes isolated from different parts of the field-grown plants. In *S. nigrum*, the root growth promotion was correlated with ACC deaminase activity and IAA production while root growth in *N. attenuata* was associated with IAA but not ACC deaminase activity (Long et al. 2008).

VOCs produced by rhizosphere microorganisms have received more and more attention over the past few years. Some VOCs were able to either modulate plant growth or to mediate the intricate dialogs between microorganisms and plants. Dimethyl sulfide can interfere with quorum sensing of different bacteria by suppressing the transcription of the N-acyl-homoserine lactone synthase genes (Chernin et al. 2011). In addition, the VOC blend mostly containing the S-containing compound dimethyl disulfide (DMDS) emitted by *Bacillus* sp. B55 promotes *N. attenuata* seedling growth. DMDS production is one of the plant growth promotion mechanisms which is particularly beneficial for the wild-type plants growing in S-deficient soils and for the ethylene-insensitive (35)S-ethylene response1 plant(35S-etr1) due to their impaired S uptake/assimilation/metabolism

(Meldau et al. 2013). Finally, VOCs can also induce systemic resistance in plants and promote plant growth (Ryu et al. 2003, 2004).

Leaf-associated microbes are well-known for symbiosis and mutualism that can influence host plant growth by the production of growth-promoting substances and nutrients and prevent the hosts from pathogens. Plant defense chemicals are detoxified by plant foliage-associated bacteria which reduce the adverse effects on insect defoliators (Mason et al. 2014). In addition, living in a sunlight-exposed habitat, phyllospheric microbiome lives can benefit from organic sources as nutrients produced by the host plant. Study of metagenomic information has shown that bacterial rhodopsin genes are present in phyllospheric communities. Microbial retinal-dependent rhodopsin proton pumps found in some epiphytic microorganisms which participate in photosynthesis and making of plant sugars, are finally accessible to epiphytic microorganisms (Atamna-Ismaeel et al. 2012).

13.3.2 Effects of Microorganisms on Nutrient Acquisition by Plants

Members of the rhizospheric microbiome can dramatically influence the plant's nutrient status. The nitrogen-fixing rhizobia and the mycorrhizal fungi that facilitate phosphorus uptake belong to this group of microbiomes. The notable role of symbionts such as mycorrhizal fungi for nutrient and mineral movement, physical structuring in soil and generating stable soil aggregates is well-recognized and documented as reviewed by (Rosenberg and Zilber-Rosenberg 2013).

Rhizospheric microorganisms are also involved in facilitating the plant's uptake of specific trace elements such as iron. Although iron is abundant in soil, it exists primarily in the insoluble ferric oxide form which is limited to microbes under neutral to alkaline conditions. Bacteria thereby utilize mechanisms to regulate intracellular iron concentrations by secretion of siderophores under such conditions (Lindsay and Schwab 1982). Various studies have proposed that plants employ an additional process for iron acquisition involving the use of iron chelated to microbial siderophores. Siderophores produced by fluorescent pseudomonads promoted iron nutrition in Graminaceous plants and some dicotyledonous plant species (Lemanceau et al. 2007). A fungal siderophore, so-called rhizoferrin, produced by *Rhizopus arrhizus* was found to be an efficient carrier of iron to plants. Its efficiency was comparable to that of the synthetic chelates (Yehuda et al. 2000).

13.3.3 Impacts of the Microbiome on Plant Health

Plant diseases can be caused by diverse microbial agents including fungi, bacteria, and viruses. In nature, plant-associated beneficial microorganisms can protect plants from various pathogens either through direct inhibition with the pathogen growth or by inducing systemic resistance in plant hosts. For example, *Pseudomonas*, *Flavobacteria*, *Bacillus,* and *Trichoderma* are some of beneficial microbes that can improve or help plant defense against microbe infections. Interestingly, the inoculation of beneficial bacteria to the roots of plants can help to prevent leaf diseases (Cytryn and Kolton 2012). One of the mechanisms is that the antimicrobial products produced by these rhizospheric bacteria can be transported to upper parts of the plant.

Most rhizobacteria produce antimicrobial metabolites that might inhibit the growth or activity of other microorganisms. Seventy-three endophytic bacteria were isolated from native wild eggplant *Solanum* sp. of 73 isolates, 40 were found to have *in vitro* antagonistic activity against various plant pathogenic bacteria (Long et al. 2003) and 13 showed *in vitro* inhibitory activity against *R. solanacearum*, the causal agent of bacterial wilt in tobacco (Long et al. 2004). Rhizosphere fungi are also prolific producers of antibiotic metabolites. *Trichoderma* species producing antimicrobial compounds have been widely acknowledged (Elad et al. 1983; Harman et al. 2004; Mathivanan et al. 2005). Fungal and bacterial strains can produce antimicrobial compounds. In addition, the antimicrobial compounds could have overlapping or different degrees of antimicrobial activity and therefore can be used as biocontrol agents (Kerr 1980). For example, *Agrobacterium radiobacter* produces agrocin 84 which exhibits antimicrobial activities against closely related genera. Many lipopeptides from *Bacillus* and *Pseudomonas* species have been involved in competitive interactions with their coexisting organisms such as bacteria, fungi, oomycetes, protozoa, nematodes, and plants (Raaijmakers et al. 2010).

Produced by rhizobacteria, quorum-sensing molecules can induce a range of plant responses, including the activation of various defense-related genes in plants such as *MPK3, MPK6, WRKY22, WRKY29,* and *Pdf1.2* (reviewed in Hartmann and Schikora 2012). Several studies investigated the transcriptional and metabolic changes in plants by rhizobacteria over the past few years. Relatively few transcriptional changes were observed in *Arabidopsis* when bacterial strains induce the host plant resistance via JA/ET pathways (Pieterse et al. 1998; Verhagen et al. 2004). However, substantial, transcriptional, and metabolic changes were also observed when rhizobacterial strains induce resistance in *Arabidopsis* via SA pathway (van de Mortel et al. 2012).

Microbes, which can produce quorum quenching enzymes, could be used to suppress the virulence expression of pathogens and stop bacterial infection. It has been shown that quorum quenching enzymes produced by endophytes have a more limited selection pressure for microbial survival than biocide treatments (Cirou et al. 2012). Under *in vitro* conditions, quorum quenching enzymes produced by endophytic bacteria can be applied to have a priming effect in defense activation. More investigations on this aspect should be carried out to efficiently control the virulent pathogens (Alagarasan et al. 2017).

13.3.4 Impacts of Microbiome on Plant Fitness

Abiotic stresses to plants include elevated temperatures, cold stress, polluted soils, and salinity stress. Some bacteria can help plants to withstand or resist against abiotic stresses by releasing compounds into the soil or triggering plants to synthesize compounds such as osmoprotectants and heat shock proteins (reviewed in Grover et al. 2011). In heavy metal contaminated soils, bacterial siderophores can protect plants against heavy metal toxicity (Burd et al. 2000). The plant stress tolerance conferred by some endophytes involves habitat-specific fungal adaptations. In the geothermal soils of Yellowstone National Park, a fungal endophyte, *Curvularia protuberata*, has been found to dominantly colonize in *Dichanthelium lanuginosum* (Panic grass). It is notable that while neither the fungus nor the plant can survive separately when exposed to heat stress >38°C, only after the colonization of this fungus, the host plant and the fungus are able to resist the heat stress >38°C (Redman et al. 2002).

The role of endophytic fungi in heat tolerance in desert plants is another example of the importance of microbiota in helping plants survive under extreme conditions (Rodriguez and Redman 2008). All plants in nature are thought to be in close interactions with mycorrhizal and/or endophytic fungi. The fungi expressing mutualistic lifestyles could promote plant fitness by conferring the temperature stress tolerance. The plants with no fungal endophytes are unable to survive in their native habitats. Heat shock proteins are found to be involved in this tolerance mechanism (Turbyville et al. 2006). *A. thaliana* became more thermotolerant when one of the inhibitors of heat shock protein HSP-90, originally isolated from the fungi, was applied (McLellan et al. 2007).

Other abiotic factors like soil pH, high concentrations of toxic compounds, etc. may adversely influence plant growth, development and yield. These factors are major challenges in many cultivation systems worldwide. Under a low-pH soil condition, foliar lesions were induced on corn plants. When these plants were inoculated with a 2,4-diacetylphloroglucinol (DAPG)-producing *P. fluorescens* strain, the foliar lesions were dramatically reduced. This was the first report that DAPG producers can act to alleviate the abiotic stress factors (Raudales et al. 2009).

There are 11 pathways involved in osmolyte utilization in mangrove rhizosphere microbiome which were discovered by metagenomic analysis. Metagenomic library was constructed from *Arthrocnemum macrostachyum* rhizosphere grown in moderate salinity (Mirete et al. 2015). By expressing the metagenomic DNA in an osmosensitive strain MKH13 of *Escherichia coli* and screening for salt resistance, 11 genes conferring salt resistance were identified. Some genes are known to encode for the proteins related to osmoadaptation such as a glycerol transporter and a proton pump. In another study, the functional capability of the plant-associated microbes to enhance salt tolerance in *Suaeda salsa* was investigated (Yuan et al. 2016). The core microbes and abundance of genes contributing to salt stress acclimatization, nutrient solubilization and competitive root colonization were identified by pyrosequencing the *S. salsa*-associated microbial metagenome. Interestingly, the inoculation of the core microbial species to the rhizosphere of cucumber (*Cucumis sativus*) and rice (*Oryza sativa*) had significantly increased the salt tolerance in these non-host plants.

13.4 Bioengineering Microbial Communities

Ecologically sustainable agricultural practices are essential and plant-associated microbiomes are playing a potential role in sustainable crop production. However, the efficacy of microbial inoculants after their application in the field is still unpredictable since the performance of inoculants is affected by various host-driven factors such as native microbial communities, soil, plant genotype, agricultural practices, the application of fertilizers or pesticides, and other factors as well. In addition, the knowledge on mechanisms of how microbiomes affect host plants is not enough to explain the observed results (Sessitsch and Mitter 2015; Singh et al. 2016). Nevertheless, the increased awareness of the microbiota roles and their potential roles in sustainable crop production lead to increased investment in this area and one approach of developing a good microbial product is to manipulate and engineer microbial communities.

The effects of plant-associated microbiomes on host plants are not required for the whole microbiome since not all of them have positive effects on plants. In addition, only some of the microbiomes are the host preference, since they were selected during plant evolutionary history and through the domestication activity of mankind (Santhanam et al. 2015; Richardson 2017). Therefore, only a part of the microbiome could be considered as a target for bioengineering to improve the effectiveness on host plants (Orozco-Mosqueda et al. 2018).

Engineering of microbiota to optimize plant growth, health and fitness is a valuable target of sustainable agricultural productivity (Mueller and Sachs 2015; Singh et al. 2016). Using RNA-based meta-transcriptomics and metabolic approaches to understand the rhizosphere colonization, the active microbiomes in soil and rhizospheres of wheat, oat, pea, and an oat plants previously grown in similar bulk soil were analyzed. Not surprisingly, the microbiomes were found to be different from each other, indicating that the symbiotic interactions between plants and microbes is highly host-specific (Turner et al. 2013a). In rice, a single gene mutation has significant impact not only on the plant growth under both paddy field and upland conditions but also on the community structure of rhizosphere microbiome (Ikeda et al. 2011). In *Arabidopsis*, while the microbial community at the seedling stage was different from the other developmental stages, plant development stages such as seedling, vegetative, bolting, and flowering showed no significant changes in bacterial community structure. These observations indicate that there are many factors influencing the plant-associated microbe. Therefore, in order to manipulate, engineer, and apply microbes in agriculture, it is necessary to understand the mechanisms behind the decision of colonization of specific microbes and the interactions that occur between plants and their associated microorganisms (Rosier et al. 2016). However, it is hard to point out the specific genes/alleles that regulate microbial communities and the exact host factors that contribute to the regulation of the microbial populations (Lakshmanan et al. 2014). In addition, to successfully apply plant-associated microbes in sustainable agriculture, it is also necessary to investigate how microbial communities respond to fertilizers, to maximize crop yields and minimize nutrient pollution caused by fertilizer application (Yeoh et al. 2016). The plant-associated microbiomes are hugely diverse and complex, which makes the task of engineering the structures of the plant-associated microbiomes a real challenge. Fortunately, although the core microbiome is less complex than the whole microbial communities of a plant, it may give a stable, consistent effect when applied to hosts. A core microbiome has shared members among microbial consortia from similar habitats (Shade and Handelsman 2012). Hence, manipulating and engineering the core microbiome which influence plant growth and fitness could be a feasible approach to achieve desired outcomes (Shade and Handelsman 2012; Foo et al. 2017). Moreover, there is an overlap of core-microbiome between sugarcane, rice, and *Arabidopsis* (Lundberg et al. 2012; Edwards et al. 2015; Yeoh et al. 2016). This observation prompts a strategy of producing a wide host range application of microbial products on unrelated plant species (Gopal and Gupta 2016). Moreover, there are methods for developing engineered "modular microbiomes" – synthetic microbial consortia that are able to orchestrate with the host to improve host health and fitness have been reviewed (Oyserman et al. 2017).

Up to now, there are many strategies for engineering microbiomes (for review see Orozco-Mosqueda et al. 2018); however, it can be classified into three strategies including: the microbiome-based strategy, the plant-based strategy, and the combination of both microbiome and plant-based methods when microbiomes and plants are recognized as parts of a whole meta-organism (Quiza et al. 2015).

The microbiome-based strategy: This strategy basically is the direct inoculation of microorganisms (Quiza et al. 2015). Since the soil microbial communities are rich, the formulations of inoculants are expected to have multiple species or strains of beneficial microbes to obtain a better performance. Nevertheless, single strain inoculants are often used which could be the reason for the undesired effectiveness under field conditions (Trabelsi and Mhamdi 2013). Since inoculation with only intended beneficial microbes could disturb the equilibrium between the native microbe population, it is not always the case that multiple species or strains in inoculants can improve the efficacy of inoculants. Yet, in many cases, using a combination of strains or even at the levelof microbiome or core-microbiome that may have different mechanisms of action.However, as long as they are compatible, they could improve performance due toan additive or synergic effect. However, because of the complex nature of the interactions between microbes, preparing appropriate consortia remains challenging (Gopal and Gupta 2016). Nevertheless, Teruo Higa, University of the Ryukyus, Okinawa, Japan successfully developed Effective Microorganisms® (EM®), originally for organic farming purposes. It consists of over 80 strains of both aerobic and anaerobic microorganisms, from 10 genera belonging to 5 different families (Ahn et al. 2014). Moreover, using recombinant strains is an option; however, because of the potential risk caused by the release of recombinant strains into the environment, this option must be carefully evaluated and considered (Barac et al. 2004; Ryan et al. 2009).

The plant-based strategy: Plants can affect the soil microbiome by producing exudates and signaling compounds (Downie 1994; Bais et al. 2006; Carney and Matson 2006). Root exudates contain primary and secondary metabolites such as sugars, amino acids, proteins, fatty acids, flavonoids, and strigolactones. These substances can be used by microbes as growth substrates or signal compounds for symbiotic and pathogenic plant-microbe interactions (Steinkellner et al. 2007; Broeckling et al. 2008; Bakker et al. 2012). These observations suggest an option to manipulate soil microbiomes by engineering plants. In fact, plant breeding or engineering focusing on improving mutual symbiosis, exudates production, and plant resistance to disease have been done (Hohmann and Messmer 2017). For example, genetically modified *A. thaliana* exuding the xenotopic octopine can recruit more bacteria, and the effect is dose dependent. Opine-producing transgenic Lotus plants recruit more opine-utilizers in the rhizosphere (Oger et al. 2004; Mondy et al. 2014).

The evolution of plants has resulted in the complexity of the root system for multiple purposes and one of them facilitates the interaction with the surrounding microorganisms in soil to promote plant growth, fitness, and survival under diverse environmental conditions. We can manipulate root architecture or function through genetic engineering, conventional breeding, or root symbionts (Shrivastava et al. 2014).

In addition, bacterial quorum sensing compounds are bacterial intercellular communication mechanisms; however, they are also important in microbe–plant interactions. Thus, quorum sensing has been targeted for engineering to mimic or interrupt bacterial quorum sensing signals. This approach could result in the improvement of plant defenses against some pathogenic bacteria (Braeken et al. 2008; Hartmann et al. 2014; Quiza et al. 2015).

The meta-organism strategy: There are complex relationships between diverse microbial consortia and plants and the microbiome is regarded as the plant's second genome, hence, the plant and its microbiome should be considered as a meta-organism or holobiont. Because of this, plants and microbes should be studied and manipulated in the context of the whole system (Farrar et al. 2014; Lakshmanan et al. 2014). In *A. thaliana*, using a core synthetic community approach, it is not only possible to assemble the synthetic communities mimicking the initial microbial structure, but also possible to point out the host genetic factors that affect community composition and/or bacterial abundance. This approach would therefore be feasible to apply in manipulating the holobiont (Bodenhausen et al. 2014; Bai et al. 2015; Sánchez-Cañizares et al. 2017). The idea of engineering both plants and microbes in improvement of plant traits, however, is still hard to explore since the mechanisms of the interaction between plant and microbe in a complex ecological context are largely unknown. Nevertheless, progress in system biology, modeling of biological networks, multi-omics approach, and efficient genetic engineering tools will enable us in further understanding of interactions (Palsson and Zengler 2010; Kumar et al. 2016; Hasin et al. 2017; Yan et al. 2017).

13.5 Implication of Plant-Associated Microbiomes in Sustainable Agriculture

All life is part of the ecosystem, and the interactions between organisms make life on Earth stable. Green plants are organisms that are able to produce their own food by absorbing the energy from sunlight to build up life ingredients (Stapleton 1992; Caldwell et al. 2007). Plants are the lowest part of the food chain which contributes to a multitude of food webs of huge complexity. Because plants cannot move around from one place to another, they solely depend on soil and its complex material that supports plant life including microbiomes and other environmental factors. Microbiomes have attracted enormous attention because of the potential applications in agriculture with significant economic effects (Aberle and Malzahn 2007; Wackers et al. 2007; Meyer et al. 2012). Therefore, managing factors such as plant microbiome that affect natural ecosystems will not only benefit plants but also secure sustainable agriculture (Lakshmanan et al. 2014).

The plant microbiome is a complex of microorganisms that could be harmful or beneficial. It is now clear that microbiota affect plant health and fitness by affecting soil fertility, soil remediation, disease/pest management, waste management, and also the plant immune system (Crotti et al. 2012; Schulz et al. 2013; Selosse et al. 2014; Veera Gnaneswar 2015; Rashid et al. 2016; Trivedi et al. 2016). In addition, it is also indicated that plants can shape the soil microbiome through their own activities such as the secretion of root exudates or actively releasing signals and cues for plant–microbe communication (Ortíz-Castro et al. 2009; Padje et al. 2016). Moreover, plant-associated soil microbes can rapidly shift in less than two years, hence, applying the plant-associated microbiomes approach in sustainable agriculture could be highly efficient in a reasonable duration (Herzberger et al. 2014). Therefore, finding the key molecular mechanism, key microorganism of the whole microbial community structure, and supporting their growth by creating the ideal conditions while suppressing the competitor or harmful species, could result in sustainable agriculture.

13.5.1 Soil Management

Soil fertility is the capacity to receive, store and transmit energy to sustain plant growth. Three main components of soil fertility are physical, chemical and biological, and their interactions result in the level of soil fertility (Abbott and Murphy 2007). Misuse of synthetic fertilizers, insecticides, fungicides, and herbicides can disrupt the delicate balance between the three components and have a negative impact soil fertility or even causing degraded and/or contaminated soil. In sustainable agriculture, it is widely accepted that healthy soil is a key factor to make plants less susceptible to disease, and insect pests. In addition, healthy soil produces healthy crop plants and could deliver high yield, and good quality. Thus, keeping soil healthy is the key component of sustainable agriculture. Therefore, it is necessary to find out efficient methods for improving or maintaining the quality of the soil to ensure its long-term productivity and stability.

Contaminated soil can be remediated and recovered by using different strategies such as: isolation, immobilization, toxicity and/or mobility reduction, physical separation, and extraction. While physical and chemical strategies are costly and non-sustainable, biological treatment such as microbe-assisted phytoremediation is sustainable and has become one of the most important approaches (Evanko and Dzombak 1997; Karimi et al. 2018). In addition, manipulating the soil microbiome can affect soil health and plant fertility (Chaparro et al. 2012). Moreover, efficient methods for recycling nutrients (for example by balanced nutrient management) to prevent soil degradation may be done by beneficial microorganisms (Nie et al. 2011; Zolla et al. 2013; Gatheru Waigi et al. 2017). Phytoremediation is the use of plants and their microbial association for remediation of contaminated soils (Ma et al. 2016) or polluted air (Weyens et al. 2015). During phytoremediation, pollutants can be accumulated, degraded, detoxified, or sequestrated by plants and microbes. These activities can promote plant growth and therefore, increase plant biomass (Weyens et al. 2015). Phytoremediation is an effective, eco-friendly and low-cost approach, and likely to be safe to human health and the environment, while resulting in good soil quality when compared with other strategies (Jagtap et al. 2016).

Arbuscular mycorrhizal fungi (AMF) have been widely used as natural biofertilizers (Berruti et al. 2015; Igiehon and Babalola 2017). In addition, AMF have been considered as the most important tool for phytoremediation (Cabral et al. 2015). Therefore, having both characteristics makes AMF simple and perfectly fit for phytoremediation application. In fact, inoculated plants with AMF and PGP rhizobacteria improve not only plant growth but also lead (Pb) uptake in *Centaurea cyanus*. In addition, both *Pseudomonas* spp. and *Glomus* spp. enhanced the efficiency of Pb phytoextraction by *C. cyanus* (Karimi et al. 2018).

13.5.2 Crop Management

Plants have coevolved with microorganisms and use them to counter biotic and abiotic stresses. Plant-associated microorganisms protect plants from pathogens via multiple mechanisms. For example, plant-associated microorganisms themselves compete with pathogens for niches and nutrients, antibiosis activities, production of lytic enzymes, and inhibition of pathogen virulence. In addition, plant-associated microorganisms can activate plant defense mechanisms to fight against pathogens or insect herbivores (Gopal and Gupta 2016; Muller et al. 2016). Moreover, the plant-associated microbiome can affect nutrient acquisition, benefit to plant fitness by shaping plant response to abiotic and biotic stresses, phytohormone production, increase in plant health, and enhanced competition. Therefore, applying beneficial microbes to manipulate the bacterial community structure for use as biological controls of aerial or soil-borne plant pathogens could be one approach for sustainable agriculture. In fact, many microbial products have been developed and commercially sold as crop protection tools or crop fertilizers (Rout et al. 2013; Souza et al. 2015; Maroniche et al. 2016; Muller et al. 2016; Paredes-Paliz et al. 2016; Verma et al. 2016; Finkel et al. 2017).

Compared to the soil microbials, the most species-rich microbial ecosystem in nature, the plant-associated microbes contain only a small number of taxa which indicates that either plants can actively select a microbial niche or only a small number of microbial species have the capacity to successfully colonize plants(Lundberg et al. 2012; Bulgarelli et al. 2013). Although plant-associated microbial taxonomy is limited, up to now, it is a challenge to achieve reliable and consistent results with agricultural microbial products. It could be that the introduced microbes require specific hosts, are easily affected by the indigenous soil microbes or abiotic, soil conditions, climatic conditions, proper inoculation methods and other factors that make them unable to colonize plants (Trabelsi and Mhamdi 2013; Bashan et al. 2014; Agler et al. 2016). In addition, the soil microbial community is complex and dynamic and the synergistic/antagonistic interactions between the introduced microbes with the native soil microbes always occurs. Inoculants with biocontrol microbes could disturb the original ecological balance of the soil microbial community. In some cases, the inoculated plants even showed more negative results than expected. Moreover, plants are able to select and shape microbe communities themselves. These effects could cause unpredictable, unreliable and inconsistent results when applying inoculants (Johnson et al. 2008; Ballesteros-Almanza et al. 2010; Janos et al. 2013; Trabelsi and Mhamdi 2013).

As previously discussed, there are many factors affecting the survival and performance of the inoculants, therefore limiting the effectiveness of introduced microorganisms on growth and fitness of field-grown plants. In general, the microbial heterogeneity is one of the most important factors affecting inoculant efficacy. To support the microbial heterogeneity, besides the use of broad host range strains and the good production of microbe-assisted products, it is important to develop inoculation technologies to create suitable conditions such as a favorable microenvironment, physical protection, healthy microbial cells at required threshold concentrations. They also need to be left for a required duration in order to support the survival, competition and performance of introduced microbes (Finkel et al. 2017). In addition, there are many methods for microbial inoculants and formulations of inoculants which have been applied (Stockwell et al. 2011; Bashan et al. 2014; Muller et al. 2016).

13.6 Conclusion and Future Perspectives

It is estimated that, by 2020, 30 per cent of global cultivated soil will be converted into degraded land due to the excessive use of chemical fertilizers, pesticides, insecticides, intensive agricultural practices and cultivation of exhaustive crops (Rashid et al. 2016).

Therefore, it is critical to increase the sustainability of agricultural practices such as crop rotation, soil preparation practices, and fertilizer or pesticide treatments to meet sustainable agriculture without compromising yields. In this context, harnessing the plant microbiome is a key and effective approach towards sustainable agriculture (Sessitsch and Mitter 2015; Kaul et al. 2017).

Although the plant-associated microbiome can positively influence plant growth and fitness, successful implementation of microbial inoculants is still in its early stage due to multiple limitations such as the unstable efficacy when applied to different environments and plant species. To overcome these constraints, it is necessary to study further several limiting factors such as communication between associated microorganisms with the host plant, metabolic and molecular changes in plants in response to microorganisms, the interactions between microorganisms and plants, and how these interactions shape the microorganism population dynamics and its structures. In addition, the microbiomes of wild relatives of economically important food crops could potentially be useful when combined with the associated microorganisms of crop plants to improve plant performance due to an additive and/or synergic effect. Moreover, plant breeding programs, especially for sustainable agriculture purposes, need to consider traits which are essential for hosting beneficial microorganisms in order to improve stress resistance, plant nutrition, and plant health, which can finally positively influence plant fitness.

References

Abbott L, and Murphy D, eds. 2007. Soil biological fertility: A key to sustainable land use in agriculture. Springer.

Aberle, N. and Malzahn, A.M. (2007). Interspecific and nutrient-dependent variations in stable isotope fractionation: experimental studies simulating pelagic multitrophic systems. *Oecologia* 154: 291–303.

Agler, M.T., Ruhe, J., Kroll, S. et al. (2016). Microbial hub taxa link host and abiotic factors to plant microbiome variation. *PLoS Biology* 14: e1002352.

Ahn, K., Lee, K.-B., Kim, Y.-J., and Koo, Y.-M. (2014). Quantitative analysis of the three main genera in effective microorganisms using qPCR. *Korean Journal of Chemical Engineering* 31: 849–854.

Alagarasan, G., Aswathy, K.S., and Madhaiyan, M. (2017). Shoot the message, not the messenger - combating pathogenic virulence in plants by inhibiting quorum sensing mediated signaling molecules. *Frontiers in Plant Science* 8: 556.

Alcaraz L.D., Peimbert M., Barajas H.R., Dorantes-Acosta A.E., Bowman J.L. & Arteaga-Vázquez M.A. (2018). Marchantia liverworts as a proxy toplants' basal microbiomes. Sci Rep. Vol 8. pp. 12712.

Andersohn, C. (2004). Does tree height determine epiphyte diversity? *Selbyana* 25: 101–117.

Andreote, F.D., Gumiere, T., and Durrer, A. (2014). Exploring interactions of plant microbiomes. *Scientia Agricola* 71: 528–539.

Andrews, J.H. and Harris, R.F. (2000). The ecology and biogeography of microorganisms on plant surfaces. *Annual Review of Phytopathology* 38: 145–180.

Antoniou, A., Tsolakidou, M.D., Stringlis, I.A., and Pantelides, I.S. (2017). Rhizosphere microbiome recruited from a suppressive compost improves plant fitness and increases. *Frontiers in Plant Science* 8: 2022.

Atamna-Ismaeel, N., Finkel, O.M., Glaser, F. et al. (2012). Microbial rhodopsins on leaf surfaces of terrestrial plants. *Environmental Microbiology* 14: 140–146.

Bae, H., Sicher, R.C., Kim, M.S. et al. (2009). The beneficial endophyte *Trichoderma hamatum* isolate DIS 219b promotes growth and delays the onset of the drought response in *Theobroma cacao*. *Journal of Experimental Botany* 60: 3279–3295.

Bai, Y., Muller, D.B., Srinivas, G. et al. (2015). Functional overlap of the Arabidopsis leaf and root microbiota. *Nature* 528: 364–369.

Bais, H.P., Weir, T.L., Perry, L.G. et al. (2006). The role of root exudates in rhizosphere interactions with plants and other organisms. *Annual Review of Plant Biology* 57: 233–266.

Bakker, M.G., Manter, D.K., Sheflin, A.M. et al. (2012). Harnessing the rhizosphere microbiome through plant breeding and agricultural management. *Plant and Soil* 360: 1–13.

Ballesteros-Almanza, L., Altamirano-Hernandez, J., Pena-Cabriales, J.J. et al. (2010). Effect of co-inoculation with mycorrhiza and rhizobia on the nodule trehalose content of different bean genotypes. *The Open Microbiology Journal* 4: 83–92.

Barac, T., Taghavi, S., Borremans, B. et al. (2004). Engineered endophytic bacteria improve phytoremediation of water-soluble, volatile, organic pollutants. *Nature Biotechnology* 22: 583.

Bashan, Y., de Bashan, L.E., Prabhu, S.R., and Hernandez, J.P. (2014). Advances in plant growth-promoting bacterial inoculant technology: formulations and practical perspectives (1998-2013). *Plant and Soil* 378: 1–33.

Beattie, G.A. and Lindow, S.E. (1995). The secret life of foliar bacterial pathogens on leaves. *Annual Review of Phytopathology* 33: 145–172.

Benizri, E., Nguyen, C., Piutti, S. et al. (2007). Additions of maize root mucilage to soil changed the structure of the bacterial community. *Soil Biology and Biochemistry* 39: 1230–1233.

Berendsen, R.L., Pieterse, C.M., and Bakker, P.A. (2012). The rhizosphere microbiome and plant health. *Trends in Plant Science* 17: 478–486.

Berruti, A., Lumini, E., Balestrini, R., and Bianciotto, V. (2015). Arbuscular mycorrhizal fungi as natural biofertilizers: let's benefit from past successes. *Frontiers in Microbiology* 6: 1559.

Bodenhausen, N., Bortfeld-Miller, M., Ackermann, M., and Vorholt, J.A. (2014). A synthetic community approach reveals plant genotypes affecting the phyllosphere microbiota. *PLoS Genetics* 10: e1004283.

Boon, E., Meehan, C.J., Whidden, C. et al. (2014). Interactions in the microbiome: communities of organisms and communities of genes. *FEMS Microbiology Reviews* 38: 90–118.

Bouffaud, M.L., Poirier, M.A., Muller, D., and Moenne-Loccoz, Y. (2014). Root microbiome relates to plant host evolution in maize and other Poaceae. *Environmental Microbiology* 16: 2804–2814.

Braeken, K., Daniels, R., Ndayizeye, M. et al. (2008). Quorum sensing in bacteria-plant interactions. In: *Molecular Mechanisms of Plant and Microbe Coexistence* (eds. C.S. Nautiyal and P. Dion), 265–289. Berlin, Heidelberg: Springer.

Bressan, M., Roncato, M.A., Bellvert, F. et al. (2009). Exogenous glucosinolate produced by *Arabidopsis thaliana* has an impact on microbes in the rhizosphere and plant roots. *The ISME Journal* 3: 1243–1257.

Brigham, L.A., Woo, H.H., Nicoll, S.M., and Hawes, M.C. (1995). Differential expression of proteins and mRNAs from border cells and root tips of pea. *Plant Physiology* 109: 457–463.

Broeckling, C.D., Broz, A.K., Bergelson, J. et al. (2008). Root exudates regulate soil fungal community composition and diversity. *Applied and Environmental Microbiology* 74: 738–744.

Buddrus-Schiemann, K., Schmid, M., Schreiner, K. et al. (2010). Root colonization by *Pseudomonas* sp. DSMZ 13134 and impact on the indigenous rhizosphere bacterial community of barley. *Microbial Ecology* 60: 381–393.

Bulgarelli, D., Rott, M., Schlaeppi, K. et al. (2012). Revealing structure and assembly cues for Arabidopsis root-inhabiting bacterial microbiota. *Nature* 488: 91–95.

Bulgarelli, D., Schlaeppi, K., Spaepen, S. et al. (2013). Structure and functions of the bacterial microbiota of plants. *Annual Review of Plant Biology* 64: 807–838.

Bulgarelli, D., Garrido-Oter, R., Munch, P.C. et al. (2015). Structure and function of the bacterial root microbiota in wild and domesticated barley. *Cell Host & Microbe* 17: 392–403.

Burd, G.I., Dixon, D.G., and Glick, B.R. (2000). Plant growth-promoting bacteria that decrease heavy metal toxicity in plants. *Canadian Journal of Microbiology* 46: 237–245.

Cabral, L., Soares, C., Giachini, A., and Siqueira, O. (2015). Arbuscular mycorrhizal fungi in phytoremediation of contaminated areas by trace elements: mechanisms and major benefits of their applications. *World Journal of Microbiology and Biotechnology* 31 (11): 1655–1664.

Cakmak, S., Gülüt, K.Y., Marschner, H., and Graham, R.D. (1994). Effect of zinc and iron deficiency on phytos1derophore release in wheat genotypes differing in zinc efficiency. *Journal of Plant Nutrition* 17: 1–17.

Caldwell, M.M., Bornman, J.F., Ballare, C.L. et al. (2007). Terrestrial ecosystems, increased solar ultraviolet radiation, and interactions with other climate change factors. *Photochemical & Photobiological Sciences* 6: 252–266.

Camehl, I., Sherameti, I., Venus, Y. et al. (2010). Ethylene signalling and ethylene-targeted transcription factors are required to balance beneficial and nonbeneficial traits in the symbiosis between the endophytic fungus *Piriformospora indica* and *Arabidopsis thaliana*. *New Phytologist* 185: 1062–1073.

Cankar, K., Kraigher, H., Ravnikar, M., and Rupnik, M. (2005). Bacterial endophytes from seeds of Norway spruce (*Picea abies* L. Karst). *FEMS Microbiology Letters* 244: 341–345.

Cardinale, M., Grube, M., Erlacher, A. et al. (2015). Bacterial networks and co-occurrence relationships in the lettuce root microbiota. *Environmental Microbiology* 17: 239–252.

Carney, K.M. and Matson, P.A. (2006). The influence of tropical plant diversity and composition on soil microbial communities. *Microbial Ecology* 52: 226–238.

Carter, J.P., Spink, J., Cannon, P.F. et al. (1999). Isolation, characterization, and avenacin sensitivity of a diverse collection of cereal-root-colonizing fungi. *Applied and Environmental Microbiology* 65: 3364–3372.

Carvalhais, L.C., Dennis, P.G., Fan, B. et al. (2013). Linking plant nutritional status to plant-microbe interactions. *PLoS One* 8 (7): e68555.

Chaparro, J.M., Sheflin, A.M., Manter, D.K., and Vivanco, J.M. (2012). Manipulating the soil microbiome to increase soil health and plant fertility. *Biology and Fertility of Soils* 48: 489–499.

Chaparro, J.M., Badri, D.V., Bakker, M.G. et al. (2013). Root exudation of phytochemicals in Arabidopsis follows specific patterns that are developmentally programmed and correlate with soil microbial functions. *PLoS One* 8: e55731.

Chaparro, J.M., Badri, D.V., and Vivanco, J.M. (2014). Rhizosphere microbiome assemblage is affected by plant development. *The ISME Journal* 8: 790–803.

Chernin, L., Toklikishvili, N., Ovadis, M. et al. (2011). Quorum sensing quenching by rhizobacterial volatiles. *Environmental Microbiology Reports* 3: 698–704.

Cirou, A., Mondy, S., An, S. et al. (2012). Efficient biostimulation of native and introduced quorum-quenching *Rhodococcus erythropolis* populations is revealed by a combination of analytical chemistry, microbiology, and pyrosequencing. *Applied and Environmental Microbiology* 78: 481–492.

Coleman-Derr, D., Desgarennes, D., Fonseca-Garcia, C. et al. (2016). Plant compartment and biogeography affect microbiome composition in cultivated and native Agave species. *The New Phytologist* 209: 798–811.

Cook, R.J., Thomashow, L.S., Weller, D.M. et al. (1995). Molecular mechanisms of defense by rhizobacteria against root disease. *Proceedings of the National Academy of Sciences of the United States of America* 92: 4197–4201.

Costa, J.M. and Loper, J.E. (1994). Characterization of siderophore production by the biological control agent *Enterobacter cloacae. Molecular Plant Microbe Interactions: MPMI* 7: 440–448.

Cregger, M.A., Veach, A.M., Yang, Z.K. et al. (2018). The Populus holobiont: dissecting the effects of plant niches and genotype on the microbiome. *Microbiome* 6: 31.

Crotti, E., Balloi, A., Hamdi, C. et al. (2012). Microbial symbionts: a resource for the management of insect-related problems. *Microbial Biotechnology* 5: 307–317.

Cytryn, E. and Kolton, M. (2012). Microbial protection against plant disease. In: *Beneficial Microorganisms in Multicellular Life Forms* (eds. E. Rosenberg and U. Gophna), 123–136. Springer.

de Souza, R.S., Okura, V.K., Armanhi, J.S. et al. (2016). Unlocking the bacterial and fungal communities assemblages of sugarcane microbiome. *Scientific Reports* 6: 28774.

Deveau, A. (2016). How does the tree root microbiome assemble? Influence of ectomycorrhizal species on *Pinus sylvestris* root bacterial communities. *Environmental Microbiology* 18: 1303–1305.

Dias, A.C.F., Costa, F.E.C., Andreote, F.D. et al. (2009). Isolation of micropropagated strawberry endophytic bacteria and assessment of their potential for plant growth promotion. *World Journal of Microbiology and Biotechnology* 25: 189–195.

Doty, S.L. (2017). *Functional Importance of the Plant Microbiome - Implications for Agriculture, Forestry and Bioenergy*. Springer International Publishing.

Downie, J.A. (1994). Signalling strategies for nodulation of legumes by rhizobia. *Trends in Microbiology* 2: 318–324.

Edwards, J., Johnson, C., Santos-Medellin, C. et al. (2015). Structure, variation, and assembly of the root-associated microbiomes of rice. *Proceedings of the National Academy of Sciences of the United States of America* 112: E911–E920.

Elad, Y., Barak, R., Chet, I., and Henis, Y. (1983). Ultrastructural studies of the interaction between *Trichoderma* spp. and plant pathogenic fungi. *Journal of Phytopathology* 107: 168–175.

Emmett, B.D., Youngblut, N.D., Buckley, D.H., and Drinkwater, L.E. (2017). Plant phylogeny and life history shape rhizosphere bacterial microbiome of summer annuals in an agricultural field. *Frontiers in Microbiology* 8: 2414.

Estendorfer, J., Stempfhuber, B., Haury, P. et al. (2017). The influence of land use intensity on the plant-associated microbiome of *Dactylis glomerata* L. *Frontiers in Plant Science* 8: 930.

Evanko, R.C. and Dzombak, A.D. (1997). *Remediation of Metals-Contaminated Soils and Groundwater*. Pittsburgh: Ground-Water Remediation Technologies Analysis Center.

Farrar, K., Bryant, D., and Cope-Selby, N. (2014). Understanding and engineering beneficial plant–microbe interactions: plant growth promotion in energy crops. *Plant Biotechnology Journal* 12: 1193–1206.

Finkel, O.M., Castrillo, G., Herrera Paredes, S. et al. (2017). Understanding and exploiting plant beneficial microbes. *Current Opinion in Plant Biology* 38: 155–163.

Fitzpatrick, C.R., Copeland, J., Wang, P.W. et al. (2018). Assembly and ecological function of the root microbiome across angiosperm plant species. *Proceedings of the National Academy of Sciences of the United States of America* 115: E1157–e1165.

Foo, J.L., Ling, H., Lee, Y.S., and Chang, M.W. (2017). Microbiome engineering: current applications and its future. *Biotechnology Journal* 12 (3): 1600099. https://doi.org/10.1002/biot.201600099.

Gagne-Bourgue, F., Aliferis, K.A., Seguin, P. et al. (2013). Isolation and characterization of indigenous endophytic bacteria associated with leaves of switchgrass (*Panicum virgatum* L.) cultivars. *Journal of Applied Microbiology* 114: 836–853.

Gaiero, J.R., McCall, C.A., Thompson, K.A. et al. (2013). Inside the root microbiome: bacterial root endophytes and plant growth promotion. *American Journal of Botany* 100: 1738–1750.

Gatheru Waigi, M., Sun, K., and Gao, Y. (2017). Sphingomonads in microbe-assisted phytoremediation: tackling soil pollution. *Trends in Biotechnology* 35: 883–899.

Goh, C.H., Vallejos, D.F.V., Nicotra, A.B., and Mathesius, U. (2013). The impact of beneficial plant-associated microbes on plant phenotypic plasticity. *Journal of Chemical Ecology* 39: 826–839.

Gopal, M. and Gupta, A. (2016). Microbiome selection could spur next-generation plant breeding strategies. *Frontiers in Microbiology* 7: 1971.

Grover, M., Ali, S.Z., Sandhya, V. et al. (2011). Role of microorganisms in adaptation of agriculture crops to abiotic stresses. *World Journal of Microbiology and Biotechnology* 27: 1231–1240.

Grube, M., Schmid, F., and Berg, G. (2011). Black fungi and associated bacterial communities in the phyllosphere of grapevine. *Fungal Biology* 115: 978–986.

Haichar, F.Z., Heulin, T., Guyonnet, J.P., and Achouak, W. (2016). Stable isotope probing of carbon flow in the plant holobiont. *Current Opinion in Biotechnology* 41: 9–13.

Hallmann, J., QuadtHallmann, A., Mahaffee, W.F., and Kloepper, J.W. (1997). Bacterial endophytes in agricultural crops. *Canadian Journal of Microbiology* 43: 895–914.

Hardoim, P.R., van Overbeek, L.S., and van Elsas, J.D. (2008). Properties of bacterial endophytes and their proposed role in plant growth. *Trends in Microbiology* 16: 463–471.

Harman, G.E., Howell, C.R., Viterbo, A. et al. (2004). *Trichoderma* species-opportunistic, avirulent plant symbionts. *Nature Reviews Microbiology* 2: 43.

Hartmann, A., Rothballer, M., and Schmid, M. (2008). Lorenz Hiltner, a pioneer in rhizosphere microbial ecology and soil bacteriology research. *Plant and Soil* 312: 7–14.

Hartmann, A. and Schikora, A. (2012). Quorum sensing of bacteria and trans-kingdom interactions of *N*-acyl homoserine lactones with eukaryotes. *Journal of Chemical Ecology* 38: 704–713.

Hartmann, A., Rothballer, M., Hense, B.A., and Schröder, P. (2014). Bacterial quorum sensing compounds are important modulators of microbe-plant interactions. *Frontiers in Plant Science* 5: 131.

Hasin, Y., Seldin, M., and Lusis, A. (2017). Multi-omics approaches to disease. *Genome Biology* 18: 83.

van der Heijden, M.G. and Schlaeppi, K. (2015). Root surface as a frontier for plant microbiome research. *Proceedings of the National Academy of Sciences of the United States of America* 112: 2299–2300.

Hein, J.W., Wolfe, G.V., and Blee, K.A. (2008). Comparison of rhizosphere bacterial communities in *Arabidopsis thaliana* mutants for systemic acquired resistance. *Microbial Ecology* 55: 333–343.

Herrera, S.D., Grossi, C., Zawoznik, M., and Groppa, M.D. (2016). Wheat seeds harbour bacterial endophytes with potential as plant growth promoters and biocontrol agents of *Fusarium graminearum*. *Microbiological Research* 186-187: 37–43.

Herzberger, A.J., Duncan, D.S., and Jackson, R.D. (2014). Bouncing back: plant-associated soil microbes respond rapidly to prairie establishment. *PLoS One* 9: e115775.

Hiltner, L. and Hinsinger, P. (2006). *Rhizosphere: Perspective and Challenges; a Tribute to Lorenz Hiltner*. Springer.

Hinsinger, P., Bengough, A.G., Vetterlein, D., and Young, I.M. (2009). Rhizosphere: biophysics, biogeochemistry and ecological relevance. *Plant and Soil* 321: 117–152.

Hirsch, P.R. and Mauchline, T.H. (2012). Who's who in the plant root microbiome? *Nature Biotechnology* 30: 961–962.

Hohmann, P. and Messmer, M.M. (2017). Breeding for mycorrhizal symbiosis: focus on disease resistance. *Euphytica* 213: 113.

Igiehon, N.O. and Babalola, O.O. (2017). Biofertilizers and sustainable agriculture: exploring arbuscular mycorrhizal fungi. *Applied Microbiology and Biotechnology* 101: 4871–4881.

Ikeda, S., Okubo, T., Takeda, N. et al. (2011). The genotype of the calcium/calmodulin-dependent protein kinase gene (*CCaMK*) determines bacterial community diversity in rice roots under paddy and upland field conditions. *Applied and Environmental Microbiology* 77: 4399–4405.

Jagtap U, Bapat V, Saladin G, Chudzińska E, Krzesłowska M, Pawlaczyk E, Komal T, Gul A, Sherameti I, Ali Z 2016. Role of microbes and plants in phytoremediation: Potential of genetic engineering, In: (ed. V.R. Squires) Ecological Restoration ISBN: 978-1-63484-611-0. Nova Science Publishers.

Janos, D.P., Scott, J., Aristizabal, C., and Bowman, D.M. (2013). Arbuscular-mycorrhizal networks inhibit *Eucalyptus tetrodonta* seedlings in rain forest soil microcosms. *PLoS One* 8: e57716.

Johnson, N.C., Graham, J.H., and Smith, F.A. (2008). Functioning of mycorrhizal associations along the mutualism–parasitism continuum. *New Phytologist* 135: 575–585.

Jones, D.L. and Hinsinger, P. (2008). The rhizosphere: complex by design. *Plant and Soil* 312: 1–6.

Kaga, H., Mano, H., Tanaka, F. et al. (2009). Rice seeds as sources of endophytic bacteria. *Microbes and Environments* 24: 154–162.

Kaki, A.A., Kacem, N.C., Dehimat, L. et al. (2013). Biocontrol and plant growth promotion characterization of *Bacillus* species isolated from *Calendula officinalis* rhizosphere. *Indian Journal of Microbiology* 53: 447–452.

Karimi, A., Khodaverdiloo, H., and Rasouli-Sadaghiani Mir, H. (2018). Microbial-enhanced phytoremediation of lead contaminated calcareous soil by *Centaurea cyanus* L. *CLEAN Soil, Air, Water* 46: 1700665.

Kasim, W.A., Gaafar, R.M., Abou-Ali, R.M. et al. (2016). Effect of biofilm forming plant growth promoting rhizobacteria on salinity tolerance in barley. *Annals of Agricultural Sciences* 61: 217–227.

Kaul, S., Choudhary, M., Sharma, T., and Dhar, M.K. (2017). Harnessing the plant microbiome: a key towards sustainable agriculture. In: *Plant-Microbe Interactions in Agro-Ecological Perspectives: Volume 2: Microbial Interactions and Agro-Ecological Impacts* (eds. D.P. Singh, H.B. Singh and R. Prabha), 307–322. Singapore: Springer.

Kawasaki, A., Donn, S., Ryan, P.R. et al. (2016). Microbiome and exudates of the root and rhizosphere of *Brachypodium distachyon*, a model for wheat. *PLoS One* 11: e0164533.

Kerr, A. (1980). Biological control of crown gall through production of agrocin 84. *Plant Disease* 64: 25–30.

Knee, E.M., Gong, F.C., Gao, M. et al. (2001). Root mucilage from pea and its utilization by rhizosphere bacteria as a sole carbon source. *Molecular Plant-Microbe Interactions* 14: 775–784.

Koenig, J.E., Spor, A., Scalfone, N. et al. (2011). Succession of microbial consortia in the developing infant gut microbiome. *Proceedings of the National Academy of Sciences of the United States of America* 108 (Suppl 1): 4578–4585.

Kretzschmar, T., Kohlen, W., Sasse, J. et al. (2012). A petunia ABC protein controls strigolactone-dependent symbiotic signalling and branching. *Nature* 483: 341–344.

Kumar, V., Baweja, M., Singh, P.K., and Shukla, P. (2016). Recent developments in systems biology and metabolic engineering of plant-microbe interactions. *Frontiers in Plant Science* 7: 1421.

Lakshmanan, V., Selvaraj, G., and Bais, H.P. (2014). Functional soil microbiome: belowground solutions to an aboveground problem. *Plant Physiology* 166: 689–700.

Lambais, M.R., Crowley, D.E., Cury, J.C. et al. (2006). Bacterial diversity in tree canopies of the Atlantic forest. *Science* 312: 1917.

Lebeis, S.L., Paredes, S.H., Lundberg, D.S. et al. (2015). Plant Microbiome. Salicylic acid modulates colonization of the root microbiome by specific bacterial taxa. *Science* 349: 860–864.

Lederberg, J. and McCray, A.T. (2001). 'Ome sweet 'Omics - a genealogical treasury of words. *The Scientist* 15: 8.

Lee, B., Lee, S., and Ryu, C.M. (2012). Foliar aphid feeding recruits rhizosphere bacteria and primes plant immunity against pathogenic and non-pathogenic bacteria in pepper. *Annals of Botany* 110: 281–290.

Leite, J., Fischer, D., Rouws, L.F. et al. (2016). Cowpea nodules harbor non-rhizobial bacterial communities that are shaped by soil type rather than plant genotype. *Frontiers in Plant Science* 7: 2064.

Lemanceau, P., Robin, A., Mazurier, S., and Vansuyt, G. (2007). Implication of pyoverdines in the interactions of fluorescent *Pseudomonads* with soil microflora and plant in the rhizosphere. In: *Microbial Siderophores* (eds. A. Varma and S.B. Chincholkar), 165–192. Springer.

Lindow, S.E. and Brandl, M.T. (2003). Microbiology of the phyllosphere. *Applied and Environmental Microbiology* 69: 1875–1883.

Lindsay, W.L. and Schwab, A.P. (1982). The chemistry of iron in soils and its availability to plants. *Journal of Plant Nutrition* 5: 821–840.

Long, H.H., Furuya, N., Kurose, D. et al. (2003). Isolation of endophytic bacteria from *Solanum* sp. and their antibacterial activity against plant pathogenic bacteria. Journal of the Faculty of Agriculture. *Journal of the Faculty of Agriculture, Kyushu University* 48: 21–28.

Long, H.H., Furuya, N., Kurose, D. et al. (2004). Identification of the endophytic bacterial isolates and their *in vitro* and *in vivo* antagonism against *Ralstonia solanacearum*. *Journal of the Faculty of Agriculture, Kyushu University* 49: 233–241.

Long, H.H., Schmidt, D.D., and Baldwin, I.T. (2008). Native bacterial endophytes promote host growth in a species-specific manner; phytohormone manipulations do not result in common growth responses. *PLoS One* 3.

Long, H.H., Sonntag, D.G., Schmidt, D.D., and Baldwin, I.T. (2010). The structure of the culturable root bacterial endophyte community of *Nicotiana attenuata* is organized by soil composition and host plant ethylene production and perception. *New Phytologist* 185: 554–567.

Lopez-Raez, J.A., Verhage, A., Fernandez, I. et al. (2010). Hormonal and transcriptional profiles highlight common and differential host responses to arbuscular mycorrhizal fungi and the regulation of the oxylipin pathway. *Journal of Experimental Botany* 61: 2589–2601.

Lundberg, D.S., Lebeis, S.L., Paredes, S.H. et al. (2012). Defining the core *Arabidopsis thaliana* root microbiome. *Nature* 488 86-+.

Ma, Y., Rajkumar, M., Zhang, C., and Freitas, H. (2016). Beneficial role of bacterial endophytes in heavy metal phytoremediation. *Journal of Environmental Management* 174: 14–25.

Madhaiyan, M., Poonguzhali, S., Ryu, J., and Sa, T. (2006). Regulation of ethylene levels in canola (*Brassica campestris*) by 1-aminocyclopropane-1-carboxylate deaminase-containing *Methylobacterium fujisawaense*. *Planta* 224: 268–278.

Maheshwari, D.K. (2011). *Bacteria in Agrobiology: Crop Ecosystems*. Springer Science & Business Media.

Maroniche, G.A., Rubio, E.J., Consiglio, A., and Perticari, A. (2016). Plant-associated fluorescent *Pseudomonas* from red lateritic soil: beneficial characteristics and their impact on lettuce growth. *The Journal of General and Applied Microbiology* 62: 248–257.

Marques, J.M., da Silva, T.F., Vollu, R.E. et al. (2014). Plant age and genotype affect the bacterial community composition in the tuber rhizosphere of field-grown sweet potato plants. *FEMS Microbiology Ecology* 88: 424–435.

Mason, C.J., Couture, J.J., and Raffa, K.F. (2014). Plant-associated bacteria degrade defense chemicals and reduce their adverse effects on an insect defoliator. *Oecologia* 175: 901–910.

Massalha, H., Korenblum, E., Malitsky, S. et al. (2017). Live imaging of root-bacteria interactions in a microfluidics setup. *Proceedings of the National Academy of Sciences of the United States of America* 114: 4549–4554.

Mathivanan, N., Prabavathy, V.R., and Vijayanandraj, V.R. (2005). Application of talc formulations of *Pseudomonas fluorescens* Migula and *Trichoderma viride* Pers. ex SF Gray decrease the sheath blight disease and enhance the plant growth and yield in rice. *Journal of Phytopathology* 153: 697–701.

McLellan, C.A., Turbyville, T.J., Wijeratne, E.M.K. et al. (2007). A rhizosphere fungus enhances *Arabidopsis* thermotolerance through production of an HSP90 inhibitor. *Plant Physiology* 145: 174–182.

Meldau, D.G., Meldau, S., Hoang, L.H. et al. (2013). Dimethyl disulfide produced by the naturally associated bacterium *Bacillus* sp B55 promotes *Nicotiana attenuata* growth by enhancing sulfur nutrition. *The Plant Cell* 25: 2731–2747.

Mendes, R., Kruijt, M., de Bruijn, I. et al. (2011). Deciphering the rhizosphere microbiome for disease-suppressive bacteria. *Science* 332: 1097–1100.

Meyer, K.M., Vos, M., Mooij, W.M. et al. (2012). Testing the paradox of enrichment along a land use gradient in a multitrophic aboveground and belowground community. *PLoS One* 7: e49034.

Mirete, S., Mora-Ruiz, M.R., Lamprecht-Grandio, M. et al. (2015). Salt resistance genes revealed by functional metagenomics from brines and moderate-salinity rhizosphere within a hypersaline environment. *Frontiers in Microbiology* 6: 1121.

Monchgesang, S., Strehmel, N., Schmidt, S. et al. (2016). Natural variation of root exudates in *Arabidopsis thaliana*-linking metabolomic and genomic data. *Scientific Reports* 6: 29033.

Mondy, S., Lenglet, A., Beury-Cirou, A. et al. (2014). An increasing opine carbon bias in artificial exudation systems and genetically modified plant rhizospheres leads to an increasing reshaping of bacterial populations. *Molecular Ecology* 23: 4846–4861.

van de Mortel, J.E., de Vos, R.C.H., Dekkers, E. et al. (2012). Metabolic and transcriptomic changes induced in Arabidopsis by the rhizobacterium *Pseudomonas fluorescens* SS101. *Plant Physiology* 112: 207324.

Mueller, U.G. and Sachs, J.L. (2015). Engineering microbiomes to improve plant and animal health. *Trends in Microbiology* 23: 606–617.

Muller, D.B., Vogel, C., Bai, Y., and Vorholt, J.A. (2016). The plant microbiota: systems-level insights and perspectives. *Annual Review of Genetics* 50: 211–234.

Naylor, D., DeGraaf, S., Purdom, E., and Coleman-Derr, D. (2017). Drought and host selection influence bacterial community dynamics in the grass root microbiome. *The ISME Journal* 11: 2691–2704.

Neal, A.L., Ahmad, S., Gordon-Weeks, R., and Ton, J. (2012). Benzoxazinoids in root exudates of maize attract *Pseudomonas putida* to the rhizosphere. *PLoS One* 7: e35498.

Neumann, G. and Martinoia, E. (2002). Cluster roots-an underground adaptation for survival in extreme environments. *Trends in Plant Science* 7: 162–167.

Nie, M., Wang, Y., Yu, J. et al. (2011). Understanding plant-microbe interactions for phytoremediation of petroleum-polluted soil. *PLoS One* 6: e17961.

Oburger, E., Gruber, B., Schindlegger, Y. et al. (2014). Root exudation of phytosiderophores from soil-grown wheat. *The New Phytologist* 203: 1161–1174.

Ofek, M., Voronov-Goldman, M., Hadar, Y., and Minz, D. (2014). Host signature effect on plant root-associated microbiomes revealed through analyses of resident vs. active communities. *Environmental Microbiology* 16: 2157–2167.

Oger, P.M., Mansouri, H., Nesme, X., and Dessaux, Y. (2004). Engineering root exudation of Lotus toward the production of two novel carbon compounds leads to the selection of distinct microbial populations in the rhizosphere. *Microbial Ecology* 47: 96–103.

Orozco-Mosqueda, M.C., Rocha-Granados, M.C., Glick, B.R., and Santoyo, G. (2018). Microbiome engineering to improve biocontrol and plant growth-promoting mechanisms. *Microbiological Research* 208: 25–31.

Ortíz-Castro, R., Contreras-Cornejo, H.A., Macías-Rodríguez, L., and López-Bucio, J. (2009). The role of microbial signals in plant growth and development. *Plant Signaling & Behavior* 4: 701–712.

Oyserman, B.O., Medema, M.H., and Raaijmakers, J.M. (2017). Road MAPs to engineer host microbiomes. *Current Opinion in Microbiology* 43: 46–54.

Padje, A.V., Whiteside, M., and Kiers, E.T. (2016). Signals and cues in the evolution of plant-microbe communication. *Current Opinion in Plant Biology* 32: 47–52.

Palsson, B. and Zengler, K. (2010). The challenges of integrating multi-omic data sets. *Nature Chemical Biology* 6: 787–789.

Paredes-Paliz, K.I., Pajuelo, E., Doukkali, B. et al. (2016). Bacterial inoculants for enhanced seed germination of *Spartina densiflora*: implications for restoration of metal polluted areas. *Marine Pollution Bulletin* 110: 396–400.

Peay, K.G., Bruns, T.D., Kennedy, P.G. et al. (2007). A strong species-area relationship for eukaryotic soil microbes: island size matters for ectomycorrhizal fungi. *Ecology Letters* 10: 470–480.

Peiffer, J.A., Spor, A., Koren, O. et al. (2013). Diversity and heritability of the maize rhizosphere microbiome under field conditions. *Proceedings of the National Academy of Sciences of the United States of America* 110: 6548–6553.

Penmetsa, R.V. and Cook, D.R. (1997). A legume ethylene-insensitive mutant hyperinfected by its rhizobial symbiont. *Science* 275: 527–530.

Perez-Jaramillo, J.E., Mendes, R., and Raaijmakers, J.M. (2016). Impact of plant domestication on rhizosphere microbiome assembly and functions. *Plant Molecular Biology* 90: 635–644.

Pfeiffer, S., Mitter, B., Oswald, A. et al. (2017). Rhizosphere microbiomes of potato cultivated in the High Andes show stable and dynamic core microbiomes with different responses to plant development. *FEMS Microbiology Ecology* 93.

Philippot, L., Raaijmakers, J.M., Lemanceau, P., and van der Putten, W.H. (2013). Going back to the roots: the microbial ecology of the rhizosphere. *Nature Reviews Microbiology* 11: 789–799.

Pierret, A., Doussan, C., Capowiez, Y. et al. (2007). Root functional architecture: a framework for modeling the interplay between roots and soil. *Vadose Zone Journal* 6 (2): 269–281.

Pieterse, C.M.J., Van Wees, S.C.M., Van Pelt, J.A. et al. (1998). A novel signaling pathway controlling induced systemic resistance in Arabidopsis. *The Plant Cell* 10: 1571–1580.

Pieterse, C.M.J., Zamioudis, C., Berendsen, R.L. et al. (2014). Induced systemic resistance by beneficial microbes. *Annual Review of Phytopathology* 52 (52): 347–375.

Pii, Y., Borruso, L., Brusetti, L. et al. (2016). The interaction between iron nutrition, plant species and soil type shapes the rhizosphere microbiome. *Plant Physiology and Biochemistry* 99: 39–48.

Quiza, L., St-Arnaud, M., and Yergeau, E. (2015). Harnessing phytomicrobiome signaling for rhizosphere microbiome engineering. *Frontiers in Plant Science* 6: 507.

Raaijmakers, J.M., Paulitz, T.C., Steinberg, C. et al. (2009). The rhizosphere: a playground and battlefield for soilborne pathogens and beneficial microorganisms. *Plant and Soil* 321: 341–361.

Raaijmakers, J.M., De Bruijn, I., Nybroe, O., and Ongena, M. (2010). Natural functions of lipopeptides from Bacillus and Pseudomonas: more than surfactants and antibiotics. *FEMS Microbiology Reviews* 34: 1037–1062.

Rashid, M.I., Mujawar, L.H., Shahzad, T. et al. (2016). Bacteria and fungi can contribute to nutrients bioavailability and aggregate formation in degraded soils. *Microbiological Research* 183: 26–41.

Raudales, R.E., Stone, E., and McSpadden Gardener, B.B. (2009). Seed treatment with 2, 4-diacetylphloroglucinol-producing pseudomonads improves crop health in low-pH soils by altering patterns of nutrient uptake. *Phytopathology* 99: 506–511.

Redman, R.S., Sheehan, K.B., Stout, R.G. et al. (2002). Thermotolerance generated by plant/fungal symbiosis. *Science* 298: 1581–1581.

Richardson, L.A. (2017). Evolving as a holobiont. *PLoS Biology* 15: e2002168.

Rodriguez, R. and Redman, R. (2008). More than 400 million years of evolution and some plants still can't make it on their own: plant stress tolerance via fungal symbiosis. *Journal of Experimental Botany* 59: 1109–1114.

Rosenberg, E. and Zilber-Rosenberg, I. (2013). Microbiotas are part of holobiont fitness. In: *The Hologenome Concept: Human, Animal and Plant Microbiota* (eds. E. Rosenberg and I. Zilber-Rosenberg), 55–80. Springer.

Rosier, A., Bishnoi, U., Lakshmanan, V. et al. (2016). A perspective on inter-kingdom signaling in plant–beneficial microbe interactions. *Plant Molecular Biology* 90: 537–548.

Rout, M.E. (2014). Chapter eleven - the plant microbiome. In: *Advances in Botanical Research*, vol. 69 (ed. A.H. Paterson), 279–309. Academic Press.

Rout, M.E., Chrzanowski, T.H., Westlie, T.K. et al. (2013). Bacterial endophytes enhance competition by invasive plants. *American Journal of Botany* 100: 1726–1737.

Rudrappa, T., Czymmek, K.J., Pare, P.W., and Bais, H.P. (2008). Root-secreted malic acid recruits beneficial soil bacteria. *Plant Physiology* 148: 1547–1556.

Ryan, P.R., Dessaux, Y., Thomashow, L.S., and Weller, D.M. (2009). Rhizosphere engineering and management for sustainable agriculture. *Plant and Soil* 321: 363–383.

Ryu, C.-M., Farag, M.A., Hu, C.-H. et al. (2003). Bacterial volatiles promote growth in *Arabidopsis*. *Proceedings of the National Academy of Sciences of the United States of America* 100: 4927–4932.

Ryu, C.-M., Farag, M.A., Hu, C.-H. et al. (2004). Bacterial volatiles induce systemic resistance in Arabidopsis. *Plant Physiology* 134: 1017–1026.

Rzanny, M. and Voigt, W. (2012). Complexity of multitrophic interactions in a grassland ecosystem depends on plant species diversity. *The Journal of Animal Ecology* 81: 614–627.

Saleem, M., Asghar, H.N., Zahir, Z.A., and Shahid, M. (2018). Impact of lead tolerant plant growth promoting rhizobacteria on growth, physiology, antioxidant activities, yield and lead content in sunflower in lead contaminated soil. *Chemosphere* 195: 606–614.

Sánchez-Cañizares, C., Jorrín, B., Poole, P.S., and Tkacz, A. (2017). Understanding the holobiont: the interdependence of plants and their microbiome. *Current Opinion in Microbiology* 38: 188–196.

Santhanam, R., Luu, V.T., Weinhold, A. et al. (2015). Native root-associated bacteria rescue a plant from a sudden-wilt disease that emerged during continuous cropping. *Proceedings of the National Academy of Sciences of the United States of America* 112: E5013–E5020.

Santhanam, R., Oh, Y., Kumar, R. et al. (2017). Specificity of root microbiomes in native-grown *Nicotiana attenuata* and plant responses to UVB increase *Deinococcus* colonization. *Molecular Ecology* 26: 2543–2562.

Santos-Medellin, C., Edwards, J., Liechty, Z. et al. (2017). Drought stress results in a compartment-specific restructuring of the rice root-associated microbiomes. *MBio* 8.

Sasse, J., Martinoia, E., and Northen, T. (2017). Feed your friends: do plant exudates shape the root microbiome? *Trends in Plant Science* 23 (1): 25–41.

Schafer, P., Pfiffi, S., Voll, L.M. et al. (2009). Manipulation of plant innate immunity and gibberellin as factor of compatibility in the mutualistic association of barley roots with *Piriformospora indica*. *The Plant Journal* 59: 461–474.

Schulz, S., Brankatschk, R., Dümig, A. et al. (2013). The role of microorganisms at different stages of ecosystem development for soil formation. *Biogeosciences* 10: 3983–3996.

Selosse, M.A., Bessis, A., and Pozo, M.J. (2014). Microbial priming of plant and animal immunity: symbionts as developmental signals. *Trends in Microbiology* 22: 607–613.

Sessitsch, A. and Mitter, B. (2015). 21st century agriculture: integration of plant microbiomes for improved crop production and food security. *Microbial Biotechnology* 8: 32–33.

Shade, A. and Handelsman, J. (2012). Beyond the Venn diagram: the hunt for a core microbiome. *Environmental Microbiology* 14: 4–12.

Shakya, M., Gottel, N., Castro, H. et al. (2013). A multifactor analysis of fungal and bacterial community structure in the root microbiome of mature *Populus deltoides* trees. *PLoS One* 8: e76382.

Shrivastava, S., Prasad, R., and Varma, A. (2014). *Anatomy of Root from Eyes of a Microbiologist*. Springer.

Siddiqui, I.A. and Shaukat, S.S. (2002). Rhizobacteria-mediated induction of systemic resistance (ISR) in tomato against *Meloidogyne javanica*. *Journal of Phytopathology-Phytopathologische Zeitschrift* 150: 469–473.

Singh, J.S., Koushal, S., Kumar, A. et al. (2016). *Microbial Inoculants in Sustainable Agricultural Productivity- Vol. II: Functional Application*, vol. 7. Frontiers Media S.A.

Smalla, K., Wieland, G., Buchner, A. et al. (2001). Bulk and rhizosphere soil bacterial communities studied by denaturing gradient gel electrophoresis: plant-dependent enrichment and seasonal shifts revealed. *Applied and Environmental Microbiology* 67: 4742–4751.

Souza, R., Ambrosini, A., and Passaglia, L.M. (2015). Plant growth-promoting bacteria as inoculants in agricultural soils. *Genetics and Molecular Biology* 38: 401–419.

Stapleton, A.E. (1992). Ultraviolet radiation and plants: burning questions. *Plant Cell* 4: 1353–1358.

Steinkellner, S., Lendzemo, V., Langer, I. et al. (2007). Flavonoids and strigolactones in root exudates as signals in symbiotic and pathogenic plant-fungus interactions. *Molecules* 12: 1290–1306.

Stockwell, V.O., Johnson, K.B., Sugar, D., and Loper, J.E. (2011). Mechanistically compatible mixtures of bacterial antagonists improve biological control of fire blight of pear. *Phytopathology* 101: 113–123.

Strong, D.R. (1992). Are trophic cascades all wet? Differentiation and donor-control in speciose ecosystems. *Ecology* 73: 747–754.

Sugiyama, A., Ueda, Y., Zushi, T. et al. (2014). Changes in the bacterial community of soybean rhizospheres during growth in the field. *PLoS One* 9 (6): e100709.

Szoboszlay, M., Lambers, J., Chappell, J. et al. (2015). Comparison of root system architecture and rhizosphere microbial communities of *Balsas teosinte* and domesticated corn cultivars. *Soil Biology and Biochemistry* 80: 34–44.

Thapa, S., Prasanna, R., Ranjan, K. et al. (2017). Nutrients and host attributes modulate the abundance and functional traits of phyllosphere microbiome in rice. *Microbiological Research* 204: 55–64.

Tkacz, A., Cheema, J., Chandra, G. et al. (2015). Stability and succession of the rhizosphere microbiota depends upon plant type and soil composition. *The ISME Journal* 9: 2349–2359.

Trabelsi, D. and Mhamdi, R. (2013). Microbial inoculants and their impact on soil microbial communities: a review. *BioMed Research International* 2013: 863240.

Trivedi, P., Trivedi, C., Grinyer, J. et al. (2016). Harnessing host-vector microbiome for sustainable plant disease management of phloem-limited bacteria. *Frontiers in Plant Science* 7: 1423.

Truchado, P., Gil, M.I., Reboleiro, P. et al. (2017). Impact of solar radiation exposure on phyllosphere bacterial community of red-pigmented baby leaf lettuce. *Food Microbiology* 66: 77–85.

Turbyville, T.J., Wijeratne, E.M.K., Liu, M.X. et al. (2006). Search for Hsp90 inhibitors with potential anticancer activity: isolation and SAR studies of radicicol and monocillin I from two plant-associated fungi of the Sonoran desert. *Journal of Natural Products* 69: 178–184.

Turner, T.R., James, E.K., and Poole, P.S. (2013a). The plant microbiome. *Genome Biology* 14 (6): 209.

Turner, T.R., Ramakrishnan, K., Walshaw, J. et al. (2013b). Comparative metatranscriptomics reveals kingdom level changes in the rhizosphere microbiome of plants. *The ISME Journal* 7: 2248–2258.

Vaishampayan, P.A., Kuehl, J.V., Froula, J.L. et al. (2010). Comparative metagenomics and population dynamics of the gut microbiota in mother and infant. *Genome Biology and Evolution* 2: 53–66.

Van der Ent, S., Van Wees, S.C.M., and Pieterse, C.M.J. (2009). Jasmonate signaling in plant interactions with resistance-inducing beneficial microbes. *Phytochemistry* 70: 1581–1588.

Varkey, S., Anith, K.N., Narayana, R., and Aswini, S. (2018). A consortium of rhizobacteria and fungal endophyte suppress the root-knot nematode parasite in tomato. *Rhizosphere* 5: 38–42.

Veera Gnaneswar G 2015. A new perspective on microbiome and resource management in wastewater systems. doi: 10.4172/2155-952X.1000184.

Venturi, V. and Keel, C. (2016). Signaling in the rhizosphere. *Trends in Plant Science* 21: 187–198.

Verhagen, B.W.M., Glazebrook, J., Zhu, T. et al. (2004). The transcriptome of rhizobacteria-induced systemic resistance in *Arabidopsis*. *Molecular Plant-Microbe Interactions* 17: 895–908.

Verma, P., Yadav, A.N., Khannam, K.S. et al. (2016). Molecular diversity and multifarious plant growth promoting attributes of Bacilli associated with wheat (*Triticum aestivum* L.) rhizosphere from six diverse agro-ecological zones of India. *Journal of Basic Microbiology* 56: 44–58.

Vicre, M., Santaella, C., Blanchet, S. et al. (2005). Root border-like cells of Arabidopsis. Microscopical characterization and role in the interaction with rhizobacteria. *Plant Physiology* 138: 998–1008.

Vokou, D., Vareli, K., Zarali, E. et al. (2012). Exploring biodiversity in the bacterial community of the Mediterranean phyllosphere and its relationship with airborne bacteria. *Microbial Ecology* 64: 714–724.

van der Voort, M., Kempenaar, M., van Driel, M. et al. (2016). Impact of soil heat on reassembly of bacterial communities in the rhizosphere microbiome and plant disease suppression. *Ecology Letters* 19: 375–382.

Wackers, F.L., Romeis, J., and van Rijn, P. (2007). Nectar and pollen feeding by insect herbivores and implications for multitrophic interactions. *Annual Review of Entomology* 52: 301–323.

Wakelin, S.A., Warren, R.A., Harvey, P.R., and Ryder, M.H. (2004). Phosphate solubilization by *Penicillium* spp. closely associated with wheat roots. *Biology and Fertility of Soils* 40: 36–43.

Wassermann, B., Rybakova, D., Muller, C., and Berg, G. (2017). Harnessing the microbiomes of Brassica vegetables for health issues. *Scientific Reports* 7: 17649.

Watson, B.S., Bedair, M.F., Urbanczyk-Wochniak, E. et al. (2015). Integrated metabolomics and transcriptomics reveal enhanced specialized metabolism in *Medicago truncatula* root border cells. *Plant Physiology* 167: 1699–1716.

Weisskopf, L., Heller, S., and Eberl, L. (2011). *Burkholderia* species are major inhabitants of white lupin cluster roots. *Applied and Environmental Microbiology* 77: 7715–7720.

Weller, D.M., Raaijmakers, J.M., Gardener, B.B., and Thomashow, L.S. (2002). Microbial populations responsible for specific soil suppressiveness to plant pathogens. *Annual Review of Phytopathology* 40: 309–348.

Wen, F., White, G.J., VanEtten, H.D. et al. (2009). Extracellular DNA is required for root tip resistance to fungal infection. *Plant Physiology* 151: 820–829.

Weyens, N., Thijs, S., Popek, R. et al. (2015). The role of plant-microbe interactions and their exploitation for phytoremediation of air pollutants. *International Journal of Molecular Sciences* 16: 25576–25604.

Xu, Y., Wang, G., Jin, J. et al. (2009). Bacterial communities in soybean rhizosphere in response to soil type, soybean genotype, and their growth stage. *Soil Biology and Biochemistry* 41: 919–925.

Yan, J., Risacher, S.L., Shen, L., and Saykin, A.J. (2018). Network approaches to systems biology analysis of complex disease: integrative methods for multi-omics data. *Briefings in Bioinformatics* 19 (6): 1370–1381. https://doi.org/10.1093/bib/bbx066.

Yehuda, Z., Shenker, M., Hadar, Y., and Chen, Y. (2000). Remedy of chlorosis induced by iron deficiency in plants with the fungal siderophore rhizoferrin. *Journal of Plant Nutrition* 23: 1991–2006.

Yeoh, Y.K., Paungfoo-Lonhienne, C., Dennis, P.G. et al. (2016). The core root microbiome of sugarcanes cultivated under varying nitrogen fertilizer application. *Environmental Microbiology* 18: 1338–1351.

Yuan, Z., Druzhinina, I.S., Labbe, J. et al. (2016). Specialized microbiome of a halophyte and its role in helping non-host plants to withstand salinity. *Scientific Reports* 6: 32467.

Zachow, C., Muller, H., Tilcher, R., and Berg, G. (2014). Differences between the rhizosphere microbiome of *Beta vulgaris* ssp. *maritima*-ancestor of all beet crops-and modern sugar beets. *Frontiers in Microbiology* 5: 415.

Zamioudis, C. and Pieterse, C.M.J. (2012). Modulation of host immunity by beneficial microbes. *Molecular Plant-Microbe Interactions* 25: 139–150.

Zgadzaj, R., Garrido-Oter, R., Jensen, D.B. et al. (2016). Root nodule symbiosis in *Lotus japonicus* drives the establishment of distinctive rhizosphere, root, and nodule bacterial communities. *Proceedings of the National Academy of Sciences of the United States of America* 113: E7996–E8005.

Zinniel, D.K., Lambrecht, P., Harris, N.B. et al. (2002). Isolation and characterization of endophytic colonizing bacteria from agronomic crops and prairie plants. *Applied and Environmental Microbiology* 68: 2198–2208.

Zolla, G., Bakker, M.G., Badri, D.V. et al. (2013). Understanding root–microbiome interactions. In: *Molecular Microbial Ecology of the Rhizosphere* (ed. F.J. de Bruijn). Heidelberg: Springer-Verlag, Berlin https://doi.org/10.1002/9781118297674.ch70.

14

Microbiome Genomics and Functional Traits for Agricultural Sustainability

Amy Novinscak, Antoine Zboralski, Roxane Roquigny, and Martin Filion

Université de Moncton, Moncton, New Brunswick, Canada

14.1 Introduction

A sustainable approach to agriculture is required to meet the ever increasing demands in food supply caused by the growth of the human population. One area of interest in sustainable agriculture is the use of microorganisms' abilities to increase plant yields (Godfray et al. 2010). Studies of plant microbiomes have increased greatly and the importance of the plant's microbiome to its survival has led researchers to consider the microbiome to be part of the plant's pan-genome (Berg et al. 2014; Tkacz and Poole 2015) or consider the plant to be a holobiont (Vandenkoornhuyse et al. 2015), similar to humans and their microbiome (Bulgarelli et al. 2012; Hirsch and Maucheline 2012; Lundberg et al. 2012). Studies have found that the plant microbiome is involved in many areas, including plant defense against pathogens, plant growth, and nutrient supply as well as contributing to stress tolerance due to environmental conditions, proving that it is crucial to the plant's health and productivity (Berendsen et al. 2012; Berg 2009; Berg et al. 2013; Bragina et al. 2014; Grube et al. 2015). A better understanding of the plant microbiome could lead to efficient management practices that could decrease the effect of plant diseases and increase crop yields, reducing the reliance on chemical pesticides and fertilizers (García-Fraile et al. 2015).

It is well-known that plant roots and the surrounding microbial diversity are physically and chemically connected. Many signals are secreted by the plants, also called roots exudates, which influence the composition of the microbiome (Bais et al. 2006; el Zahar Haichar et al. 2014; Philippot et al. 2013; Tkacz et al. 2015). It has been demonstrated that the microbial communities' structure is significantly different between plant species, as demonstrated for example by Garbeva et al. (2008) for maize, oat, and barley. In return, microbiome composition also interferes with plant health in ensuring functions like plant growth promotion, disease suppression, denitrification, or carbon metabolism (Babalola 2010; Olanrewaju et al. 2017; Raaijmakers et al. 2009; Rodríguez and Fraga 1999). With up to 10^{11} microbial cells per gram of root (Berendsen et al. 2012) and more than 30,000 bacterial and archaeal operational taxonomic units detected in the sugar beet rhizosphere soil

The Plant Microbiome in Sustainable Agriculture, First Edition. Edited by Alok Kumar Srivastava, Prem Lal Kashyap and Madhumita Srivastava.

(Mendes et al. 2011), the rhizosphere is considered as the second genome of plants and as a hotspot for plant roots–microbe interactions (Berendsen et al. 2012). Density-dependent behavior, quorum sensing (QS), is one example of microbial communication shown to be used by rhizosphere microorganisms for ecological functions (competence, biofilm formation, antibiotics, and extracellular enzymes production) (DeAngelis 2016).

Plants can be colonized by numerous microbial species, with most being unable to be cultivated in the laboratory (Müller et al. 2016). The use of metagenomics approaches have been crucial to understand the composition and the function of the plant microbiome (Tkacz and Poole 2015). These approaches are based on the amplification of targeted genes (often 16S rRNA for bacteria and ITS region for fungi) directly in soil samples, circumventing the need to cultivate microorganisms. They have helped elucidate the composition of plant microbiomes. Plant microbiome composition and diversity have been documented for multiple plant species and types of soil (Zhang et al. 2017). However, given the complexity of plant–microbiome interactions, most of the underlying mechanisms of their relationships remain cryptic to date. Plant microbiome functions are also scarcely described in the literature, especially regarding the cornerstone of plant–microbiome interactions: community-scale metabolomics. The deciphering of microbiome functions from the metagenome to the metaphenome is ongoing, especially through the use of meta-omics techniques. The functional knowledge of the plant microbiome will help to fully harness the abilities of the microbiome to improve plant health and yields (Massart et al. 2015; Tkacz and Poole 2015). This chapter will thus focus on known microbiome gene-function links involved in: (i) production/improved availability of plant nutrients; and (ii) plant disease suppression leading to improved plant growth.

14.2 Production/Improved Availability of Plant Nutrients

14.2.1 Biological Nitrogen Fixation

Nitrogen is one of the most important elements for plant development because it is essential to nucleic acids, enzymes, and proteins (Inui Kishi et al. 2017), however it is not always bioavailable for plants in soils. To remedy the nitrogen limitations, nitrogen fertilizers are added during crop growth, and the application amounts have been estimated at 10^{11} kg per year (Glass 2003; Schmer et al. 2014). It is also estimated that only 40 percent of the applied nitrogen fertilizer is absorbed by plants, while the remaining is lost in the environment and can be leached into groundwater. Biological nitrogen fixation (BNF) is the process of reducing atmospheric nitrogen (N_2) to ammonia (NH_3) by diazotrophs, bacteria, and archaea that possess the nitrogenase enzyme encoded by the *nifH* gene (Hoffman et al. 2014; Krapp 2015). This gene is a commonly used genetic marker to examine diazotroph communities and nitrogen fixation in the environment (Boyd and Peters 2013; Coelho et al. 2009) because it is highly conserved and its phylogeny is correlated to 16S rRNA phylogeny (Zehr et al. 2003). Microorganisms capable of BNF are important as they may decrease the dependency on synthetic fertilizers. Potential research avenues include examining optimal use of nitrogen-fixing systems, searching for new plant–microorganism fixing associations and transferring the ability of BNF to non-nitrogen fixing microbes (Olivares et al. 2013).

BNF can be separated into two main types, (i) symbiotic nitrogen fixation by *Rhizobium* and *Frankia* in leguminous and non-leguminous actinorhizal plants, respectively and (ii) non-symbiotic nitrogen fixation by free-living and endophytic bacteria. BNF contributes to the production of more than 2×10^{10} kg of nitrogen annually, worldwide. Of this, 80 per cent is contributed by symbiotic fixation, while the remaining 20 per cent is made available by non-symbiotic fixation (Falkowski 1997). Nodulation is a highly effective method of nitrogen assimilation, however it is restricted to a subset of legumes and actinorhizal plants (Oldroyd et al. 2011; Udvardi and Poole 2013). Symbiotic bacteria forming nodules on leguminous plant roots include species of *Rhizobium, Bradyrhizobium, Allorhizobium, Mesorhizobium, Azorhizobium*, and *Sinorhizobium* (Hungria and Vargas 2000; Sy et al. 2001). Inside the nodules, they convert atmospheric N_2 into NH_3, a form available for the plants, and in return, receive carbon compounds released by the plants (Gyaneshwar et al. 2011; Oldroyd et al. 2011). *Frankia* species have been reported to form nodules in non-leguminous actinorhizal plants for N_2 fixation (Zhang et al. 2012). Non-symbiotic bacteria able to fix N_2 include species of the genus *Azotobacter, Azospirillum, Azoarcus, Bacillus*, and *Pseudomonas* among others (Bhattacharyya and Jha 2012; Saharan and Nehan 2011). Certain species of endophytic bacteria are also known to fix significant amounts of N_2 and transfer it to the plant (Turner et al. 2013).

A limited number of metagenomics studies have been conducted that specifically target nitrogen assimilation in the microbiome. In the 2000s, metagenome sequencing revealed new phyla of nitrogen-cycling microbes that were not able to be cultivated in laboratories (Stein and Klotz 2016). Bouffaud et al. (2016) examined the diazotroph community in Poaceae plants. They concluded that there was no relationship between Poaceae evolutionary history and the size, diversity, or expression of the *nifH* community. They suggest that Poaceae select for this functional group through optimization of nitrogen fixation in the rhizosphere. The use of dual host-microbe transcriptomics demonstrated that the capacity of a nitrogen-fixing *Burkholderia* strain to form microaerobic biofilms on sugarcane roots is preceded by reduced motility and immunogenicity, followed by metabolic adaptation to the sugar-rich plant environment. The sugarcane does not activate an immune response and changes its root morphology, supplying the bacterium with photosynthates (Paungfoo-Lonhienne et al. 2016).

Genome analysis of diazotrophs have demonstrated the presence of novel genes and associations between genetic operons. For example, a transcriptome study revealed 166 genes upregulated during nitrogen fixation conditions in *Pseudomonas stutzeri* A1501, however, only 95 of the up-regulated genes have orthologs in other diazotrophs. These results indicate a number of unidentified genes involved in nitrogen fixation including a newly characterized gene, *pnfA*, which is chromosomally linked to and regulated by the same sigma factor as *nifHDK* (Yan et al. 2010). Bonato et al. (2016) found that in *Herbaspirillum seropedicae*, the nitrate reductase (NAR) is not required for nitrate assimilation, but is required for the production of high levels of nitrite and nitric oxide and dissipation of redox power, leading to increased carbon consumption. Bruto et al. (2014) examined 25 plant growth promoting rhizobacterial (PGPR) genomes and found that the *nifHDK* genes occurred together with the *pqqBCDE* genes (encoding a co-factor potentially implicated in several cellular processes and contributing to phosphate solubilization). Clearly, there are still many unknown genes and gene functions that are associated with BNF in microorganisms.

14.2.2 Phosphate Nutrition

Phosphorus is the second most important nutrient for plants, following nitrogen. It is often abundant in soils, but is insoluble for plants because in acidic conditions it is normally bound to aluminum and iron and in alkaline conditions it is bound to calcium (Tkacz and Poole 2015). The unavailability of phosphorus in soil has been recognized as a major growth limiting factor in agricultural systems (Daniels et al. 2009). Phosphorus fertilizers are often applied to agricultural crops, however their efficiency rarely exceeds 30% due to fixation in soil (Sharma et al. 2013). Reports have indicated that different bacterial species are able to release organic acids that can chelate the cations bound to phosphate, releasing it into the soil (Richardson et al. 2009; Vassilev et al. 2006). However, this process is insufficient for plants to obtain all the required phosphate, making the phosphate acquisition through a plant's association with mycorrhizal fungi important to their survival (Smith et al. 2003). Current research suggests that inoculating crops with phosphate solubilizing microbes has the potential to reduce application rates of phosphate fertilizer by 50 per cent without reducing crop yield (Jilani et al. 2007; Yazdani et al. 2009).

Phosphate solubilizers are microorganisms that are able to release organic acids that lower the pH of the rhizosphere. The organic acid production causes the chelation of H^+ ions in the root microenvironment and insoluble phosphates are transformed into soluble forms (Mullen 2005). Mycorrhiza, on the other hand, are known as phosphate mobilizers, as they enhance the surface area of the roots through colonization, thus mobilizing the phosphate from areas that the roots are unable to reach (Hilda and Fraga 1999). It has been estimated that from 1 to 50 per cent of all the microbial population of soil are phosphate solubilizing bacteria (Chen et al. 2006).

One of the best known phosphate utilization examples is the interaction between mycorrhizal fungi and plant roots, which is important in providing water and phosphorus for the plant, while the mycorrhizal fungi obtains carbon (Augé 2001; Harrison 2005). Estimates include that up to 300,000 plant species have been found to interact with arbuscular mycorrhizal (AM) fungi (Bouwmeester et al. 2007). The AM fungi–plant host cross-talk is similar to nitrogen assimilation through nodulation (Oldroyd 2013). AM are the most common type of mycorrhizal fungi inhabiting the microbiome of crop plants (Baum et al. 2015). Due to their important effects on the growth and health of the plant, the exploration of the diversity and structure of AM fungal communities in agricultural ecosystems together with the understanding of how environmental variables affect them is of high relevance (De Beenhouwer et al. 2015). Because the traditional identification methods were problematic and not accurate, the use of next generation sequencing techniques has proved to be an ideal approach for studying the AM communities in soil (Higo et al. 2014). Metagenomic studies examining the structure and diversity of AM fungi have been applied in various plants, including olive crops (Montes-Borrego et al. 2014), coffee plants (De Beenhouwer et al. 2015), cover crops (Higo et al. 2014), potatoes (Senés-Guerrero and Schüßler 2015), maize (Turrini et al. 2016), and wheat (Dai et al. 2014). These results provide valuable information for understanding the effects of AM communities inhabiting the rhizosphere and their role in ultimately improving crop yields.

The genomic basis for phosphorus acquisition by bacteria has also been addressed. Unno and Shinano (2013) analyzed the rhizosphere metagenome of plants that showed changed growth in the presence of phytic acid and detected changes in the relative abundance of classes that include strains known for phytic acid utilization. They also found changes in the abundance of genes encoding enzymes related to phytic acid utilization such as alkaline phosphatase or citrate synthase. Chhabra et al. (2013) applied a functional metagenomics approach to examine mineral phosphate solubilizing bacteria in barley rhizosphere by constructing a fosmid library in *Escherichia coli*, which was screened in an assay for mineral phosphate solubilization activity. Genes and operons with homology to phosphorous uptake systems, regulatory, and solubilization mechanisms were identified. These studies provide a basic understanding of the phylogenetic and metabolic features of mineral phosphate solubilizing bacteria in the microbiome.

14.2.3 Cadaverine: A Neglected Plant and Microbial Polyamine

Cadaverine, or pentane-1,5-diamine, is a polyamine produced by different bacteria of the phyllosphere and rhizosphere (Cassán et al. 2009; Hamana et al. 1990; Lavizzari et al. 2010), and by the plant itself (Bagni et al. 1986; Rodríguez et al. 2000; Smith and Wilshire 1975) taking its name from its putrid smell. It is a key metabolic compound in plants, especially involved in cell wall architecture, membrane integrity, and gene expression regulation (Jancewicz et al. 2016). It has been shown to positively or negatively affect the development of different plant species when externally applied, but remains poorly investigated for its plant growth-promoting effect (Jancewicz et al. 2016). This compound is directly produced from the reaction of decarboxylation of the common amino acid lysine, absorbing a proton to form carbon dioxide, and leading to an increase in cytosolic pH (Meng and Bennett 1992; Takayama et al. 1994). The bacterial genes responsible for its biosynthesis and transport, respectively *cadA* and *cadB*, have been characterized in *E. coli* and are located on the *cadBA* operon (Meng and Bennett 1992; Tabor et al. 1980). They respectively encode a lysine decarboxylase and a lysine-cadaverine antiporter. Their expression is regulated by the product of the *cadC* gene, which is a transcription factor bound to the bacterial inner membrane (Neely and Olson 1996; Watson et al. 1992). It is activated by a low external pH and lysine (Neely and Olson 1996), leading to cadaverine excretion and pH regulation and providing survival in acidic condition (Soksawatmaekhin et al. 2004). The effect of a cadaverine-producing microbe on a plant has been assessed by Cassán et al. (2009). They inoculated the plant growth-promoting bacteria *Azospirillum brasilense* Az39 on leaves ofrice seedlings. They also separately applied cadaverine in similar amounts as the ones produced by *A. brasilense* Az39. Results showed a better osmotic stress tolerance for the plants and an increase in root weight to a similar extent in both settings. However, this bacterial strain is known to promote plant growth through other mechanisms than cadaverine excretion, especially phytohormones production (Perrig et al. 2007). Whether cadaverine plays a role in the rice growth promotion effect of this strain or not remains to be further investigated. Other polyamines like putrescine, spermidine, or spermine have been scarcely investigated regarding their potential roles in plant-beneficial microbes interactions (Jancewicz et al. 2016; Jiménez-Bremont et al. 2014). Further research could open the way for new applications for polyamines to develop sustainable agricultural practices.

14.3 Disease Suppression Leading to Improved Plant Growth

14.3.1 Quorum Sensing Regulation in Antibiotic Production

Since the first QS system discovered in the bioluminescent marine Gram-negative bacterium *Vibrio fischeri* (Nealson and Hastings 1979) and defined by Fuqua et al. (1994), it is well recognized that this two-component system is widely distributed among many plant-associated bacteria (and also in some eukaryotes), involved in symbiotic, pathogenic, or commensal relationships with plants (Loh et al. 2002; Miller and Bassler 2001; Whitehead et al. 2001). Typically, a two-component QS system consists of a small weight diffusible molecule (also called autoinducers, *luxI*-like) and a cognate response regulator (*luxR*-like). As a bacterial population is increasing, the extracellular concentration of autoinducers is also increasing. Upon reaching an autoinducer threshold concentration, the QS system is integrated into a complex, multilayered signal transduction network which regulates the expression of a diverse array of genes (Harjai and Sabharwal 2017; Loh et al. 2002).

Chemically, these autoinducers are mainly N-acyl-homoserine lactones (AHL) and small secreted peptides, for Gram-negative and Gram-positive bacteria, respectively. AHL and their cognate regulators are fully described in excellent reviews (González and Venturi 2013; Harjai and Sabharwal 2017; Heeb and Haas 2001; Loh et al. 2002; Miller and Bassler 2001; Rosier et al. 2016; Schauder and Bassler 2001; Waters and Bassler 2005; Williams 2007). A myriad of AHL signals exist, such as 3OH-PAME (3OH palmitic acid methyl ester), PQS (3,4-dihydroxy-2-heptylquinoline), and AI-2 (autoinducer-2), just to name a few. Due to this existing diversity, signaling within and across species in soil is pervasive (DeAngelis 2016). More work is however needed to better define the functional control of signaling in food web interactions.

With a homologous two-component system, the GacA/GacS (global antibiotics and cyanide control) regulon is the best studied QS example in several fluorescent *Pseudomonas* species. Its presence is reported by Heeb and Haas (2001) describing the phenotypes controlled by the GacA/GacS system in various Gram-negative bacteria. GacA was first described as a mediator of antibiotic and cyanide production in the biological control bacterium *Pseudomonas fluorescens* strain CHA0 (Laville et al. 1992). In this strain, the GacA function is required for antifungal activity (Heeb and Haas 2001; Laville et al. 1992). Furthermore, this antifungal activity is insured by a combination of several secondary metabolites that are essential to the biocontrol of fungal root pathogens (Haas et al. 2000). *Pseudomonas chlororaphis* 30–84 is a biological control bacterium that inhibits the fungal pathogen *Gaeumannomyces graminis* var. *Tritici* (Ggt), the causal agent of the take-all disease in wheat. In *P. chlororaphis* 30–84, two separate QS systems (PhzR/PhzI and CsaR/CsaI) have been identified, one responsible for phenazines synthesis and the other is suggested to be involved in regulating biosynthesis of cell surface components and in rhizo-competence (Zhang and Pierson 2001). Among the secondary metabolites that *P. chlororaphis* is producing, there are three phenazine antibiotics responsible for the control of Ggt (Pierson III and Pierson 1996). Phenazine production is regulated by PhzI/PhzR, where *phzI* codes for an AHL synthase and *phzR* for a regulator involved in the transcription of the phenazine biosynthetic operon, *phzABCDEFG* (Chin-A-Woeng et al. 2003;

Pierson III and Pierson 1996). It has been demonstrated that the GacA/GacS system controls the production of AHL required for the phenazine biosynthesis by mediating the transcription of *phzI* (Chancey et al. 1999). Other biosynthesis pathways of secondary metabolites with antimicrobial activities are also regulated by the GacA/GacS system, like the operons *phlABCD* and *hcnABC* responsible for the synthesis of 2,4-diacetylphloroglucinol (DAPG) and hydrogen cyanide, respectively. Several studies reported the implication of AHL production and QS regulation on antibiotic production in *Pseudomonas* species (Morohoshi et al. 2013; Peng et al. 2018; Selin et al. 2012). Metabolites biosynthesis and their implication in control of plant diseases have been studied intensively. Here are excellent reviews on this topic: (Blankenfeldt 2013; Chin-A-Woeng et al. 2003; Haas and Defago 2005; Lugtenberg and Kamilova 2009; Lugtenberg et al. 2002; Mavrodi et al. 2013).

14.3.2 Induction of Plant Defense Mechanisms

In addition to being involved in antibiotics production, AHL have been suggested as capable of eliciting an induced systemic resistance response in tomato (Schuhegger et al. 2006). Salicylic acid (SA) levels were increased in leaves when AHL-producing bacteria (*Serratia liquefaciens* MG1 and *Pseudomonas putida* IsoF) colonized the rhizosphere. It was also revealed that AHL molecules systemically induce SA- and ethylene-dependent plant defense gene expression.

More and more studies are also reporting microbial perception of AHL produced by eukaryotes such as plants, which in turn often specifically respond to the same microbial-produced signals. AHL-mimicking compounds and AHL inhibitory activities were reported in exudates from pea seedlings and from seedlings of the model legume *Medicago truncatula* (Gao et al. 2003; Teplitski et al. 2000). The different responses of AHL reporters to the different plant extracts suggested that the plants may have adapted AHL mimics to communicate with specific bacteria (Loh et al. 2002).

Given the pivotal role QS plays in pathogenic interactions (DeAngelis 2016; Loh et al. 2002), inactivation or degradation of AHLs present an attractive target for designing innovative disease control strategies. The first application of QS disruption for disease control was the study of the *aiiA* gene from *Bacillus* sp. 240B1, which is encoding a 250 amino acid protein capable of AHL inactivation (AiiA, AHL-lactonase). The gene has been cloned in the pathogen *Erwinia carotovora*, resulting in reduced AHL production and attenuation of *E. carotovora* pathogenicity on potato, eggplant, Chinese cabbage, carrot, celery, cauliflower, and tobacco (Dong et al. 2000). Expression of this gene and production of AiiA in the genetically modified leaves of tobacco and tubers of potato plants paralyzed the quorum-sensing system of *E. carotovora*, resulting in increased plant disease resistance (Dong et al. 2001). Transformation of *P. fluorescens* P3, an ineffective biological control agent against plant pathogens, with a plasmid carrying the *aiiA* gene, providing the ability to degrade AHLs, significantly reduced potato soft rot caused by *E. carotovora* and crown gall of tomato caused by *Agrobacterium tumefaciens*. Results demonstrated that AHL degradation can be used as a novel biocontrol mechanism, but also demonstrated the potential for non-target interactions that can interfere with the biocontrol efficacy of other strains (Molina et al. 2003). A similar approach has also been used to generate transgenic tobacco lines expressing *expI*, a gene responsible for AHL biosynthesis in *E. carotovora*. The AHL-producing (*expI*-positive) tobacco lines exhibited enhanced resistance to infection by wild-type *E. carotovora* (Mäe et al. 2001). More examples on the importance of the communication occurring in the rhizosphere between plant-beneficial bacteria, pathogenic agents and plants are also discussed in these reviews: Bakker et al. 2013; el Zahar Haichar et al. 2014; González and Venturi 2013; Huang et al. 2014; Lemanceau et al. 2017; Zhang et al. 2017.

14.3.3 Type III Secretion System: Coping with the Plant Immune System

Microbes living in the vicinity of plants, such as in the phyllosphere or the rhizosphere, have to cope with, and even take advantage of their neighbors in order to thrive. They especially need to overcome the defense reaction of the plant. This has been observed in the case of plant pathogens: to infect their host plant and retrieve its nutrients, they have to remain unnoticed by the plant immune system (Jones and Dangl 2006), especially by injecting effectors into the host cells. Some vital compounds produced by microbes, plant pathogens, or not, serve as first signals of microbial presence for the plant: the microbial-associated molecular patterns (MAMPs) (Bittel and Robatzek 2007). MAMPs, like the bacterial elongation factor Tu and flagellin, or fungal chitin, trigger a plant immunity response leading to an oxidative burst, cell wall disorganization, production of reactive nitrogen species (especially nitric oxide), production of other antimicrobial compounds, and pathogenesis-related (PR) proteins (Newman et al. 2013). Such reactions aim at preventing microbes from thriving, but they also inhibit plant growth (Hacquard et al. 2017). The flagelline protein, the main constituent of the bacterial flagellum, has been extensively studied for its role in bacterial recognition by plants (Newman et al. 2013). Its detection is mediated by the plant receptor kinase flagellin-sensing 2 (FLS2) (Gómez-Gómez and Boller 2000; Zipfel 2014) and occurs through a particular epitope called flg22, which is a 22-amino acid highly conserved N-terminal domain of the flagellin (Felix et al. 1999). In a recent transcriptomic study, Stringlis et al. (2018) have inoculated either the PGPR *Pseudomonas simiae* WCS417 or its flg22 peptide alone ($flg22^{417}$) on *Arabidopsis thaliana* roots. They then looked for the transcriptional response of the plant. The authors noticed that 74 percent of the genes differentially expressed after the WCS417 treatment were also differentially expressed after the $flg22^{417}$ treatment. This means that the root response to this bacterium was mostly due to the flagellin or one or more other WCS417 MAMPs activating the same pathways. Results also showed that 50 per cent of the genes differentially expressed after the $flg22^{417}$ treatment were not affected by the WCS417 treatment, especially defense-associated genes, suggesting a mitigating effect from the bacterium on the plant immune system in order to establish a beneficial interaction with the plant. This hypothesis is supported by Millet et al. (2010) who inoculated WCS417 first, followed by flg22. WCS417 inhibited the activation of the plant defense-related gene *CYP71A12* in presence of flg22 alone.

In the model plant pathogen *Pseudomonas syringae*, the suppression of the MAMPs-triggered immunity is mainly mediated by the type III secretion systems (T3SS) (Guo et al. 2009). These systems are membrane protein complexes able to inject small proteins called effectors into plant cells to affect cell signaling (Cornelis and Van Gijsegem 2000). Many phytobeneficial bacteria also have T3SS (Zamioudis and Pieterse 2012), including the WCS417 strain (Stringlis 2018). T3SS-injected effectors found in another *Pseudomonas* sp., *P. fluorescens* Q8r1-96, have been shown to suppress the plant oxidative burst as well as the hypersensitive response, another plant defense mechanism (Mavrodi et al. 2011). However, the T3SS-defective mutant of this strain was still able to colonize its usual habitat, the rhizosphere. In the WCS417 strain, the T3SS are encoded by a cluster of nine structural and two regulatory genes named *rsc* for "rhizosphere-expressed secretion protein conserved" (Stringlis 2018). In the same strain, 11 genes were identified as putative secreted effectors, mostly unrelated to known effector families. Further investigation will be needed to characterize their potential effect on plant physiology.

The role of T3SS extends beyond the *Pseudomonas* genus in the bacterial microbiomes of plants. Some rhizobia are indeed producing T3SS, such as *Sinorhizobium fredii* NGR234 (Viprey et al. 1998), *Mesorhizobium loti* MAFF303099 (Okazaki et al. 2010), or *Bradyrhizobium elkanii* USDA61 (Okazaki et al. 2009). These rhizobia are able to secrete effectors via their T3SS to help them establish symbiotic relationships with plants (Deakin and Broughton 2009). These effectors, called Nop for "nodulation outer protein," are homologous to the effectors found in pathogens (Gourion et al. 2015; Kambara et al. 2009). Plant-rhizobia symbioses are usually initiated by the secretion of lipochito oligosaccharides by rhizobia, called Nod factors (NF), responding to plant root secretion of flavonoids (Dénarié et al. 1996; Subramanian et al. 2007). Okazaki et al. (2013) demonstrated the ability of a NF-defective mutant derived from the USDA61 strain to induce nodule formation in soybean (*Glycine max*) roots. When the T3SS genes were also knocked-out, the mutant could not induce nodule formation anymore. It appeared that the presence of the T3SS allowed the rhizobia to bypass the plant NF recognition in order to initiate the symbiosis. Effectors secretion through the T3SS by the same strain can also lead to a plant immune response blocking the symbiotic process, depending on the plant genotype (Yasuda et al. 2016).

Even though only bacteria belonging to the *Proteobacteria* phylum seem to encode T3SS, such as rhizobia or *Pseudomonas* spp. (Hacquard et al. 2017), other plant-growth promoting microbes like mycorrhizal fungi can benefit from these plant signaling-disruption systems. Warmink and Van Elsas (2008) showed a five-time increased abundance of homologous genes of the highly conserved T3SS *hcrR* gene in the mycosphere soil of the mycorrhizal fungi *Laccariaproxima* compared to the bulk soil. Two bacterial strains were identified as the main *hcrR* carriers: *P. fluorescens* BS053 and *Burkholderia terrae* BS110. Another study found similar results in the mycorrhizosphere of the plant *Medicago trunculata* compared to a nonmycorrhizal rhizosphere (Viollet et al. 2011). In both studies, the authors hypothesized that this bacterial enrichment could reflect potentially beneficial fungi–bacteria interactions. This hypothesis was supported by a study of Cusano et al. (2011). They sequenced the genome of *P. fluorescens* BBc6R8, a known mycorrhiza helper bacteria (MHB, see Deveau and Labbé 2017 for a review) for *Lacarria bicolor* S238N (Deveau et al. 2007; Duponnois and Garbaye 1991), and identified T3SS genes highly homologous to others known in the *Pseudomonas* genus. Then, they observed that a T3SS-defective mutant of this strain was unable to promote mycorrhization. A similar study conducted on another MHB, *P. fluorescens* strain C7R12, also demonstrated the necessary presence of the T3SS for mycorrhization promotion (Viollet et al. 2017). The T3SS may disrupt the plant immune system, thus promoting mycorrhizal symbiosis (Cusano et al. 2011). Many other identified MHB belong to the *Proteobacteria* phylum (Frey-Klett et al. 2007) and could potentially produce T3SS. Genome sequencing should elucidate this question and lead to the discovery of unknown effectors.

Bacteria potentially helping mycorrhizal fungi can also be found inside them, as illustrated by the fungal–beneficial endobacterium *Candidatus Glomeri bactergigasporarum* in the mycorrhizal fungi *Gigaspora margarita* (Salvioli et al. 2016). This bacteria possess a T3SS that may be more synthesized when the fungal host is in symbiosis with the plant (Bonfante and Anca 2009; Ghignone et al. 2012).

T3SS has a role in many inter-kingdom multipartite interactions in the plant microbiome, involving plants, mycorrhizal fungi, MHB, endobacteria, and the other microorganisms of the plant microbiome. These interactions can be pathogenic, as illustrated by some plant pathogens such as *P. syringae*, or beneficial, as shown in many plant- and fungi-associated bacteria promoting directly or indirectly plant growth. At the microbiome level, one study reports the use of metagenomics datasets to investigate the abundance of T3SS genes (Barret et al. 2013). The use of metatranscriptomic and metaproteomic approaches will allow a better understanding of the involvement of the T3SS and thus paving the way to the development of new techniques for sustainable agriculture.

14.4 Future Prospects

One important limitation of meta-omics approaches is the presence of a high number of sequences coding for genes with unknown function coupled with sequences from unknown microorganisms for which no homologous sequences are found in public databases (Knief 2014). Genome sequencing of strain collections (Bai et al. 2015; Lebeis et al. 2015) might provide a better screening tool for sets of plant growth promoting traits that could be readily detected in genomes (Bruto et al. 2014). The identification of appropriate genomic markers and screening of genome collections might provide a less laborious alternative, while also providing the opportunity for the discovery of correlated and novel plant growth promotion-associated genes.

The complementation of metagenomics data with other meta-omics data will be one of the future goals to obtain a more complete view of the activities and the physiological potential of plant associated microbial communities under given conditions (Knief et al. 2011; Segata et al. 2013). This information is required to construct models that can explain and predict microbial mediated processes and interactions under different environmental conditions.

14.5 Conclusion

The great biodiversity in plant microbiomes is far from being uncovered as many microorganisms in these complex communities remain unculturable. Despite the emergence of next-generation sequencing tools, complete genome assembly and new protein characterizations remain difficult for these organisms. Even though the availability of large high-quality databases in meta-omics is rising, knowledge is still lacking to link plant microbiome metagenome to its phenome. Future research should thus focus on further characterizing microbiome metagenomes, metatranscriptomes, metaproteomes, and community-scale metabolomes. This information will allow the construction of a holistic scheme of the microbiome functions in relation to its metagenome.

References

Augé, R.M. (2001). Water relations, drought and vesicular-arbuscular mycorrhizal symbiosis. *Mycorrhiza* 11: 3–42.

Babalola, O.O. (2010). Beneficial bacteria of agricultural importance. *Biotechnology Letters* 32: 1559–1570.

Bagni, N., Creus, J., and Pistocchi, R. (1986). Distribution of cadaverine and lysine decarboxylase activity in *Nicotiana glauca* plants. *Journal of plant physiology* 125: 9–15.

Bai, Y., Müller, D.B., Srinivas, G. et al. (2015). Functional overlap of the Arabidopsis leaf and root microbiota. *Nature* 528: 364–369.

Bais, H.P., Weir, T.L., Perry, L.G. et al. (2006). The role of root exudates in rhizosphere interactions with plants and other organisms. *Annual Reviews in Plant Biology* 57: 233–266.

Bakker, P., Doornbos, R.F., Zamioudis, C. et al. (2013). Induced systemic resistance and the rhizosphere microbiome. *The Plant Pathology Journal* 29: 136–143.

Barret, M., Egan, F., and O'Gara, F. (2013). Distribution and diversity of bacterial secretion systems across metagenomic datasets: secretion systems in metagenomes. *Environmental Microbiology Reports* 5: 117–126.

Baum, C., El-Tohamy, W., and Gruda, N. (2015). Increasing the productivity and product quality of vegetable crops using arbuscular mycorrhizal fungi: A review. *Scientia Horticulturae* 187: 131–141.

Berendsen, R.L., Pieterse, C.M., and Bakker, P.A. (2012). The rhizosphere microbiome and plant health. *Trends in Plant Science* 17: 478–486.

Berg, G. (2009). Plant-microbe interactions promoting plant growth and health: perspectives for controlled use of microorganisms in agriculture. *Applied Microbiology and Biotechnology* 84: 11–18.

Berg, G., Zachow, C., Müller, H. et al. (2013). Next-generation bio-products sowing the seeds of success for sustainable agriculture. *Agronomy* 3: 648–656.

Berg, G., Grube, M., Schloter, M., and Smalla, K. (2014). Unraveling the plant microbiome: looking back and future perspectives. *Frontiers in Microbiology* 5: 148.

Bhattacharyya, P.N. and Jha, D.K. (2012). Plant growth-promoting rhizobacteria (PGPR): emergence in agriculture. *World Journal of Microbiology and Biotechnology* 28: 1327–1250.

Bittel, P. and Robatzek, S. (2007). Microbe-associated molecular patterns (MAMPs) probe plant immunity. *Current Opinion in Plant Biology* 10: 335–341.

Blankenfeldt, W. (2013). The biosynthesis of phenazines. In: *Microbial Phenazines: Biosynthesis, Agriculture and Health* (eds. S. Chincholkar and L. Thomashow), 1–17. Springer.

Bonato, P., Batista, M.B., Camilios-Neto, D. et al. (2016). RNA-seq analyses reveal insights into the function of respiratory nitrate reductase of the diazotroph *Herbaspirillum seropedicae*. *Environmental Microbiology* 18: 2677–2688.

Bonfante, P. and Anca, I.-A. (2009). Plants, mycorrhizal fungi, and bacteria: A network of interactions. *Annual Review of Microbiology* 63: 363–383.

Bouffaud, M.-L., Renoud, S., Moënne-Loccoz, Y., and Muller, D. (2016). Is plant evolutionary history impacting recruitement of diazotrophs and *nifH* expression in the rhizosphere? *Scientific Reports* 6: 21690.

Bouwmeester, H.J., Eouz, X., Lopez-Raez, J.A., and Becard, G. (2007). Rhizosphere communication of plants, parasitic plants and AM fungi. *Trends in Plant Science* 12: 224–230.

Boyd, E.S. and Peters, J.W. (2013). New insights into the evolutionary history of biological nitrogen fixation. *Frontiers in Microbiology* 4: 201.

Bragina, A., Oberauner-Wappis, L., Zachow, C. et al. (2014). The Sphagnum microbiome supports bog ecosystem functioning under extreme conditions. *Molecular Ecology* 23: 4498–4510.

Bruto, M., Prigent-Combaret, C., Muller, D., and Moënne-Loccoz, Y. (2014). Analysis of genes contributing to plant-beneficial functions in plant growth-promoting rhizobacteria and related Proteobacteria. *Scientific Reports* 4: 6261.

Bulgarelli, D., Rott, M., Schlaeppi, K. et al. (2012). Revealing structure and assembly cues for *Arabidopsis* root-inhabiting bacterial microbiota. *Nature* 488: 91–95.

Cassán, F., Maiale, S., Masciarelli, O. et al. (2009). Cadaverine production by *Azospirillum brasilense* and its possible role in plant growth promotion and osmotic stress mitigation. *European Journal of Soil Biology* 45: 12–19.

Chancey, S., Wood, D., and Pierson, L.R. (1999). Two-component transcriptional regulation of N-acyl-homoserine lactone production in *Pseudomonas aureofaciens*. *Applied and Environmental Microbiology* 65: 2294–2299.

Chen, Y.P., Rekha, P.D., Arun, A.B. et al. (2006). Phosphate solubilizing bacteria from subtropical soil and their tricalcium phosphate solubilizing abilities. *Applied Soil Ecology* 34: 33–41.

Chhabra, S., Brazil, D., Morrisey, J. et al. (2013). Characterization of mineral phosphate solubilisation traits from a barley rhizosphere soil functional metagenome. *Microbiology Open* 2: 717–724.

Chin-A-Woeng, T.F.C., Bloemberg, G.V., and Lugtenberg, B.J.J. (2003). Phenazines and their role in biocontrol by *Pseudomonas* bacteria. *New Phytologist* 157: 503–523.

Coelho, M.R.R., Marriel, I.E., Jenkins, S.N. et al. (2009). Molecular detection and quantification of *nifH* gene sequences in the rhizosphere of sorghum (*Sorghum bicolor*) sown with two levels of nitrogen fertilizer. *Applied Soil Ecology* 42: 48–53.

Cornelis, G.R. and Van Gijsegem, F. (2000). Assembly and function of type III secretory systems. *Annual Reviews in Microbiology* 54: 735–774.

Cusano, A.M., Burlinson, P., Deveau, A. et al. (2011). *Pseudomonas fluorescens* BBc6R8 type III secretion mutants no longer promote ectomycorrhizal symbiosis: MHB effect lost in T3SS mutants. *Environmental Microbiology Reports* 3: 203–210.

Dai, M., Hamel, C., Bainard, L.D. et al. (2014). Negative and positive contributions of arbuscular mycorrhizal fungal taxa to wheat production and nutrient uptake efficiency in organic and conventional systems in the Canadian prairie. *Soil Biology and Biochemistry* 74: 156–166.

Daniels, C., Michan, C., and Ramos, J.L. (2009). New molecular tools for enhancing methane production, explaining thermodynamically limited lifestyles and other important biotechnological issues. *Microbial Biotechnology* 2: 533–536.

De Beenhouwer, M., Van Geel, M., Ceulemans, T. et al. (2015). Changing soil characteristics alter the arbuscular mycorrhizal fungi communities of Arabica coffee (*Coffea arabica*) in Ethiopia across a management intensity gradient. *Soil Biology and Biochemistry* 9: 133–139.

Deakin, W.J. and Broughton, W.J. (2009). Symbiotic use of pathogenic strategies: rhizobial protein secretion systems. *Nature Reviews Microbiology* 7: 312–320.

DeAngelis, K.M. (2016). Chemical communication connects soil food webs. *Soil Biology and Biochemistry* 102: 48–51.

Dénarié, J., Debelle, F., and Prome, J.-C. (1996). *Rhizobium* lipo-chitooligosaccharide nodulation factors: signaling molecules mediating recognition and morphogenesis. *Annual Review of Biochemistry* 65: 503–535.

Deveau, A. and Labbé, J. (2017). Mycorrhiza helper bacteria. In: *Molecular Mycorrhizal Symbiosis* (ed. F. Martin), 437–450. Hoboken, New Jersey: Wiley.

Deveau, A., Palin, B., Delaruelle, C. et al. (2007). The mycorrhiza helper *Pseudomonas fluorescens* BBc6R8 has a specific priming effect on the growth, morphology and gene expression of the ectomycorrhizal fungus *Laccaria bicolor* S238N. *New Phytologist* 175: 743–755.

Dong, Y.-H., Xu, J.-L., Li, X.-Z., and Zhang, L.-H. (2000). AiiA, an enzyme that inactivates the acylhomoserine lactone quorum-sensing signal and attenuates the virulence of *Erwinia carotovora*. *Proceedings of the National Academy of Sciences of the United States of America* 93: 3526–3531.

Dong, Y.-H., Wang, L.-H., Xu, J.-L. et al. (2001). Quenching quorum-sensing-dependent bacterial infection by an N-acyl homoserine lactonase. *Nature* 411: 813–817.

Duponnois, R. and Garbaye, J. (1991). Mycorrhization helper bacteria associated with the Douglas fir-*Laccaria laccata* symbiosis: effects in aseptic and in glasshouse conditions. *Annales des Sciences Forestières (EDP Sciences)* 48 (3): 239–251.

Falkowski, P.G. (1997). Evolution of the nitrogen cycle and its influence on the biological sequestration of CO2 in the ocean. *Nature* 387: 272–275.

Felix, G., Duran, J.D., Volko, S., and Boller, T. (1999). Plants have a sensitive perception system for the most conserved domain of bacterial flagellin. *The Plant Journal* 18: 265–276.

Frey-Klett, P., Garbaye, J., and Tarkka, M. (2007). The mycorrhiza helper bacteria revisited. *New Phytologist* 176: 22–36.

Fuqua, W.C., Winans, S.C., and Greenberg, E.P. (1994). Quorum sensing in bacteria: the LuxR-LuxI family of cell density-responsive transcriptional regulators. *Journal of Bacteriology* 176: 269–275.

Gao, M., Teplitski, M., Robinson, J.B., and Bauer, W.D. (2003). Production of substances by *Medicago truncatula* that affect bacterial quorum sensing. *Molecular Plant-Microbe Interactions* 16: 827–834.

Garbeva, P., Van Elsas, J., and Van Veen, J. (2008). Rhizosphere microbial community and its response to plant species and soil history. *Plant and Soil* 302: 19–32.

García-Fraile, P., Menéndez, E., and Rivas, R. (2015). Role of bacterial biofertilizers in agriculture and forestry. *AIMS Bioengineering* 2: 133–205.

Ghignone, S., Salvioli, A., Anca, I. et al. (2012). The genome of the obligate endobacterium of an AM fungus reveals an interphylum network of nutritional interactions. *The ISME Journal* 6: 136–145.

Glass, A.D.M. (2003). Nitrogen use efficiency of crop plants: physiological constraints upon nitrogen absorption. *Critical Reviews in Plant Science* 22: 453–470.

Godfray, H.C., Beddiington, J.R., Crute, I.R. et al. (2010). Food security: the challenge of feeding 9 billion people. *Science* 327: 812–818.

Gómez-Gómez, L. and Boller, T. (2000). FLS2: an LRR receptor–like kinase involved in the perception of the bacterial elicitor flagellin in *Arabidopsis*. *Molecular Cell* 5: 1003–1011.

González, J.F. and Venturi, V. (2013). A novel widespread interkingdom signaling circuit. *Trends in Plant Science* 18: 167–174.

Gourion, B., Berrabah, F., Ratet, P., and Stacey, G. (2015). *Rhizobium*–legume symbioses: the crucial role of plant immunity. *Trends in Plant Science* 20: 186–194.

Grube, M., Cernava, T., Soh, J. et al. (2015). Exploring functional contexts of symbiotic sustain within lichen-associated bacteria by comparative omics. *The ISME Journal* 9: 412–424.

Guo, M., Tian, F., Wamboldt, Y., and Alfano, J.R. (2009). The majority of the type III effector inventory of *Pseudomonas syringae* pv. tomato DC3000 can suppress plant immunity. *Molecular Plant-Microbe Interactions* 22: 1069–1080.

Gyaneshwar, P., Hirsch, A.M., Moulin, L. et al. (2011). Legume-nodulating betaproteobacteria: diversity, host range, and future prospects. *Molecular Plant-Microbe Interactions* 24: 1276–1288.

Haas, D. and Defago, G. (2005). Biological control of soil-borne pathogens by fluorescent Pseudomonads. *Nature Reviews Microbiology* 3: 307–319.

Haas, D., Blumer, C., and Keel, C. (2000). Biocontrol ability of fluorescent pseudomonads genetically dissected: importance of positive feedback regulation. *Current Opinion in Biotechnology* 11: 290–297.

Hacquard, S., Spaepen, S., Garrido-Oter, R., and Schulze-Lefert, P. (2017). Interplay between innate immunity and the plant microbiota. *Annual Review of Phytopathology* 55: 565–589.

Hamana, K., Minamisawa, K., and Matsuzaki, S. (1990). Polyamines in *Rhizobium*, *Bradyrhizobium*, *Azorhizobium* and *Argobacterium*. *FEMS Microbiology Letters* 71: 71–76.

Harjai, K. and Sabharwal, N. (2017). Biofilm formation and quorum sensing in Rhizosphere. In: *Biofilms in Plant and Soil Health* (eds. I. Ahmad and F.M. Husain), 17–29. Hoboken, NJ: Wiley.

Harrison, M.J. (2005). Signaling in the arbuscular mycorrhizal symbiosis. *Annuual Reviews in Microbiology* 59: 19–42.

Heeb, S. and Haas, D. (2001). Regulatory roles of the GacS/GacA two-component system in plant-associated and other gram-negative bacteria. *Molecular Plant-Microbe Interactions* 14: 1351–1363.

Higo, M., Isobe, K., Drijber, R.A. et al. (2014). Impact of a 5-year winter cover crop rotational system on the molecular diversity of arbuscular mycorrhizal fungi colonizing roots of subsequent soybean. *Biology and Fertility of Soils* 50: 913–926.

Hilda, R. and Fraga, R. (1999). Phosphate solubilizing bacteria and their role in plant growth promotion. *Biotechnology Advances* 17: 319–359.

Hirsch, P.R. and Maucheline, T.H. (2012). Who's who in the plant root microbiome? *Nature Biotechnology* 30: 961–962.

Hoffman, B.M., Lukoyanov, D., Yang, Z.-Y. et al. (2014). Mechanism of nitrogen fixation by nitrogenase: the next stage. *Chemical Reviews* 114: 4041–4062.

Huang, X.-F., Chaparro, J.M., Reardon, K.F. et al. (2014). Rhizosphere interactions: root exudates, microbes, and microbial communities. *Botany* 92: 267–275.

Hungria, M. and Vargas, M.A.T. (2000). Environmental factors affecting N_2 fixation in grain legumes in the tropics, with an emphasis on Brazil. *Field Crops Research* 65: 151–164.

Inui Kishi, R.N., Galdiano Júnior, R.F., Val-Moraes, S.P., and Kishi, L.T. (2017). Soil microbiome and their effects on nutrient management for plants. In: *Probiotics in Agroecosystems* (eds. V. Kumar et al.), 117–143. Singapore: Springer Nature.

Jancewicz, A.L., Gibbs, N.M., and Masson, P.H. (2016). Cadaverine's functional role in plant development and environmental response. *Frontiers in Plant Science* 7 https://doi.org/10.3389/fpls.2016.00870.

Jilani, G., Akram, A., Ali, R.M. et al. (2007). Enhancing crop growth, nutrients availability, economics and beneficial rhizosphere microflora through organic and biofertilizers. *Annals of Microbiology* 57: 177–183.

Jiménez-Bremont, J.F., Marina, M., Guerrero-González, M.d.l.L. et al. (2014). Physiological and molecular implications of plant polyamine metabolism during biotic interactions. *Frontiers in Plant Science* 5 https://doi.org/10.3389/fpls.2014.00095.

Jones, J.D.G. and Dangl, J.L. (2006). The plant immune system. *Nature* 444: 323–329.

Kambara, K., Ardissone, S., Kobayashi, H. et al. (2009). Rhizobia utilize pathogen-like effector proteins during symbiosis. *Molecular Microbiology* 71: 92–106.

Knief, C. (2014). Analysis of plant microbiome interactions in the era of next generation sequencing technologies. *Frontiers in Plant Science* 5: 216.

Knief, C., Delmotte, N., and Vorholt, J.A. (2011). Bacterial adaptation to life in associations with plants – a proteomic perspective from culture to *in situ* conditions. *Proteomics* 11: 3086–3105.

Krapp, A. (2015). Plant nitrogen assimilation and its regulation: a complex puzzle with missing pieces. *Current Opinion in Plant Biology* 25: 115–122.

Laville, J., Voisard, C., Keel, C. et al. (1992). Global control in *Pseudomonas fluorescens* mediating antibiotic synthesis and suppression of black root rot of tobacco. *Proceedings of the National Academy of Sciences of the United States of America* 89: 1562–1566.

Lavizzari, T., Breccia, M., Bover-Cid, S. et al. (2010). Histamine, cadaverine, and putrescine produced in vitro by *Enterobacteriaceae* and *Pseudomonadaceae* isolated from spinach. *Journal of food protection* 73: 385–389.

Lebeis, S.L., Herrera Paredes, S., Lundberg, D.S. et al. (2015). Salicylic acid modulates colonization of the root microbiome by specific bacterial taxa. *Science* 349: 860–864.

Lemanceau, P., Blouin, M., Muller, D., and Moënne-Loccoz, Y. (2017). Let the core microbiota be functional. *Trends in Plant Science* 22: 583–595.

Loh, J., Pierson, E.A., Pierson, L.S. et al. (2002). Quorum sensing in plant-associated bacteria. *Current Opinion in Plant Biology* 5: 285–290.

Lugtenberg, B.J.J. and Kamilova, F. (2009). Plant-growth-promoting rhizobacteria. *Annual Review of Microbiology* 63: 541–556.

Lugtenberg, B.J.J., Chin-A-Woeng, T.F.C., and Bloemberg, G.V. (2002). Microbe-plant interactions: principles and mechanisms. *Antonie Van Leeuwenhoek International Journal of General and Molecular Microbiology* 81: 373–383.

Lundberg, D.S., Lebeis, S.L., Paredes, S.H. et al. (2012). Defining the core *Arabidopsis thaliana* root microbiome. *Nature* 488: 86–90.

Mäe, A., Montesano, M., Koiv, V., and Palva, E.T. (2001). Transgenic plants producing the bacterial pheromone N-acyl-homoserine lactone exhibit enhanced resistance to the bacterial phytopathogen *Erwinia carotovora*. *Molecular Plant-Microbe Interactions* 14: 1035–1042.

Massart, S., Martinez-Medina, M., and Jijakli, M.H. (2015). Biological control in the microbiome era: challenges and opportunities. *Biological Control* 89: 98–108.

Mavrodi, D.V., Joe, A., Mavrodi, O.V. et al. (2011). Structural and functional analysis of the type III secretion system from *Pseudomonas fluorescens* Q8r1-96. *Journal of Bacteriology* 193: 177–189.

Mavrodi, D.V., Parejko, J.A., Mavrodi, O.V. et al. (2013). Recent insights into the diversity, frequency and ecological roles of phenazines in fluorescent *Pseudomonas* spp. *Environmental Microbiology* 15: 675–686.

Mendes, R., Kruijt, M., de Bruijn, I. et al. (2011). Deciphering the rhizosphere microbiome for disease-suppressive bacteria. *Science* 332: 1097–1100.

Meng, S.-Y. and Bennett, G.N. (1992). Nucleotide sequence of the *Escherichia coli cad* operon: a system for neutralization of low extracellular pH. *Journal of Bacteriology* 174: 2659–2669.

Miller, M.B. and Bassler, B.L. (2001). Quorum sensing in bacteria. *Annual Reviews in Microbiology* 55: 165–199.

Millet, Y.A., Danna, C.H., Clay, N.K. et al. (2010). Innate immune responses activated in *Arabidopsis* roots by microbe-associated molecular patterns. *The Plant Cell* 22: 973–990.

Molina, L., Constantinescu, F., Michel, L. et al. (2003). Degradation of pathogen quorum-sensing molecules by soil bacteria: a preventive and curative biological control mechanism. *FEMS Microbiology Ecology* 45: 71–81.

Montes-Borrego, M., Metsis, M., and Landa, B.B. (2014). Arbuscular mycorrhizal fungi associated with the olive crop across the Andalusian landscape: factors driving community differentiation. *PLoS One* 9: e96397.

Morohoshi, T., Wang, W.Z., Suto, T. et al. (2013). Phenazine antibiotic production and antifungal activity are regulated by multiple quorum-sensing systems in *Pseudomonas chlororaphis* subsp. *aurantiaca* StFRB508. *Journal of Bioscience and Bioengineering* 116: 580–584.

Mullen, M.D. (2005). Phosphorus in soils: biological interactions. In: *Encyclopedia of Soils in the Environment* (ed. D. Hillel), 210–215. Oxford: Elsevier.

Müller, C.A., Obermeier, M.M., and Berg, G. (2016). Bioprospecting plant-associated microbiomes. *Journal of Biotechnology* 235: 171–180.

Nealson, K. and Hastings, J.W. (1979). Bacterial bioluminescence: its control and ecological significance. *Microbiological Reviews* 43: 496–518.

Neely, M.N. and Olson, E.R. (1996). Kinetics of expression of the *Escherichia coli cad* operon as a function of pH and lysine. *Journal of Bacteriology* 178: 5522–5528.

Newman, M.-A., Sundelin, T., Nielsen, J.T., and Erbs, G. (2013). MAMP (microbe-associated molecular pattern) triggered immunity in plants. *Frontiers in Plant Science* 4: 139.

Okazaki, S., Zehner, S., Hempel, J. et al. (2009). Genetic organization and functional analysis of the type III secretion system of *Bradyrhizobium elkanii*. *FEMS Microbiology Letters* 295: 88–95.

Okazaki, S., Okabe, S., Higashi, M. et al. (2010). Identification and functional analysis of type III effector proteins in *Mesorhizobium loti*. *Molecular Plant-Microbe Interactions* 23: 223–234.

Okazaki, S., Kaneko, T., Sato, S., and Saeki, K. (2013). Hijacking of leguminous nodulation signaling by the rhizobial type III secretion system. *Proceedings of the National Academy of Sciences of the United States of America* 110: 17131–17136.

Olanrewaju, O.S., Glick, B.R., and Babalola, O.O. (2017). Mechanisms of action of plant growth promoting bacteria. *World Journal of Microbiology and Biotechnology* 33: 197. https://doi.org/10.1007/s11274-017-2364-9.

Oldroyd, G.E. (2013). Speak, friend, and enter: signalling systems that promote beneficial symbiotic associations in plants. *Nature Reviews in Microbiology* 11: 252–263.

Oldroyd, G.E.D., Murray, J.D., Poole, P.S., and Downie, J.A. (2011). The rules of engagement in the legume-rhizobial symbiosis. *Annual Reviews in Genetics* 45: 119–144.

Olivares, J., Bedmar, E.J., and Sanjuán, J. (2013). Biological nitrogen fixation in the context of global change. *Molecular Plant-Microbe Interactions* 26: 486–494.

Paungfoo-Lonhienne, C., Lonhienne, T.G.A., Yeoh, Y.K. et al. (2016). Crosstalk between sugarcane and a plant-growth promoting *Burkholderia* species. *Scientific Reports* 6: 37389.

Peng, H., Ouyang, Y., Bilal, M. et al. (2018). Identification, synthesis and regulatory function of the N-acylated homoserine lactone signals produced by *Pseudomonas chlororaphis* HT66. *Microbial Cell Factories* 17: 9. https://doi.org/10.1186/s12934-017-0854-y.

Perrig, D., Boiero, M.L., Masciarelli, O.A. et al. (2007). Plant-growth-promoting compounds produced by two agronomically important strains of *Azospirillum brasilense*, and implications for inoculant formulation. *Applied Microbiology and Biotechnology* 75: 1143–1150.

Philippot, L., Raaijmakers, J.M., Lemanceau, P., and Van Der Putten, W.H. (2013). Going back to the roots: the microbial ecology of the rhizosphere. *Nature Reviews Microbiology* 11: 789–799.

Pierson, L.S. III and Pierson, E.A. (1996). Phenazine antibiotic production in *Pseudomonas aureofaciens*: role in rhizosphere ecology and pathogen suppression. *FEMS Microbiology Letters* 136: 101–108.

Raaijmakers, J.M., Paulitz, T.C., Steinberg, C. et al. (2009). The rhizosphere: a playground and battlefield for soilborne pathogens and beneficial microorganisms. *Plant and Soil* 321: 341–361.

Richardson, A., Barea, J.-M., McNeill, A., and Prigent-Combaret, C. (2009). Acquisition of phosphorus and nitrogen in the rhizosphere and plant growth promotion by microorganisms. *Plant and Soil* 321: 305–339.

Rodríguez, H. and Fraga, R. (1999). Phosphate solubilizing bacteria and their role in plant growth promotion. *Biotechnology advances* 17: 319–339.

Rodríguez, C.W., de Armas, R., Vicente, C., and Legaz, M.-E. (2000). Changes in free and conjugated polyamines during starvation of sugarcane juices as analyzed by high-performance liquid chromatography. *Journal of Chromatography A* 881: 531–541.

Rosier, A., Bishnoi, U., Lakshmanan, V. et al. (2016). A perspective on inter-kingdom signaling in plant–beneficial microbe interactions. *Plant Molecular Biology* 90: 537–548.

Saharan, B.S. and Nehan, V. (2011). Plant growth promoting rhizobacteria: a critical review. *Life Science and Medical Research* 21: 1–30.

Salvioli, A., Ghignone, S., Novero, M. et al. (2016). Symbiosis with an endobacterium increases the fitness of a mycorrhizal fungus, raising its bioenergetic potential. *The ISME Journal* 10: 130–144.

Schauder, S. and Bassler, B.L. (2001). The languages of bacteria. *Genes and Development* 15: 1468–1480.

Schmer, M.R., Vogel, K.P., Varvel, G.E. et al. (2014). Energy potential and greenhouse gas emissions from bioenergy cropping systems on marginally productive cropland. *PLoS One* 9: e89501.

Schuhegger, R., Ihring, A., Gantner, S. et al. (2006). Induction of systemic resistance in tomato by N-acyl-L-homoserine lactone-producing rhizosphere bacteria. *Plant, Cell and Environment* 29: 909–918.

Segata, N., Boernigen, D., Tickle, T.L. et al. (2013). Computational meta'omics for microbial community studies. *Molecular and Systems Biology* 9: 666.

Selin, C., Fernando, W.G., and de Kievit, T. (2012). The PhzI/PhzR quorum-sensing system is required for pyrrolnitrin and phenazine production, and exhibits cross-regulation with RpoS in *Pseudomonas chlororaphis* PA23. *Microbiology* 158: 896–907.

Senés-Guerrero, C. and Schüßler, A. (2015). A conserved arbuscular mycorrhizal fungal core-species community colonizes potato roots in the Andes. *Fungal Diversity* 77: 317–333.

Sharma, S.B., Sayyed, R.Z., Trivedi, M.H., and Gobi, T.A. (2013). Phosphate solubilising microbes: sustainable approach for managing phosphorus deficiency in agricultural soils. *Springerplus* 2: 587–601.

Smith, T.A. and Wilshire, G. (1975). Distribution of cadaverine and other amines in higher plants. *Phytochemistry* 14: 2341–2346.

Smith, S.E., Smith, F.A., and Jakobsen, I. (2003). Mycorrhizal fungi can dominate phosphate supply to plants irrespective of growth responses. *Plant Physiology* 133: 16–20.

Soksawatmaekhin, W., Kuraishi, A., Sakata, K. et al. (2004). Excretion and uptake of cadaverine by CadB and its physiological functions in *Escherichia coli*. *Molecular Microbiology* 51: 1401–1412.

Stein, L.Y. and Klotz, M.G. (2016). The nitrogen cycle. *Current Biology* 26: R83–R101.

Stringlis, I. 2018. Signals from the underground and their interplay with plant immunity. Available at: https://dspace.library.uu.nl/handle/1874/359193.

Stringlis, I.A., Proietti, S., Hickman, R. et al. (2018). Root transcriptional dynamics induced by beneficial rhizobacteria and microbial immune elicitors reveal signatures of adaptation to mutualists. *The Plant Journal* 93: 166–180.

Subramanian, S., Stacey, G., and Yu, O. (2007). Distinct, crucial roles of flavonoids during legume nodulation. *Trends in Plant Science* 12: 282–285.

Sy, A., Giraud, E., Jourand, P. et al. (2001). Methylotrophic *Methylobacterium* bacteria nodulate and fix nitrogen in symbiosis with legumes. *Journal of Bacteriology* 183: 214–220.

Tabor, H., Hafner, E.W., and Tabor, C.W. (1980). Construction of an *Escherichia coli* strain unable to synthesize putrescine, spermidine, or cadaverine: characterization of two genes controlling lysine decarboxylase. *Journal of Bacteriology* 144: 952–956.

Takayama, M., Ohyama, T., Igarashi, K., and Kobayashi, H. (1994). *Escherichia coli cad* operon functions as a supplier of carbon dioxide. *Molecular Microbiology* 11: 913–918.

Teplitski, M., Robinson, J.B., and Bauer, W.D. (2000). Plants secrete substances that mimic bacterial N-acyl homoserine lactone signal activities and affect population density-dependent behaviors in associated bacteria. *Molecular Plant-Microbe Interactions* 13: 637–648.

Tkacz, A. and Poole, P. (2015). Role of root microbiota in plant productivity. *Journal of Experimental Botany* 66: 2167–2175.

Tkacz, A., Cheema, J., Chandra, G. et al. (2015). Stability and succession of the rhizosphere microbiota depends upon plant type and soil composition. *The ISME Journal* 9: 2349–2359.

Turner, T.R., James, E.J., and Poole, P.S. (2013). The plant microbiome. *Genome Biology* 14: 1–10.

Turrini, A., Sbrana, C., Avio, L. et al. (2016). Changes in the composition of native root arbuscular mycorrhizal fungal communities during a short-term cover crop-maize succession. *Biology and Fertility of Soils* 52: 643–653.

Udvardi, M. and Poole, P.S. (2013). Transport and metabolism in legume-rhizobia symbioses. *Annual Reviews in Plant Biology* 64: 781–805.

Unno, Y. and Shinano, T. (2013). Metagenomic analysis of the rhizosphere soil microbiome with respect to phytic acid utilization. *Microbes and Environments* 28: 120–127.

Vandenkoornhuyse, P., Quaiser, A., Duhamel, M. et al. (2015). The importance of the microbiome of the plant holobiont. *New Phytologist* 206: 1196–1206.

Vassilev, N., Vassileva, M., and Nikolaeva, I. (2006). Simultaneous P-solubilizing and biocontrol activity of microorganisms: potentials and future trends. *Applied Microbiology and Biotechnology* 71: 137–144.

Viollet, A., Corberand, T., Mougel, C. et al. (2011). Fluorescent pseudomonads harboring type III secretion genes are enriched in the mycorrhizosphere of *Medicago truncatula*: Pseudomonads harboring T3SS are enriched in mycorrhizosphere. *FEMS Microbiology Ecology* 75: 457–467.

Viollet, A., Pivato, B., Mougel, C. et al. (2017). *Pseudomonas fluorescens* C7R12 type III secretion system impacts mycorrhization of *Medicago truncatula* and associated microbial communities. *Mycorrhiza* 27: 23–33.

Viprey, V., Del Greco, A., Golinowski, W. et al. (1998). Symbiotic implications of type III protein secretion machinery in *Rhizobium*. *Molecular Microbiology* 28: 1381–1389.

Warmink, J.A. and Van Elsas, J.D. (2008). Selection of bacterial populations in the mycosphere of *Laccaria proxima*: is type III secretion involved? *The ISME Journal* 2: 887–900.

Waters, C.M. and Bassler, B.L. (2005). Quorum sensing: cell-to-cell communication in bacteria. *Annual Reviews in Cell Developmental Biology* 21: 319–346.

Watson, N., Dunyak, D.S., Rosey, E.L. et al. (1992). Identification of elements involved in transcriptional regulation of the *Escherichia coli cad* operon by external pH. *Journal of Bacteriology* 174: 530–540.

Whitehead, N.A., Barnard, A.M., Slater, H. et al. (2001). Quorum-sensing in Gram-negative bacteria. *FEMS Microbiology Reviews* 25: 365–404.

Williams, P. (2007). Quorum sensing, communication and cross-kingdom signalling in the bacterial world. *Microbiology* 153: 3923–3938.

Yan, Y., Ping, S., Peng, J. et al. (2010). Global transcriptional analysis of nitrogen fixation and ammonium repression in root-associated *Pseudomonas stutzeri* A1501. *BMC Genomics* 11: 11.

Yasuda, M., Miwa, H., Masuda, S. et al. (2016). Effector-triggered immunity determines host genotype-specific incompatibility in legume–*Rhizobium* symbiosis. *Plant and Cell Physiology* 57: 1791–1800.

Yazdani, M., Bahmanyar, M.A., Pirdashti, H., and Esmaili, M.A. (2009). Effect of phosphate solubilization microorganisms (PSM) and plant growth promoting rhizobacteria (PGPR) on yield and yield components of corn (*Zea mays* L.). *World Academy of Science Engineering and Technology* 49: 90–92.

el Zahar Haichar, F., Santaella, C., Heulin, T., and Achouak, W. (2014). Root exudates mediated interactions belowground. *Soil Biology and Biochemistry* 77: 69–80.

Zamioudis, C. and Pieterse, C.M. (2012). Modulation of host immunity by beneficial microbes. *Molecular Plant-Microbe Interactions* 2: 139–150.

Zehr, J.P., Jenkins, B.D., Short, S.M., and Steward, G.F. (2003). Nitrogenase gene diversity and microbial community structure: a cross-system comparison. *Environmental Microbiology* 5: 539–554.

Zhang, Z. and Pierson, L.S. (2001). A second quorum-sensing system regulates cell surface properties but not phenazine antibiotic production in *Pseudomonas aureofaciens*. *Applied and Environmental Microbiology* 67: 4305–4315.

Zhang, X., Shen, A., Wang, Q., and Chen, Y. (2012). Identification and nitrogen fixation effects of symbiotic *Frankia* isolated from *Casuarina* spp. in Zhejiang, China. *African Journal of Biotechnology* 11: 4022–4029.

Zhang, R., Vivanco, J.M., and Shen, Q. (2017). The unseen rhizosphere root–soil–microbe interactions for crop production. *Current Opinion in Microbiology* 37: 8–14.

Zipfel, C. (2014). Plant pattern-recognition receptors. *Trends in Immunology* 35: 345–351.

Index

The Plant Microbiome in Sustainable Agriculture, First Edition. Edited by Alok Kumar Srivastava, Prem Lal Kashyap and Madhumita Srivastava.